Wind Over Waves II:
Forecasting and Fundamentals of Applications

Shahrdad G Sajjadi, Professor of Oceanography,
John C Stennis Space Centre, Mississippi
and
Lord Julian Hunt, Professor of Space and Climate Physics,
University College London

Horwood Publishing
Chichester

The Institute of Mathematics
and its Applications

HORWOOD PUBLISHING LIMITED
International Publishers, Coll House, Westergate, Chichester,
West Sussex, PO20 3QL, England
First published in 2003

British Library Cataloguing in Publication Data
A catalogue record of this book is available from the British Library

ISBN 1-898563-81-0

Printed in Great Britain by Antony Rowe Ltd, Eastbourne

Preface

The study of waves on surfaces of liquids continues to throw up new phenomena and new theoretical questions that are of great interest to many scientific and engineering disciplines. Fluid mechanics, meteorology, oceanography, geology and civil, chemical and electrical engineering being some of those that feature in the contribution to this volume. Here we particularly focus on surface waves driven by the gas flow over the surface with the main emphasis being on ocean winds force by the natural wind. However many of the results are certainly applicable to waves of engineering importance near the coast or in confined or open channels.

Because of the scientific importance of these topics and the growing interest in their applications, an extended programme of lectures and seminars was held on the 'Mathematics of Surface Water Waves' at the Isaac Newton Institute (INI) for Mathematical Sciences in Cambridge. On the organising committee of the programme, held in August and September 2001, were T. Bridges, S.E. Belcher, S.G. Sajjadi. This was followed (from September 3–5) by a conference organised by the Institution of Mathematics and its Applications (IMA) of Churchill College on the subject of 'Wind over waves; fundamentals, forecasting and applications'.

This second volume in the IMA series of wind over waves is based on the excellent theoretical and experimental papers presented at the conference and on the new ideas and discussions at both meetings, (probably the longest and largest ever held on this topic). The opening remarks of Julian Hunt setting the scene for the programme and the conference are included here on pp 1–2.

During the conference the annual Theoretical and Applied Mechanics Day (August 27) was celebrated. This was recently instituted by the state of Illinois. Professor Moffatt, Director of INI composed some suitable verse after Robbie Burns (following the tradition of Maxwell who illustrated his dynamics lectures in Cambridge in the 1870's with such lines from Burns' poems, as 'when a body meets a body'!)

The first paper in this volume by Craik is based on extensive historical research into how Stokes' basic linear and non-linear theory of small amplitude waves emerged from the considerable number of earlier studies. But, as Craik, whose own work on waves is well known, remarks, none of these had the clarity and correctness of Stokes' analysis. Stokes spent some of his career (as did one of the editors) involved in the administration of the Meteorological Office which is why he developed great interest in the interaction between waves and weather. He was not always so far sighted in his views on scientific development, making the mistake that O. Reynolds work on turbulence was unimportant - as Prof. B.E. Launder FRS has recently discovered in Royal Society archives).†

† In a personal communication, Prof. Launder wrote to Dr Sajjadi: 'My remarks

Ironically the study of turbulent air flow over water waves is largely based on the work of Stokes and Reynolds, together with that of the other great 19th century scientists Kelvin and Helmholtz.

Smedman, Larsen & Hogstrom provide new measurements of the vertical wind speed profile over waves in the Baltic Sea. They conclude that this takes on a logarithmic form for certain but not all wave conditions. In special circumstances, with larger waves, there is, they deduce, a resonant coupling between the waves and the turbulence in the air flow, which changes the mean profile and thence the transfer of energy from the air flow to the waves.

The three next papers report on new studies of this transfer process. Taylor & Yelland compare different formulae that have been proposed for calculating the roughness length z_0, that determines how the drag of the water waves affects the wind profile. Charnock originally proposed that the water surface forms waves that adjust locally to the wind stress $u_\star$, so that $z_0 \sim u_\star^2/g$. Ongoing research reported here shows that this is too simplistic. More accurate, empirically based formulae are proposed to allow for the history of the waves' development and/or variations in the wave height and slope. Not only is new data presented, here but there is also a detailed analysis of the accuracy of ship and buoy measurements. At higher wind speeds the formulae predict changes in the variation of wave drag with wind speed but there is not yet sufficient data to confirm this prediction. Makin & Kudryavtsov have developed a theoretically based model which differs in some respects from the usual 'WAM' operational model, by relating the sea drag directly to the aerodynamical properties of wind waves and to the time-history of their interactions with the wind. They combine recent quasi-steady modelling of turbulent shear flow over waves having low slopes with other models for the breaking effect of steeper waves using statistical distributions of the wave surface. The results compare well with data, and, they claim, provide a methodology for examining more complex aspects of waves in future. Wind forcing of waves is also reviewed by Papadimitrakis & Papaioannou, who assume that it is only the waves large enough to break which determine the air flow over a rough sea. This approach differs appreciably from that in the previous paper. By analysing such waves over all length scales they calculate how the wave spectrum determines the surface drag coefficient and roughness length. Using meteorological data they compare the results with new measurements in the Mediterranean.

The next paper focuses on the often overlooked situation when the direction

were made on the basis of reviews of Osborne Reynolds' Reynolds-averaging paper by Stokes and other 19th Century luminaries in Fluid Mechanics which are now made available for consultation in situ at the Royal Society. I suspect that the paper would not have been published if Reynolds had not already gone to the expense of having the paper printed privately. In the end the Editor was able to rationalize that while they didn't understand or agree with everything the paper said at least Reynolds must have a very strong conviction since he had sunk his own money into having it printed. The referees and editor also said that his earlier paper was very good (i.e. his paper on the transition of flow in a tube) so there may be merit in the later one. A typical refereeing fudge, but it's just as well that in the end they made the right decision.

of the wind differs from that of the wave propagation, such as occurs when there is a swell travelling from a distant storm.

Grachev, Fairall, Hare & Edson show that the mean shear stress in the wind has three separate components determined by the turbulent wind, by the waves driven by the wind and by the swell coming from another direction. Since the latter are often quite large, the usual relationships between wind stress and wind direction are not always applicable. They test their hypothesis against data taken on a platform in the Pacific.

The transfer of the winds' momentum into surface waves continues to be an active topic of fundamental fluid mechanics research. The general consensus is that there are two basic mechanisms; the instability and/or turbulence of the air flow drive pressure and shear stress fluctuations that initiate surface perturbations, and then the aerodynamic drag of the undulating surface that further amplifies. But there remain considerable differences of opinion about the physical nature of these mechanisms (for example whether they change appreciably in flows over unsteady or breaking waves), about their analysis and quantitative evaluation by analytical and numerical models (for example the relative importance of critical layers and turbulent shear stresses) how the mechanisms might interact over waves of finite amplitude and when vortical flow in the water is considered. These controversies are not academic; because current theories still tend to under predict the growth rate of wind driven waves. These controversies are well reflected in the papers in this volume. Sajjadi & Bettencourt analyse and compute the turbulent sheared wind flow over the sharply sloping 'Stokes waves' (whose surface has to be represented with higher harmonics). They explain why the predicted rates of growth of these waves are somewhat greater (by about 15%) than those predicted for monochromatic waves in the standard operational models (e.g. WAM).

McIntyre reviews, with the aid of meteorological examples, the broad mathematical and physical reasons why the disturbances in inviscid shear flows grow with time, including those in air flows over waves, provided the waves themselves are unsteady, as occurs in groups of waves. Whether unsteadiness is intrinsic/crucial for the applicability of Miles' inviscid theory remains one of the key controversies in the subject.

Despite these controversies over detailed questions, there is a broad consistency between the model estimates for the positive and negative forcing of waves by the wind as a function of their slope and of the relative magnitudes of wind speed and wave speed. Willemsen combines models for forcing of waves, including those where the wave speed of long waves exceeds the wind speed, with another model for the non-linear hydrodynamic interactions between waves. This leads to a prediction for how the spectrum of waves that is forced by a growing turbulent wind waves tends to generate waves at ever longer wavelengths until their speed much exceeds that of the wind. Then they are damped by drag forces. At shorter wavelengths the internal dissipation of waves and wave breaking are significant. Note that these processes appear to lead to an equilibrium wave spectrum in a finite time if the wind speed is constant!

The non-linear interactions between different Fourier components not only change their relative energies, i.e. the spectra, but also generate highly non-Gaussian inhomogeneities with patches of high amplitude waves, and areas with distinct patterns of waves such as that recently observed in the Marseille wind-wave facility by Caulliez & Collard. Badulin, Shrira, Voronovich, Kozhelupova & Yurezamkuya examine this spatio-temporal data in order to understand the non-linear dynamics involved in these well defined events. Their technique involves a direct analysis of the resonance of groups of three and four wave numbers. They find that some waves are simply 'slaves' to a dominant or 'master' wave, an idea that may simplify future dynamical studies.

Many of the most significant physics and ecologically important processes involving waves on the surfaces of lakes and oceans depend on complex motions below the air-water surface. These can often be observed in the moving patterns of the surface, marked by bubbles and floating particles. The paper by Hunt, Eames & Belcher reviews the basic mechanisms that control the vorticity in the water. For waves travelling with low to moderate slope that do not break and are not driven by the wind, the vorticity at the surface is significant but alternating and tend to be cancelled out before it is transported into the interior of the water column. This is the implicit assumption of 'numerical wave tank' models of waves now being developed in laboratories around the world. In waves at higher slope, when the water surface becomes multiply connected and dissipative breaking occurs significant, vorticity is generated not only near the surface, but tends to penetrate well below the surface through steady and unsteady processes. The distortion of this vorticity in three dimensional waves has a strong effect on the wave dynamics and mixing processes.

W.R.C. Phillips takes up the theme of three dimensional vorticity dynamics in the water with his non-linear analysis of Langmuir circulations that are familiar to marine observers as 'wind-rows' of floating particles and as streaks of still and ruffled water caused by vortices in water lying parallel to the wind. The pioneering theory of Craik & Leibovich showed how the mean transport or 'drift' of fluid by waves distorts the mainstream vorticity in the water produced by wind shear stress over the surface. The important point of the new calculations presented here is that the different patterns and scales of circulations which develop quite suddenly are shown to be bifurcations of this complex non-linear system. This leads, for example, to sudden changes in the size and spacing of the longitudinal vortices. He concludes with some interesting suggestions about future work, including the effects of higher amplitude waves.

Readers of this introduction will already have learnt that waves have many very different forms and types of motion both individually and in groups, and that those may either be quite robust or tend to change quite suddenly. The question is whether the underlying mathematical structure for wave motion can explain and perhaps systematise some of these general properties in particular their sensitivity to the initial forcing? Only mathematicians can provide these answers; since no amount of computation or experiment can do so.

In this volume Kirchgassner analyses 2-dimensional incompressible irrotational water waves travelling with a given speed. He describes some exact results for the existence and uniqueness of flows (without proofs) for particular types of initial data. This should lead to more complete theories for the stability and instability of certain types of solitary wave depending on the initial conditions, which are still major goals of wave theory.

In the paper based on his Benjamin memorial lecture, Bridges takes up other fundamental questions about the stability of solitary waves, starting from the pioneering idea of Benjamin (1972) to use Hamiltonian formulation in terms of energy and momentum integrals. This approach is extended using the 'multi-symplectic' framework which enables different components in the flow system to be examined separately, such as three dimensional growing instabilities within a basically two dimensional wave. These theoretical developments also indicate how numerical methods may be improved for computing the stability of these waves.

Solitary waves are generally assumed to be planar and exactly two dimensional. Baskar & Prasad ask the question how such a wave is affected by small variations along its crest. They use the general kinematic analysis applicable to wide class of hydrodynamic waves, including compressible waves. An important result of the analysis for water waves is that transverse waves are generated which propagate along the crest line.

A less well known but equally important aspect of wave theory predicts how the modulation of the pressure field and water surface can be measured remotely. In the first volume (edited by Sajjadi, Thomas & Hunt) the history of the pioneering studies of how waves generated by distant storms could be analysed from local observations was described by F. Ursell. (which tend to confirm 'Folk-lore' about the noise produced by stones on a beach tells you about the storms at sea - *e.g.* as in Westward Ho by C. Kingsley). In this volume Longuet-Higgins shows how such distant waves can also be detected through the elastic waves in the ground microseismis (which are regularly observed at earthquake monitoring stations, such as Eskdale Muir in Scotland) and the infrasound pressure fluctuations set up in the atmosphere (measured remotely for example in the Netherlands). His analysis shows that nonlinear interaction between *oppositely traveling* wave components in the spectrum of ocean waves (which are commonly observed) will cause large-scale pressure oscillations to be transmitted to depths large compared to the length of the surface waves. These nonlinear pressure oscillations have twice the frequency of the surface waves and generate microseisms propagating outwards from the storm area. A similar mechanism generates microbaroms (low-frequency pressure waves) in the atmosphere. Probably of even greater importance is his discovery of a triple-period instability in steep standing waves. This instability leads to the ejection of spray into the atmospheric boundary-layer, so affecting the dynamics and thermodynamics of the air-sea boundary. The great importance of sea spray in hurricane dynamics has been emphasized by Lighthill in a paper in the preceding WoW volume. Longuet-Higgins describes some laboratory experiments which demonstrate such instabilities. As we expect from this author the theory and experiment are exceptionally elegant and successful.

Another remote sensing technique of increasing importance (and minimal cost) is the use of high frequency radar back scatter. Green describes how the Doppler spectrum echoes from the sea surface can now be computed by discretising equations that Barrick introduced in the 1970s, and introducing the techniques of tomographic inversion. However, if the wave dynamics are better understood, this will also improve the radar based remote sensing.

The same conclusion can be drawn from many studies of the practical effects of waves, it shows further that recent research on wind driven waves is still vitally important.

The editors would particularly like to thank the other members of the International Organising Committee for suggesting invited speakers and topics, and the invited speakers for their contributions. We would like to thank all the speakers for their efforts in preparing and delivering their talks, and we are grateful to those who so ably Chaired the various sessions. We are grateful to both the INI and the IMA for their continuous help in the organisation of these two conferences. The 'Surface Water Waves 2001' was sponsored by the Isaac Newton Institute for Mathematical Sciences, and the 'Wind-over-Waves: Fundamentals, Forecasting and Applications' was sponsored by the Institute of Mathematics and its Applications, and we gratefully acknowledge their assistance.

For the production of this volume, we would like to thank all those who contributed, for their co-operation, for their patience, and their willingness to cope with the technicalities of achieving uniformity of style. The editors sought comments on all the papers from appropriate experts in the field, and we are grateful to them, and also to the authors, for responding so readily to suggestions which were relayed to them. The papers were checked and put into final LaTeXcamera ready copy by Mrs LuAnn Sajjadi before their final journey to Horwood Publishing Ltd. We would like to express our thanks to her for her patience and hard work. Also our sincere thanks to the individuals concerned at Horwood Publishing Ltd for their assistance throughout the production of this volume.

J.C.R. Hunt
University College London.

S.G. Sajjadi
Stennis Space Center.

Preface to the second printing

Our thanks are due to many colleagues for their kindness in informing us of errors and missprints which they had noticed in the first printing of this book; in particular, we cannot miss this opportunity of expressing our gratitude to Mrs Sajjadi for the monumental task she took to correct, re-type and proof read the entire manuscript of the second printing.

J.C.R. Hunt and S.G. Sajjadi

TAM-day musings:† A poem to celebrate TAM-day, 27th August 2001

K. Moffatt

After Tam O'Shanter With apologies to Robert Burns

When lecture rooms are growing stale,
And thoughts begin to turn to ale;
When blackboards thick in chalk are smothered,
With scarce a new truth there discovered;
When loud the arguments have railed,
But longed-for inspiration failed;
Why, that's the time to take a break,
To turn to coffee, tea and cake;
Or sip the vineyard's produce cool,
And contemplate the game of boules.

This truth was haply brought to mind,
By letter sent and duly signed,
Conveying tidings of great joy
Hot from the heart of Illinois;
There it was formally declared
What ne'er had hitherto been dared:
Mechanics as a worldwide art
Should henceforth have a day apart,
A day when man may ruminate
Upon the subject's vibrant state;
A day when blackboard toil should cease,
And staff should have a bit of peace!
Straight from the land of Al Capone,
Instruction came by telephone;
It seemed there was no time to lose;
Invited thus, one can't refuse!

And thus it was TAM-day was born,
An August Monday to adorn,
A day this year decreed by heaven
To fall on August twenty-seven;
So three times three, again times three,
A date on which TAM holds the key
To open Archimedes' door,

† This poem was written in response to the following e-mail message received from Professor Hassan Aref on 7th August 2001.

And celebrate on every shore;
What time of day? you well may ask;
The answer's plain: from dawn till dusk.

In Cambridge town we heard the call,
And rallied to the central hall
Of mighty Newton's Institute,
Irreverently called by some the Newt!
He who bestrode the pebbled shore
Of Ocean's ever-mobile floor;
Here a pebble, there a pebble,
But was the system integrable?
Said Newton "If I shed a tear
Upon this ocean wide and clear,
Will this affect the rain in Laos?"
And thus he sowed the seeds of Chaos.

But to our tale: the day dawned bright,
The weather forecast had been right;
The warming sun on boules court shimmered,
The overarching crane fair glimmered;
That day a child might understand
An awesome drama was to hand.

Well practiced in the laws of motion,
And fortified by vintage potion,
The gifted savants slow foregathered
Hard by the shed where bikes are tethered;
From every land and clime they came,
Experts of legendary fame;
From Russia, It'ly and Japan,
And every country known to man;
From Poland, Greece and USA,
A clash of titans underway!

But here my Muse her wing maun cour,
This glorious game made such a stoor;
The boules of glittering steel were round,
And sped unerring on the ground;
They rolled, they arched, they spun, they clickit,
More action here than seen in cricket!
The game might well have run till dawn
There by the side of Newton's lawn.
The cochinet brent new frae France
Was kissed with steel, and touched perchance,
Balls tossed with Zakharovian skill;
Then Kruskal clad in T-shirt still,
He who could aim a ball and roll-it-on,

Invariant as any soliton,
With concentration ever keener
Entered upon that tense arena;
And cast one ball, a crafty throw
That scattered those of every foe.
The cochinet was split asunder,
Th'encircling crowd was mute with wonder.

Now, who this tale o' truth shall read,
Mechanics of whatever creed:
When overwork becomes a grind,
And problems tangle up the mind,
Fill well the cup and fill it full,
Take refuge in a game of boules!

Contents

Contributors

S.I. BADULIN: P.P. Shirshov Institute of Oceanology 36 Nakhimovsky pr., Moscow 117851, Russia

S. BASKAR: Department of Mathematics, Indian Institute of Science, Bangalore 560 012

S.E. BELCHER: Department of Meteorology, University of Reading, Earley Gate, Reading RG6 6BB, UK

M.T. BETTENCOURT: CHL, John C. Stennis Space Center, Building 1103, Suite 103, Mississippi 39529, USA

T.J. BRIDGES: Department of Mathematics and Statistics, University of Surrey, Guildford, Surrey GU2 7XH, UK

A.D.D. CRAIK: School of Mathematics and Statistics, University of St Andrews, St Andrews, Fife KY16 9SS, Scotland, UK

I. EAMES: Department of Mechanical Engineering, University College London, Torrington Place, London WC1E 7JE, UK

J.B. EDSON: Woods Hole Oceanographic Institution, Woods Hole, Massachussetts, USA

C.W. FAIRALL: NOAA Environmental Technology Laboratory, Boulder, Colorado, USA

A.A. GRACHEV: University of Colorado CIRES/NOAA ETI, Boulder, Colorado, USA

J.J. GREEN: Department of Applied Mathematics, University of Sheffield, UK

J.E. HARE: University of Colorado CIRES/NOAA ETL, Boulder, Colorado, USA

U. HÖGSTRÖM: Department of Earth Sciences, Meteorology, Uppsala university, Uppsala, Sweden

J.C.R. HUNT: Departments of Space and Climate Physics and Geological Sciences, University College London, Gower Street, London WC1E 6BT, UK, and J.M. Burgers Centre, Delft University of Technology, Delft, Netherlands

K. KIRCHGÄSSNER: Mathematisches Institut A, Universität Stuttgart, Germany

N.G. KOZHELUPOVA: P.P.Shirshov Institute of Oceanology 36 Nakhimovsky pr., Moscow 117851, Russia

V.N. KUDRYAVTSEV: Marine Hydrophysical Institute, Sebastopol, Ukraine

X.G. LARSEN: Department of Earth Sciences, Meteorology, Uppsala university, Uppsala, Sweden

M.S. LONGUET-HIGGINS: Institute for Nonlinear Science, University of California, San Diego, La Jolla, California 92093-0402, USA

V.K. MAKIN: Netherlands Meteorological Institute (KNMI), De Bilt, The Netherlands

M.E. MCINTYRE: Centre for Atmospheric Science Department of Applied Mathematics and Theoretical Physics, University of Cambridge, Silver Street, Cambridge CB3 9EW, UK

K. MOFFATT: Isaac Newton Institute for Mathematical Sciences, 20 Clarkson Road, Cambridge, CB3 0EH, UK

I. PAPADIMITRAKIS: National Technical University of Athens Department of Civil Engineering Hydraulics, Water Resources and Maritime Engineering, Greece

A.I. PAPAIOANNOU: National Technical University of Athens Department of Civil Engineering Hydraulics, Water Resources and Maritime Engineering, Greece

W.R.C. PHILLIPS: Department of Theoretical and Applied Mechanics, University of Illinois at Urbana-Champaign, Urbana, IL 61801-2935, USA

P. PRASAD: Department of Mathematics, Indian Institute of Science Bangalore 560 012

S.G. SAJJADI: CHL, John C. Stennis Space Center, Building 1103, Suite 103, Mississippi 39529, USA

V.I. SHIRIRA: Department of Mathematics, Keele University, Keele ST5 5BG, UK

A.-S. SMEDMAN: Department of Earth Sciences, Meteorology, Uppsala university, Uppsala, Sweden

P.K. TAYLOR: James Rennell Division for Ocean Circulation and Climate, Southampton Oceanography Centre, UK

A.G. VORONOVICH: COETL NOAA, 325 Broadway, Boulder, Colorado, USA

J.F. WILLEMSEN: Rosenstiel School of Marine and Atmospheric Science, University of Miami, Miami, Florida, USA

M.J. YELLAND: James Rennell Division for Ocean Circulation and Climate, Southampton Oceanography Centre, UK

Y.S. YURENZANSKAYA: Moscow Institute of Physics and Technology, 141700, Dolgoprudny, Moscow region, Russia

PLATE 1. Conference participants in the grounds of the Isaac Newton Institute, Cambridge

INI Water Waves Conference Opening Remarks

J.C.R. Hunt, CB, FRS
University College London

As Chairman of the opening session, I should like to welcome you all very warmly to the Isaac Newton Institute and express my appreciation that such a distinguished international gathering of experts is attending this our first programme on surface water waves. We have had previous INI programmes in mathematical sciences on various aspects of fluid mechanics, including atmosphere-ocean dynamics in 1996 and turbulence in 1999. (A report on the latter, which was published this summer, is now in the INI library). Wave motions differ from other forms of fluctuations in that they persist over many periods at approximately constant frequency and wavelength; but the dynamics of interacting non-linear waves, although distinct from those of vortices and turbulent eddies, have some significant features in common. These are increasingly being recognized especially by the Russian school of wave theorists, led by Professor Zakharov, whose presence is particularly welcome at this programme.

Surface wave motion has been a paradigm problem for mathematical sciences because waves are the essential feature of so many natural and artificial phenomena whether in fluids, solids, plasmas, nerve channels, or communications systems, etc. On the other hand, research on the theory of waves has also benefited from using concepts, and analytical methods derived in other areas of mathematics and science, especially those involving heterogeneous and non-linear systems. In particular plasma physics has greatly contributed to the modern theory and practice of wave spectral modelling, which is now applied daily in operational forecasts of ocean waves and may indeed provide some lessons for the developers of practical models of turbulence spectra, which are not so well established.

In the 19th century the study of wave motion led to new mathematics on partial differential equations and asymptotic methods, which we will hear about from Alex Craik in his historical lecture. However here we will probably be focusing more on the developments of the 20th and 21st century research on non-linear interactions, solitons and the general mathematical properties of wave solutions, especially those derived using Hamiltonian methods. The organizers of this programme, Tom Bridges, Stephen Belcher and Shahrdad Sajjadi (who sadly could not be here) were pleased that the INI scientific committee also saw the need to hold this programme, both because the study of surface waves is still an exciting field of mathematical science, and because of the urgent practical need to improve our understanding and computational modelling of surface waves. With average sea levels rising by about half a metre over the next 100 years, coastal defences around the world are being re-assessed and, in many places, new construction is underway. Engineers need to improve their predictions of waves in shallow waters and how waves impact on the coastal structures. The famous constructor of Scotland's lighthouses, and grandfather of the novelist Robert Louis Stevenson, studied carefully the huge Atlantic waves approaching the west coast and estimated that the peak pressure his masonry structures would need to withstand was about three tons per square foot, which roughly corresponds to 30 metre waves!

We shall hear about the importance of predicting such 'freak' ocean waves as well as the average statistics of waves from engineers and ocean. But sometimes the greatest effects on structures and ships come from waves with imperceptible amplitude, but damaging oscillations can be produced on oil drilling barges when the wave frequencies resonate with the structure. Recent research has shown how waves affect the intensity of ocean storms over the period of several days as they develop which is another reason why predicting wave statistics is now an integral feature of weather forecasting. This spectacular application of our science will be described in the opening lecture by Peter Jansen. Exciting interdisciplinary applications of wave dynamics are emerging for interpreting remote sensing of the ocean surfaces, e.g. for coastal surveying, wave-ice interactions and improving climate models for the global transport of heat by ocean waves. Although INI is a UK and European institution, it is situated in Cambridge, where there has been a tradition of research on hydrodynamics and surface water waves since the 19th century. Here Stokes and Thomson (Kelvin) laid down the foundation of the subject. Their vigorous correspondence with lots of nice diagrams can be seen in the rare books room of the University library. Characteristically, their great achievements were recognized by the British establishment with Stokes (like Newton) becoming a knight and MP for Cambridge, and Thomson taking a seat in the House of Lords as Lord Kelvin of Largs. In the 1920's, Harold Jeffreys improved on Kelvin's theory for the generation of waves by the wind. To support this hypothesis on the 'sheltering' mechanism he photographed the waves on Newnham duck pond next to the Granta pub. He compared the feeble ripples there with Cornish's photographs of Atlantic waves.

Cambridge research on surface waves in the 20th century was greatly stimulated by laboratory wave tanks. G.I. Taylor discovered the 'peaky' form of transverse standing waves in a square shaped tank. But in his wartime consulting before the Normandy Landings in 1944 he also showed how waves can be damped by surface currents driven by the releasing of plumes of air bubbles. Probably the most remarkable application of wave tank research here was the experimental discovery by Jim Feir and Thomas Brooke Benjamin that Stokes waves are unstable and evolve into groups. Their quite unexpected and exciting observations in the summer of 1964, which I saw as a research student, then led to the famous Benjamin Feir theory of wave instability. The DAMTP laboratory in Silver Street now houses their long tank until the laboratory moves to the new Mathematical Sciences building in 2002, next to the Isaac Newton Institute. I hope you all enjoy this programme and that it will be as successful as others in stimulating new research as we work together over the next three weeks.

G.G. Stokes and His Precursors on Water Wave Theory †

A.D.D. Craik
School of Mathematics and Statistics, University of St Andrews, St Andrews, Fife KY16 9SS, Scotland, UK.

ABSTRACT

George Gabriel Stokes (born August 13 1819) died almost one hundred years ago, on 1st February 1903, and it is just over 150 years since he published his great 1847 paper on water waves. His contributions to water-wave theory are widely and deservingly acknowledged, but the context of his work is nowadays largely forgotten. Stokes' important early work, published during 1842-50, was influenced by his French predecessors, Laplace, Lagrange, Poisson and Cauchy, and by his near-contemporaries in Britain: Challis, Green, Russell, Kelland, Airy, Earnshaw and Thomson. Later contributions to water wave theory by Stokes and others are discussed more briefly.

1. BRITISH FLUID MECHANICS, 1835-1885

Before 1835, the state of knowledge of fluid mechanics in Britain was unimpressive: few seem to have read the work of Johann and Daniel Bernoulli, Jean d'Alembert and Leonhard Euler from a century earlier. Lagrange's *Mécanique Analytique*, Laplace's *Mécanique Céleste*, Poisson's *Mécanique*, and Fourier's *Théorie Analytique de la Chaleur* were rather better known; but, of the generations before Stokes, only James Ivory (1765-1842) and George Biddell Airy (1801-1892) stand out among British scientists as having mastered these analytical works sufficiently to make their own original contributions. Yet English translations, with explanatory notes, of Book 1 of Laplace's *Mécanique Céleste* were published by John Toplis (Nottingham, 1814) and Henry Harte (Dublin, 1822); Thomas Young had written anonymously his *Elementary Illustrations of the Celestial Mechanics of Laplace* [Book 1] in 1821; and the great American translation and commentary of vols. 1-4 by Nathaniel Bowditch appeared during 1829-38. All of these included descriptions of the equations of inviscid fluid mechanics and must have helped to raise awareness of this field. Of works by British authors, Samuel Vince's old-fashioned *The Principles of Hydrostatics*, first published in 1798, was reprinted in a 6th edition in 1829.

† The initial impetus for this work was an invitation from Professor Alastair Wood to lecture at the G.G. Stokes Summer School, held at Stokes' birthplace in Skreen, Co. Sligo, in August 2000. I am most grateful to him for this invitation. Further work was carried out during a visit to the Research Institute for Mathematical Sciences, Kyoto University. I am grateful to the Director and staff of the Institute for their kind hospitality; and especially to Professors H. Okamoto and M. Funakoshi for their invitation to lecture on this theme at a RIMS symposium. The resources of St Andrews University Library are also acknowledged with thanks.

Henry Moseley's 1830 *A Treatise on Hydrostatics & Hydrodynamics...* is unimpressive; Thomas Webster's 1836 The *Theory of the Equilibrium and Motion of Fluids* was "compiled principally from the writings of Poisson and Challis"; James Challis himself prepared two commissioned reports (1833, 1836) on fluid motion for the recently-founded British Association for the Advancement of Science; John Henry Pratt's 1836 *The Mathematical Principles of Mechanical Philosophy* briefly described the equations of inviscid flow; William Walton's 1847 *A Collection of Problems in Illustration of the Principles of Theoretical Hydrostatics and Hydrodynamics* and William H. Miller's *The Elements of Hydrostatics & Hydrodynamics* (1st ed. 1831) are uninspired, but the 1850 4th edition of the latter at least contained a 7-page Appendix on the 'Theory of Long Waves', showing some awareness of recent work.

Every one of the above-named authors was based, or had received his education, at Cambridge University, and these textbooks were mainly aimed at the captive audience of Cambridge undergraduates.† Vince and later Challis held the post of Plumian Professor of Astronomy, and Miller was Professor of Mineralogy, while others were or had been college fellows. The few works on fluid mechanics published in Britain between 1800 and 1840 by non-Cambridge graduates were mostly by Scottish authors and aimed at a general audience, concerned more with practical hydraulic and pneumatical devices than with mathematical theory.

George Biddell Airy's long article on *Tides & Waves* was published in the *Encyclopaedia Metropolitana* in 1841. When it appeared, Airy's scientific reputation as mathematician and astronomer was already high: he was then Astronomer Royal at Greenwich Observatory, following a period at Cambridge as Lucasian Professor of Mathematics and then as Plumian Professor of Astronomy (between Vince and Challis). Airy's article was of a different calibre from the above texts. It was studied by the young G.G. Stokes who, in its year of publication, had graduated from Cambridge as Senior Wrangler and 1st Smith's prizeman. In that year, Stokes became a fellow of Pembroke College, with ample time for private study and research; and he followed the advice of his former coach, William Hopkins, that hydrodynamics would be a fruitful field. The British Association for the Advancement of Science had been founded in 1831, largely through the initiative of the Scot, David Brewster (Howarth 1931, Ch.1). Though the Cambridge professoriate were at first suspicious of the new national body, and none attended its inaugural meeting in York, the influential William Whewell of Trinity College, and then Cambridge's Professor of Mineralogy, wrote urging the Association to commission "reports... concerning the present state of science, drawn up by competent persons" (Howarth 1931, p.25). Over the next few years, several reports on 'Hydrodynamics' and on 'Waves' were duly commissioned. The first two, by James Challis (1833, 1836),

† More on the Cambridge milieu of that time may be found in Harman (1985): of particular note are the articles by D.B. Wilson contrasting Cambridge with the Scottish universities of Edinburgh and Glasgow, and by I. Grattan-Guinness on French influences on Cambridge before 1840.

are mainly noteworthy for their pompous style and lack of substance, but they at least drew attention to the possibilities for advance:

> The foregoing review of the theory of fluid motion... may suffice to show that this department of science is in an extremely imperfect state. Possibly it may on that account be the more likely to receive improvements; and I am disposed to think that such will be the case. (Challis 1833, p.151)

James Challis (1803-1882) published no fewer than 14 papers on the equations of hydrodynamics between 1829 and 1845, mostly in *Philos. Mag. and Trans. Camb. Philos. Soc.* and mostly worthless. Stokes, who had attended at least some of Challis' lectures, soon quarreled with him, objecting to his errors both by letter (unpublished, in Cambridge University Library's Stokes Collection) and eventually in print: Stokes (1842a, 1851), Challis (1851), see also Wilson (1987, 132-3).

In 1837, the British Association set up a 'Committee on Waves' to conduct observations and experiments. This committee consisted of John Scott Russell (1808-1882) and Sir John Robison (1778-1843) (Howarth 1931, pp. 266, 271). Robison, the son of a former professor of Natural Philosophy at Edinburgh, was a well-to-do amateur scientist and Secretary of the Royal Society of Edinburgh during 1828-39 (Campbell & Smellie, 1983). Russell had briefly taught at Edinburgh University; but in 1844 he left Scotland for London, to pursue a notable and turbulent career as engineer and naval architect (Emmerson 1977).

A substantial report by Robison & Russell (1837) was followed by a brief one (1840). Robison died in 1843; and Russell alone wrote a brief supplementary report (1842), then his major 'Report on Waves' (1844). Russell paid tribute to Robison as "a kind friend", but made a point of claiming all the experimental work as his own:

> In all these researches the responsible duties were mine, and I alone accountable for them; but in forwarding the objects of the investigation I always found him a valuable counselor and a respected and cordial cooperator. (Russell 1844, p.311)

These reports constitute a remarkable series of observations, at sea, in rivers and canals, and in Russell's own wave tank constructed for the purpose. There was much for the mathematicians to think about: for, in 1837, not even the linear theory of small-amplitude plane waves was well known. Though Russell's experiments are now famous for his discovery of the nonlinear solitary wave, it is not surprising that, in his own day, this aspect of his work proved contentious and misunderstood.

Others at Cambridge who published on water waves before Stokes were George Green (1793-1841) and Samuel Earnshaw (1805-88). Philip Kelland (1808-1879) had also studied at Cambridge, and for a time was a fellow of Queen's College; but, from 1838 until his death, he was Professor of Mathematics at Edinburgh University. In Edinburgh, he met John Scott Russell and became interested in water waves. In 1840 he contributed a short paper to the British Association for the Advancement of Science. In the same year, he published the first part of a long two-part paper 'On the Theory of Waves' in the Transactions of the Royal Society of Edinburgh, the second part following in 1844.

The papers of Green, Kelland and Earnshaw, the encyclopaedia article by Airy, and the experimental reports by Russell & Robison and Russell were all read by Stokes and informed his own work on water waves.

George Gabriel Stokes was born in the rural parish of Skreen, Co. Sligo, Ireland, a son of the (Episcopal) Church of Ireland Rector of that parish. During 1835-37, he was educated at Bristol College, and entered Pembroke College, Cambridge, as an undergraduate in 1837. He was appointed to the Lucasian Chair of mathematics of Cambridge University in 1849, a post which he held until his death 53 years later.†

Stokes' collected papers (Stokes, 1880-1905) and some though not all of his voluminous correspondence (Larmor 1907; Wilson 1990) are published. Remarkably, all his significant work on fluids was written during the two periods 1842-50 and 1880-98: in other words, when he was aged around 23-30 and 60-79. Between the ages of 30 and 60, Stokes' published research concerned mainly optics, and mathematical problems deriving from optics: see Wilson (1987), Paris (1996). Also, throughout 1854-85, Stokes spent much, arguably too much, time performing onerous duties as Secretary of the Royal Society of London.

Though it might appear that Stokes' revived interest in fluid motion was prompted by the need to prepare for publication his collected *Mathematical & Physical Papers*, this is only part of the story. His correspondence from the 1860's and 1870's includes several letters about wave motion. In particular, his appointment to the Meteorological Council led him to consider practical aspects of wave propagation in open seas: how to measure wave amplitudes and periods from ships, and how such information might be used to deduce weather conditions at a distance.

The individual most closely associated with Stokes and his work throughout his long career was William Thomson (1824-1907). In 1892 Thomson became Baron Kelvin of Largs. At the age of 10, the precocious Thomson became a student of Glasgow University, where his father James was a professor of mathematics. Before enrolling at Cambridge in 1841, he already had an impressive knowledge of mathematics and natural philosophy. Shortly after his graduation, Thomson edited the *Cambridge Mathematical Journal* (later *the Cambridge & Dublin Mathematical Journal*). This was aimed at a student readership, and Thomson initiated a series of 'Notes on Hydrodynamics', some written by himself and three, published in 1848-49, by Stokes. Most of Thomson's own researches in fluids, particularly waves, instabilities and vortices, date from a later period during the 1870's and 1880's.

Stokes and Thomson maintained a frequent correspondence, both friendly and scientific, throughout their adult lives (Wilson 1987, 1990). They clearly

† Fuller biographies are in Wilson (1987, 1990), Larmor (1907), Wood (1995) and at the websites `http://www-history.mcs.standrews.ac.uk/history/References/Stokes.html` and `http://webpages.dcu.ie/~wooda/stokes/ggstokes.html`

felt a close intellectual affinity, despite their different personalities and scientific attitudes: Stokes the meticulous and narrowly-focused Cambridge scholar (described by Lord Rayleigh as "over-cautious" [Stokes' Papers 5, p. xix]), devoutly religious, and willingly burdened with administrative duties; and Thomson, the more speculative and wide-ranging Glasgow Professor of Natural Philosophy.

Stokes' main original contributions to fluid mechanics are undoubtedly his fundamental 1845 paper on the equations of viscous flow, where he re-derives from a more general hypothesis the equations first given by Navier, and gives a few exact solutions; his 1847 paper 'On the Theory of Oscillatory Waves' considered below; his 1848 work on bodies moving slowly through a viscous fluid (now called 'Stokes flow': surprisingly published as part of a paper on the 'Lumeniferous Ether'); and 'On the Effect of the Internal Friction of Fluids on the Motion of Pendulums' (1856, read 1850). These works gave a major boost to the subsequent study of fluid flow in Britain; but Stokes himself published nothing more on this topic for the next 30 years.

Though Stokes and Thomson's didactic 'Notes on Hydrodynamics' in the *Camb. & Dublin Math. Jour.* (1847-49) did not claim originality, they too played a crucial part in facilitating the study of fluid mechanics until more satisfactory monographs and textbooks eventually appeared. The best of these are W.H. Besant's *A Treatise of Hydromechanics* (1877); Horace Lamb's *A Treatise on the Mathematical Theory of the Motion of Fluids* (1879), later enlarged as *Hydrodynamics* (1st ed. 1895, and continuously in print since then through six editions); and A.B. Basset's two-volume *A Treatise on Hydrodynamics* (1888). Even more influential, but of a different kind, were William Thomson & Peter Guthrie Tait's *Treatise on Natural Philosophy*, vol.1 (all published) (1867), James Clerk Maxwell's pioneering *A Treatise of Electricity and Magnetism* (1873), and John William Strutt (Lord Rayleigh)'s *The Theory of Sound* (1877-8).

Samuel Earnshaw had complained in 1845 that:

> Though it is now about a hundred years since the general equations of fluid motion, expressed in partial differential coefficients, were first given to the world, I am not aware that any important case of fluid motion has hitherto been rigorously extracted from them. This however has not arisen from want of effort....; but rather from the peculiarly rebellious character of the equations themselves, which resist every attack, except it have reference to some case of a very simple and uninteresting nature. (Earnshaw 1847, p.326)

The remarkable progress made during the 50 years from 1835 to 1885 is made abundantly clear by comparing the textbooks of, say, Lamb and Miller; or W.M. Hicks' (1881-2) 'Report on Recent Progress in Hydrodynamics' (the last so-titled) to the British Association, and that of Challis in 1833; or by contrasting the review by A.G. Greenhill (1887) of 'Wave Motion in Hydrodynamics' with Kelland's (1840a).

2. Work on water waves before 1840

Isaac Newton was the first to attempt a theory of water waves. In *Book II, Prop. XLV of his Principia* (1687), he proposed a dubious analogy with oscillations in a U-tube, correctly deducing that the frequency of deep-water waves must be proportional to the inverse of the square root of the "breadth of the wave". Much later, after Euler's (1757a,b; 1761) derivation of the equations of hydrodynamics, Laplace (1776) and Lagrange (1781, 1786) obtained the linearized governing equations for small-amplitude waves, and Lagrange gave their approximate solution in the limiting case of long plane waves in shallow water.

A remarkable early paper by Gerstner (1802) gave the first *exact nonlinear* solution for waves of finite amplitude on deep water. But the Gerstner wave solution was for long overlooked; and even today it is usually regarded more as a curiosity than a result of practical importance, because the wave is not irrotational. Its independent rediscovery by W. J. Macquorn Rankine (1863) revived interest in it, and led Stokes to comment.

Thomas Young wrote extensively on tides, but only briefly on waves. On waves, he added nothing new but gave perhaps the first account in English of waves in shallow water (1821, pp. 318-327).

In Dec. 1813, the French Académie des Sciences announced a mathematical prize competition on surface wave propagation. In July 1815 the 25-year-old Cauchy submitted his entry; and, in August, Poisson, one of the judges, deposited a memoir of his own in order to record his independent work (Dalmedico 1988). Cauchy was awarded the prize in 1816; Poisson's memoir was published in 1818 and Cauchy's work, with an astonishing 188pp. of additional notes, eventually appeared in 1827.

The work of Cauchy and Poisson was far ahead of its time. It confronted the general initial value problem for linearized water waves: given any localized initial disturbance of the liquid surface, what is the subsequent motion? Both axisymmetric and two-dimensional cases were considered. The analysis employed *de facto* Fourier transforms, comprising superpositions of standing-wave modes; and asymptotic approximations to evaluate the resultant integrals. Not only did the methods of analysis repel most readers; the results, too, seemed baffling and contrary to intuition, for there was then no understanding of wave interference and dispersion. Though the Cauchy-Poisson analysis is now acknowledged as a milestone in the mathematical theory of initial-value problems, its influence on the development of wave theory was slight. Kelland, Airy and Stokes all thought it to have little practical relevance, despite its mathematical sophistication.

Shortly before the delayed appearance of Cauchy's memoir, a very different work on waves was published in Leipzig. This was *Wellenlehre auf Experimente gegründet...*, (1825) by the brothers Ernst Heinrich Weber (1795-1878) & Wilhelm Eduard Weber (1804-1891). This describes the Webers' laboratory experiments on plane periodic waves in a channel, and gives extracts from, and comments on, earlier works. Airy, in his article (1841*b* pp. 344-350), devoted seven pages to the observations of the Webers and of Russell. Probably through Airy's comments, The Webers' work became known to Russell in time for mention in his 1844 report: he was no doubt relieved to be able to write that "their labours and mine do not in the least degree supersede or interfere with each other" (Russell, 1844, p.332 footnote). Thus, just when Stokes was starting to work on water waves, accounts of these first-rate experiments were conveniently to hand.

3. Airy, Russell, Green, Kelland and Earnshaw

Airy's long 155pp. article on Tides and Waves for the *Encyclopaedia Metropolitana* appeared in 1841. Although its main focus was tidal phenomena, Airy also gave attention to the 'Theory of Waves in Canals' (Section IV) and an 'Account of Experiments on Waves' (Section V). In the former, he gives the now-standard linear theory for plane waves, finding the dispersion relation for the wave velocity in terms of the wavelength and the liquid depth. He also shows that particle paths are ellipses (or circles for infinite depth). He attempts a non-linear (and non-dispersive) *Theory of Long Waves in which the Elevation of the Water bears a sensible proportion to the depth of the Canal,* arriving at an approximate equation which he uses to deduce that the surface elevation is

$$V = k\left\{1 - am\sin(mvt - mx') + \tfrac{3}{4}a^2m^3.x'.\sin(2mvt - 2mx')\right\} \qquad (3.1)$$

where m is the wavenumber, and a the wave amplitude. Here, v is the linear wave speed $gk^{1/2}$ where k is the liquid depth; and x' is "measured from the point where the canal communicates with the open sea." Thus,

> (... When the wave leaves the open sea, its front slope and its rear slope are equal in length, and similar in form. But as it advances in the canal, its front slope becomes short and steep, and its rear slope becomes long and gentle....)

By considering the time intervals between successive maxima of V when the secular term in x' is small, Airy deduces that "the phase of high water has traveled along the canal with the velocity... $\sqrt{gk(1+3b)}$", where b is the maximum elevation divided by the mean depth k. Correspondingly, the minima at low water travel with the lesser velocity with $-b$ replacing b. He next continues his approximation to $O(a^3)$, finding terms which grow as x'^2, and showing his results in a figure.

Airy then turns to small-amplitude (linear) waves in canals with slowly-varying depth and cross-section. Rather similar theories had earlier been developed by Green (1837, 1838) and Kelland (1840). Airy's results for long waves agree with this earlier work, which is not cited; but his approximations seem ad hoc and unconvincing.

Airy also attempted to construct a *Theory of Waves on Canals when Friction is taken into account.* Restricting attention to long, small-amplitude (linear) waves, he reasonably models the frictional force as proportional to the horizontal velocity. Depending on the problem studied, solutions display either spatial or temporal exponential decay. Surprisingly, he leaves till last the simplest problem, the temporal damping of a uniform wavetrain.

In his *Section V. Account of Experiments on Waves*, Airy describes the experiments of the Webers, and of Russell & Robison (1837). Though he describes Russell's experiments as "upon the whole, the most important body of experimental information in regard to the motion of Waves which we possess", he is critical of "Mr. Russell's references to theory" as giving "a most erroneous notion of the extent of the Theory of Waves at the date of these experiments". Airy was unconvinced by the importance attached by Russell to "The great primary wave" or solitary wave, writing that

> We are not disposed to recognize this wave as deserving the epithets "great" or "primary"... and we conceive that, ever since it was known that the theory of shallow waves of great length was contained in the equation $\frac{d^2X}{dt^2} = gk\frac{d^2X}{dx^2}$,... the theory of the solitary wave has been perfectly well known.

Like Airy, Kelland (1840, 1844) and Earnshaw (1847) believed that, to describe a solitary wave, it was acceptable to use functions defined on a finite interval, and joined to a flat surface at the ends. But, unlike Airy, both Kelland and Earnshaw considered the wave to be nonlinear, with a wave speed that differed somewhat from linear theory. In contrast, Airy categorically stated that:

> provided it be long in proportion to the depth of the fluid... [the wave] can, when moving freely, have no other velocity than $\sqrt{gk}$... [but] Mr. Russell was not aware of the influence of the length of the wave in any case and therefore has not given it...

Russell was rightly displeased by these remarks; and, in his 1844 'Report on Waves', he expresses disappointment with Airy's recent "elaborate paper on waves". Airy's formula for the velocity of small-amplitude waves has a form "closely resembling that which Mr. Kelland had previously obtained"; and so Airy had "advanced in this direction little beyond his predecessor". He also takes issue with Airy's claim of good agreement with his [Russell's] experiments; and he unfavourably contrasts Airy's boldness with Kelland's modesty, the latter having

> not yielded to the temptation of twisting his theory to exhibit some apparent approximation to the facts, nor distorted the facts to make them appear to serve the theory, a proceeding not without precedent. (Russell 1844, p.334.)

In some respects, both Green and Kelland have legitimate claims to priority over Airy. Green's short paper "On the motion of waves in a variable canal of small depth and width" (1838), though restricted to long linear waves in shallow water, gives an exemplary analysis of the effects of slow variations. Soon after, Green published another "Note on the motion of waves in canals" (1839) Interested by Russell's "Great Primary Wave", he calculates the horizontal displacement of particles by the passage of a localized wave of elevation (with a similar result, of opposite sign, for a wave of depression). He then considers waves in a triangular channel with one side vertical, the other at an arbitrary angle, and finds that long waves of small amplitude have a velocity of propagation given by $\sqrt{(gc/2)}$ where c is the maximum liquid depth. This he compares with Russell's measured data, and finds good agreement in those cases where "the elevation above the surface of equilibrium is very small compared with the depth c". Green concludes his note with a demonstration of the now familiar, but till then unproven, result that particle paths in the presence of deep-water waves are circles, with radii which decrease exponentially with depth. With arbitrary fixed depth, the corresponding elliptical paths were deduced soon afterwards by Airy.

The clarity of Green's exposition is reminiscent of that of Stokes, who admired Green's work. Stokes gave prominence to Green's two papers in his 1846 'Report on Hydrodynamics' to the British Association. In autobiographical notes, Stokes recalls that, after graduation,

> I thought I would try my hand at original research; and, following a suggestion made to me by Mr Hopkins while reading for my degree, I took up the subject of Hydrodynamics, then at rather a low ebb in the general reading of the place, notwithstanding that George Green, who had done admirable work in this and other departments, was resident in the University till he died. (Larmor 1907, 1, *Memoir* p.8)

The work of Philip Kelland on waves is more difficult to assess. In addition to his brief report on waves to the British Association, when it met in Glasgow in 1840, he published a very long two-part paper "On the Theory of Waves" in the *Transactions of the Royal Society of Edinburgh* (1840, 1844). These papers are needlessly long-winded and contain an infuriating mixture of good and bad. But he did obtain original results, and displayed some sound intuition. Kelland begins by considering long waves in shallow water [1840, pp.501-507], eventually establishing that $c^2 = gh$ after some algebraic rambling. Here, h is the depth, c the wave speed, and g is gravitational acceleration. He next tackles waves in fluid of arbitrary depth, without assuming irrotationality but making assumptions which are in fact equivalent. Unaware that his assumed form of solution is incompatible with the *nonlinear* equations, he dubiously obtains the wave speed c as [p.513]

$$\frac{2\pi}{\lambda}c^2 = g.\frac{e^{\alpha h} - e^{-\alpha h}}{e^{\alpha h} + e^{-\alpha h}}.\frac{1}{1 - (2\pi a/\lambda)^2(e^{\alpha h} - e^{-\alpha h})^2}, \tag{3.2}$$

where a denotes the wave amplitude and $\alpha \equiv 2\pi/\lambda$ is the wavenumber. Setting $a = 0$ recovers the correct linear dispersion relation for infinitesimal waves; and it is not hard to confirm that Kelland's analysis can be justified (though it remains cumbersome) in this limit. In effect, this is the first statement of the linear dispersion relation, apart from that hidden in Poisson's long article, for it anticipates Airy's article by a year.

More of a surprise is the fact that the above result, if expanded in powers of the amplitude, yields the correct modification at $O(a^2)$ to the wave speed and frequency of deep-water waves with small but finite amplitude. That is to say, Kelland's formula correctly gives what is now commonly called the 'Stokes frequency correction', about eight years before it was derived by Stokes (1847a).

In this, Kelland was fortunate – or would have been, had he received any credit for the discovery!

Kelland next adopts a more ambitious plan, representing velocity components and vertical surface displacement as a sum of harmonics, such as

$$z = h + a_1 e^{az} \sin\theta + a_2 e^{2az} \sin 2\theta + \&\text{c.}$$
$$+ f_1 e^{-az} \sin\theta + f_2 e^{-2az} \sin 2\theta + \&\text{c.}, \qquad \theta \equiv (2\pi/\lambda)(ct - x).$$

He substitutes these expressions into the surface boundary conditions, obtaining complicated expressions representing quadratic and cubic interactions, since he allows the amplitudes of each harmonic to be of the same order of magnitude. Not surprisingly, the analysis gets bogged down and reaches no firm conclusions [pp.514-523]. Despite this failure, Kelland deserves credit for attempting what I believe to be the *first* study of finite-amplitude waves to employ an amplitude-expansion technique involving a sum of harmonics. Airy's slightly later attempt was restricted to waves in shallow water; and the definitive use of amplitude expansions appeared in Stokes (1847a).

Kelland's work is mainly remembered for his study of waves in canals with non-rectangular cross sections. For long waves of small amplitude in canals with triangular cross sections, he correctly derives the result that the wave speed is $\surd(gh/2)$ where h is the maximum depth. Though this result is also given in Green's 1839 paper, Kelland's work was independent: he, like Green, provided favourable numerical comparisons with Russell's experiments. Kelland's paper was read to the Royal Society of Edinburgh on 1 April 1839; Green's to the Cambridge Philosophical Society, just a few weeks earlier on 18 February 1839.

Kelland goes further than Green in considering long waves in canals "of any shape whatever to the vertical section" [pp.527-531], deducing that the phase speed c satisfies

$$c^2 = g.\frac{\text{area of vertical section}}{\text{breadth at surface}}. \tag{3.3}$$

But Kelland's attempt to study canals with slowly-varying breadth is wrong, where Green's is correct.

Kelland next addresses "solitary wave motion": apart from Green's brief note of 1839, this is the first theoretical attempt to model the remarkable new observations by Russell. Unlike Green, who wisely confined attention to long waves in shallow water, Kelland attempts to describe wave motion in arbitrary depth h. His analysis purports to be nonlinear, for any wave amplitude; but he assumes an incompatible form of solution. Some ill-founded working leads him to a formula for the wave velocity [p.541], reproduced by Russell (1844) when unfavourably comparing Airy's claims with Kelland's caution. Kelland warns that the associated motion is discontinuous at the ends of the finite interval occupied by the wave of elevation.

Kelland (1844) revisits many of the same topics of his 1840 paper, with minor variations and much the same mix of right and wrong. His only significant new solution is that for linear waves in a channel of arbitrary depth and triangular cross section, with one or both walls inclined at 45° to the vertical. He also addresses the Cauchy-Poisson initial-value problem, but without obtaining any new correct results.

The representation of a solitary wave by discontinuous expressions was taken up again by Samuel Earnshaw (1847). He, too, attempted a nonlinear theory, and it fails for similar reasons: he merely used a different invalid hypothesis. Eventually, Rayleigh (1876) derived the correct approximate solution, retaining both dispersion and nonlinearity. Rayleigh learned of Boussinesq's (1871) earlier version after obtaining his own; the now-famous paper by Korteweg & de Vries (1895) appeared only much later. An account of the later history of solitary waves, and a detailed description of Russell's experiments, are given by R. K. Bullough (1988) and R.K. Bullough & P.J. Caudrey (1995).

4. Stokes' early papers on waves

Stokes' first publication to address water waves was his 1846 'Report on Recent Researches in Hydrodynamics' to the British Association for the Advancement of Science (Stokes 1846). Though containing few mathematical details, this report was a great advance on the previous ones by Challis; in it, he devoted 15 pages, roughly half of the Report, to the *Theory of waves, including tides.* He mentions the "extremely difficult" work of Poisson and Cauchy, a "complicated" paper by Ostrogradsky (1832) on wave motion within a cylindrical basin, then the work on long waves by Lagrange, Green, Kelland and Airy.

On oscillatory waves in fluid of arbitrary constant depth, he cites the priority of Cauchy and Poisson in deriving the relation between frequency and wavelength for standing waves; he mentions that the full theory is given by Airy; and that the particle paths were deduced by Green and Airy. Next, he discusses Kelland's solution for waves in a canal having one or both sides inclined at 45° to the vertical, and Russell's related experiments. He notes that a limiting case of Kelland's result gives "waves propagated in deep water along the edge of a bank having a slope of 45°" [*Papers* 1, p.167]. There follows $2\frac{1}{2}$ pp. on the *Theory of Solitary Waves*, about Russell's experiments, and Green's, Airy's and Earnshaw's (but not Kelland's) theories of it. There, he cites Airy's formula for the speed of nonlinear wave crests, mentioned above. But Stokes was cautiously aware that the theory of the solitary wave was still unsatisfactory:

> It is the opinion of Mr Russell that the solitary wave is a phenomenon *sui generis*, in nowise deriving its character from the circumstances of the generation of the wave. His experiments seem to render this conclusion probable. Should it be correct, the analytical character of the solitary wave remains to be discovered.

The section concludes with a six-page review of Airy's work on the *Theory of River and Ocean Tides.*

Stokes was by then abreast of all recent developments in water-wave theory. Three years later (Stokes 1849), he gave a clear but unoriginal mathematical account of linear water-wave theory. This is *Note VI: On Waves*, one of the didactic *Notes on Hydrodynamics* which he and Thomson published in the *Cambridge and Dublin Mathematical Journal.* In contrast, Stokes' (1847a) paper 'On the Theory of Oscillatory Waves' is one of the classics of fluid mechanics, remarkable both for its clarity and its originality. Here, for the first time, we find a convincing treatment of waves of finite amplitude.†

After formulating the general nonlinear equations in two space dimensions, Stokes deduces "to a first approximation" the linear theory for waves in constant depth h. Then he carefully derives the exact nonlinear free-surface boundary conditions involving the velocity potential ϕ and the surface elevation y. These he rearranges as an equation involving ϕ-derivatives alone, evaluated at the free surface, and another for y as a function of ϕ-derivatives [his (7)-(8), p.201].

Then he supposes the motion is small, so that these equations may be approximated. Though he does not explicitly employ the amplitude a as a small parameter, this is implicit, his nonlinear results (9), (10) being correct to $O(a^2)$. Stokes observes that, when the liquid depth is infinite, his approximate results at this order simplify greatly to [p.206]

$$\phi = -ace^{-my}\sin mx, \qquad y = a\cos mx - (ma^2/2)\cos 2mx \qquad (4.1a, b)$$

where x is short for $x - ct$ and the free surface is $y = a\cos x$ at leading order. In other words, the correction to the velocity potential is zero, but the wave profile has a second-harmonic component which sharpens the crests and flattens the troughs. Stokes shows a figure of this profile, for the amplitude $a = 7\lambda/80$, where $\lambda = 2\pi/m$ is the wavelength [p.212]. Similarly, he obtains the corresponding simplification for long waves with mh is very small [p.210].

He then calculates the motion of individual fluid particles situated at $x + \xi, y + \eta$ where ξ, η are variable quantities depending on the motion. For infinite depth, the positions are

$$\xi = ae^{-my}\sin m(x - ct) + m^2a^2cte^{-2my}, \qquad \eta = ae^{-my}\cos m(x - ct). \quad (4.2)$$

> Hence the motion of the particles is the same as to a first approximation, with one important difference, which is that in addition to the motion of the oscillation the particles are transferred forwards... in the direction of propagation... [p.207].

This is Stokes' discovery of what is now called the 'Stokes drift', represented by the term in a^2t. He then extends his analysis to any constant depth h. Stokes' next section (sect. 10, pp.208-9) speculates on the likely importance of this result for navigation at sea. A ship's hull extending to a certain depth below the free surface would experience a sort of averaged Stokes drift, taken over that depth, which might significantly affect its actual velocity.

Stokes meticulously draws attention to the fact that his findings are, "at first sight, at variance with the results obtained by Mr Airy for the case of long waves": e.g. (3.3) above, where Airy's second harmonic term exhibits secular

† This paper is also summarised by Bullough (1988) and Bullough & Caudrey (1995).

growth. Impressively, he gives the correct explanation, that in his theory $a/h \ll (h/\lambda)^2$, whereas Airy's requires $a/h \gg (h/\lambda)^2$. "Thus the difference in the results obtained corresponds to a difference in the physical circumstances of the motion" [p.201].

Stokes begins his next section 12 with the startling claim: "[t]here is no difficulty in proceeding to the higher orders of approximation, except what arises from the length of the formulae." To illustrate the method, he proceeds "to the third order in the case of infinite depth, so as to find... the most important term, depending on the height of the waves, in the expression for the velocity of propagation." Retaining cubic terms in his exact boundary condition, he quickly arrives at

$$c = \left(\frac{g}{m}\right)^{1/2} \left(1 + \tfrac{1}{2}m^2a^2\right), \tag{4.3a}$$

$$y = a\cos mx - \tfrac{1}{2}ma^2\cos 2mx + \tfrac{3}{8}m^2a^3\cos 3mx \tag{4.3b}$$

for the wave velocity and surface elevation. The former gives the 'Stokes nonlinear frequency shift', in agreement with Kelland's dubiously-derived result mentioned above; and the latter is a yet closer approximation to the free surface than (4.1b).

Stokes concludes his paper with an examination of linear waves on an interface between two liquids or between liquid and air. New Appendices and a Supplement to this paper were later added in Stokes' *Mathematical & Physical Papers*, as discussed below.

Stokes made just one error in his paper, but a serious one. After describing the linear theory of wave motion as a superposition of sinusoidal modes, he wrongly claimed that this demonstrated that a solitary wave cannot be propagated without change of form, and so that the gradual "degradation" of Russell's solitary waves was "an essential characteristic".

Only much later, after Boussinesq (1871) and Rayleigh (1876) had found their approximate solution for a solitary wave, did Stokes change his view. He suggested (Stokes 1883) that his amplitude expansion failed to converge for indefinitely long waves in shallow water, "and is not therefore applicable to solitary waves". But Korteweg & de Vries (1895) eventually showed that Stokes' amplitude expansion *does* correctly represent the first few Fourier components of their periodic 'cnoidal wave' solutions represented by Jacobian elliptic functions; and these cnoidal waves approach the solitary wave solution as the periodic length increases towards infinity (see also Bullough 1988). Stokes (1891) at last came nearer to the true explanation, that there is "a relation between the height and the length in a solitary wave which can be propagated uniformly". It is unfortunate that Stokes' high reputation contributed to the delay in recognizing the importance of Russell's solitary waves.

Surprisingly, since he had himself derived the equations of viscous flow, Stokes devoted little attention to wave damping by viscosity. A brief calculation in (Stokes 1856) gives the dissipation associated with irrotational plane waves; but he did not determine the significant contributions from oscillatory viscous boundary layers – now called 'Stokes layers' – near the bottom and free surface.

A very full treatment of viscous damping of linear waves was later given by Bassett (1888).

His paper (1847a) established the twenty-seven year-old Stokes as a skilled analyst, master of rational approximations allied to sound physical insight, able to proceed systematically towards his goal unhindered by technical difficulties. His clear analytical style is reminiscent of Green's more limited but elegant papers on shallow-water waves. There is none of the algebraic clumsiness and unexplained hypotheses of Airy, nor the dubious and long-winded manipulations of Kelland. Yet Stokes owed a debt to these predecessors, and to the experimentalist Russell, who had set the stage for him. He was in the right place at the right time – and the right man in that place.

5. STOKES' LATER WORK ON WAVES

Engrossed in other matters, Stokes dropped his researches on water waves, and published nothing more on them until he prepared for publication the first volume of his collected *Mathematical & Physical Papers* (1880). But correspondence shows that he retained some interest in the topic despite his 30-year silence.

In 1862 he corresponded with William Thomson about Earnshaw's solitary-wave analysis and a paper by W.J. Macquorn Rankine (1863) recently submitted to *Phil. Trans. Roy. Soc.* for which Stokes was a referee. Rankine's exact nonlinear wave solution, obtained geometrically, turned out to be identical to Gerstner's wave of 1802. In an Appendix added to his 1847 paper (*Papers*, 1 (1880) pp.219-225) Stokes examines this solution, observing that irrotational waves, not Gerstner's rotational waves, are most likely to occur naturally; and that irrotational waves exhibit particle ('Stokes') drift while Gerstner's do not.

From 1866 to 1887, Stokes was a member of the Meteorological Council, for which he prepared several memoranda. In one of these, he examined records of wave trains at sea, deducing information about the distant storm which had originated them: see Stokes' letters to Sir Edward Sabine, a Mr Melsens of St Helena, and Capt. A.H. Toynbee (Larmor 1907, 2, 133-158). These early exercises in 'remote sensing' show Stokes' interest in usefully applying simple wave theory, and his awareness of the occurrence of groups of waves. Around this time, he also corresponded with Airy about seiches (i.e. standing-wave oscillations in closed basins) (Larmor 1907, 2, 178-185).

Though Stokes' interest in water waves may have been re-awakened by his work for the Meteorological Council, his regular family holidays on the north coast of Ireland also played their part. They normally stayed in accommodation attached to Armagh Observatory, where his father-in-law Dr Thomas Romney Robinson was Astronomer. Around 1876, Stokes there studied groups of waves (Larmor 1907, 1, pp.31, 337-8). In the same year, he set a Smith's Prize Examination question at Cambridge (*Stokes*' Papers, 5, p.362), which is credited as the first printed mention of the group velocity of dispersive waves from a theoretical standpoint. In retrospect, it seems surprising that Stokes and everyone else so long failed to notice this simple consequence of linear wave theory, developed 35 years previously. A year later, Osborne Reynolds (1877) showed that the group velocity is also the velocity of propagation of wave energy; and Rayleigh's (1878) 'Note on Progressive Waves' gave a comprehensive account of group velocity, including that of capillary waves first studied by Thomson (1871).

Stokes wrote three short Appendices and a 12-page Supplement (*Papers* 1, 1880) to his 1847 paper on 'The Theory of Oscillatory Waves'. The first, on the Gerstner-Rankine solution [pp. 219-225] was mentioned above. The second concerns the periodic irrotational wave of greatest height, showing that Rankine was wrong in claiming that this has a 90° angle at the crest. By a beautifully simple argument, Stokes shows that the angle must be 120° if a steady wave exists [pp. 225-228]. The third Appendix, a brief remark about using a stream-function rather than a velocity potential, breaks off with: "... while these sheets were going through the press I devised a totally different method... which I find possesses very substantial advantages".

This new method is described in the Supplement (*Papers* 1, 314-326). Stokes introduces the two functions ϕ and ψ (the velocity potential and stream function, but not so named); but he then recasts the equations for steady nonlinear wave propagation no longer in terms of equations for $\phi(x,y)$ and $\psi(x,y)$, but as equations for $x(\phi,\psi)$ and $y(\phi,\psi)$. Since the motion is irrotational, x and y each satisfy Lapace's equation with ϕ,ψ as coordinates, as well as the converse; and the nonlinear free-surface boundary conditions now take more convenient form. After some re-scaling and restriction to deep-water waves, x and y may be written as the expansions [p.316]

$$x = -\phi + \sum_{1}^{\infty} A_i e^{i\psi} \sin i\phi, \qquad y = -\psi + \sum_{1}^{\infty} A_i e^{i\psi} \cos i\phi. \tag{5.1}$$

The free-surface condition yields a set of relations between the undetermined coefficients A_i, B_i , and truncation of this set to the first $i+1$ relations yields an approximation to i'th-order. In this way, Stokes derives expressions for x and y which include five harmonics in $\sin n\phi$ or $\cos n\phi$ ($n = 1$ to 5). Then to 4th order only, he deduces from these the corresponding surface elevation [p.319]

$$y = a\cos mx - \left(\tfrac{1}{2}ma^2 + \tfrac{17}{24}m^3a^4\right)\cos 2mx + \tfrac{3}{8}m^2a^3\cos 3mx - \tfrac{1}{3}m^3a^4\cos 4mx \ldots.. \tag{5.2}$$

This recovers his 1847 result (4.3b), along with the new $O(a^4)$ terms of the expansion.

A similar analysis of waves in water of constant depth is pursued up to third order; but he does not derive the corresponding free-surface relation $y(x)$, remarking that "it must be allowed that the approximation is slower in the case of a finite depth". Stokes was hopeful that it would be possible to extend this method to treat waves of large amplitude, but he admitted that "[t]here can be little doubt that... the series cease to be convergent when the limiting wave, presenting an edge of 120°, is reached." This ingenious inversion of dependent and independent variables has been much employed in more recent times.

In 1879-80, Stokes exchanged more letters with Thomson about his reformulation (5.1), and about the wave of greatest height (Wilson 1990, 2, pp.462-510).

Two later papers, one published posthumously, return to the formulation (5.1): Stokes (1883) and *Stokes'Papers* 5, 146-158. Stokes' remaining contributions on waves are slight. A 3-page paper in *Philos. Mag.* (1891) responded rather testily to a criticism made by J. McCowan (1891), who had taken issue with an old assertion by Stokes (1847a) that a solitary wave could not propagate without change in form. And a single paragraph, concerning the "outskirts of the solitary wave", quoted in Lamb's *Hydrodynamics* (1895, p.421), is reproduced in *Papers* 5, p.163.

By this time, Britain had a remarkable wealth of first-rate talent working on fluid flows and wave phenomena of various kinds: William Thomson, Rayleigh, Clerk Maxwell, Reynolds, Froude, Rankine, Lamb, Bassett, G.H. Darwin, J.J. Thomson, W.M. Hicks, A.E.H. Love, A.G. Greenhill, W. Burnside, J. McCowan, J.H. Michell. Overseas, Boussinesq, St Venant, Helmholtz, Korteweg, Kirchhoff, Beltrami, Joukowsky and Poincaré were making the greatest contributions.† The foundations laid by Stokes and his precursors now supported an impressive edifice.

REFERENCES

(Excluding some works to which only passing reference is made)

AIRY, G.B. 1841 (a) art. Tides & Waves. *Encyclopaedia Metropolitana* (1817-1845), eds. H.J. Rose etc. "Mixed Sciences" **3**. Also (b) Airy, George Biddle [sic], *Trigonometry, On the Figure of the Earth, Tides and Waves* (articles from Encyclopaedia Metropolitana) 396pp. + Plates; n.d., n.p.

AIRY, G.B. 1896 *Autobiography of Sir George Biddell Airy K.C.B....* ed. Wilfrid Airy, Cambridge: Cambridge Univ. Press.

BASSET, A.B. 1888 *A Treatise on Hydrodynamics* 2v. Cambridge: Deighton, Bell & Co. Also Dover reprint, 1961.

BESANT, W.H. 1877 *A Treatise on Hydromechanics.* Cambridge: Deighton, Bell & Co.

BOUSSINESQ, J. 1871 Théorie de l'intumescence liquide appelée *onde solitaire* ou *de translation*, se propageant dans un canal rectangulaire. *Paris Acad. Sci. Comptes Rendus*, **72**, 755-759.

BULLOUGH, R.K. 1988 The Wave "par excellence", the solitary, progressive great wave of equilibrium of the fluid - an early history of the solitary wave. In *Solitons*, ed. M. Lakshmanan, Springer Series in Nonlinear Dynamics, 150-281. Springer: New York etc.

BULLOUGH, R.K. & Caudrey, P.J. 1995 Solitons and the Korteweg-de Vries Equation: Integrable Systems in 1834-1995. *Acta Applicandae Mathematicae*, **39**, 193-228.

CAMPBELL, N. & SMELLIE, R.M.S. 1983 *The Royal Society of Edinburgh* (1783-1983), *the first two hundred years.* Edinburgh: The Royal Soc. of Edinburgh.

CAUCHY, A.-L. 1827 Mémoire sur la théorie de la propagation des ondes à la surface d'un fluide pesant d'une profondeur indéfinie. *Mémoires présentés par divers Savans à l'Académie Royale des Sciences de l'Institut de France* (*Prix de l'Académie Royale des Sciences, concours de 1815 et de 1816*), **I**, 3-312.

CHALLIS, J. 1833 Report on the present State of the analytical Theory of Hydrostatics & Hydrodynamics. *Rep. Brit. Assoc. for the Advancement of Sci*, 131-151.

CHALLIS, J. 1836 Supplementary Report on the Mathematical Theory of Fluids. *Rep. Brit. Assoc. for the Advancement of Sci*, 225-252.

† A comprehensive review of hydrodynamics up to 1912 was given by Love (1912).

CHALLIS, J. 1851 (a) On the Principles of Hydrodynamics. *Philos. Mag.*, **1** (ser. 4), 26-38. (b) On the Principles of Hydrodynamics, with a Reply to the Arguments of Prof. Stokes. *ibid.* 231-241. (c) Further Discussion of the Principles of Hydrodynamics, in Reply to Prof. Stokes. *ibid.* 477-478.

DALMEDICO, A.D. 1988 La Propagation des Ondes en Eau Profonde et ses Développements Mathématiques (Poisson, Cauchy 1815-1825). *In The History of Modern Mathematics*, Vol. II, pp. 129-168, eds. David E. Rowe & John McCleary. London: Academic.

EARNSHAW, S. 1847 The Mathematical Theory of the two great Solitary Waves of the First Order. *Trans. Camb. Phil. Soc.* **8**, 326-341 (read Dec. 8, 1845).

EMMERSON, G.S. 1977 *John Scott Russell: a Great Victorian Engineer and Naval Architect.* London: John Murray.

EULER, L. 1757a Principes Géneraux du Mouvement des Fluides. *Mémoires de l'Académie des Sciences de Berlin,* **11** (1755) 271-315. Also in *Leonhardi Euleri Opera Omnia* Ser.2, XII (1954) ed. C.A. Truesdell. Lausanne: Orell Füssli.

EULER, L. 1757b Continuation des Recherches sur la Théorie du Mouvement des Fluides. *Mém. de l'Acad. des Sci. de Berlin,* **11** (1755) 316-361. Also in *Op. Omn. loc. cit.*

EULER, L. 1761 Principia Motus Fluidorum. *Novi Commentarii Acad. Sci. Petropolitanae,* **6** (1756/7) 271-311. Also in *Op. Omn. loc. cit.*

FERRERS N.M. (ed.) 1871 *Mathematical Papers of the late George Green.* London: Macmillan and Co.

GERSTNER, F.J. VON. 1802 Theorie der Wellen. *Abhand. d. kön. Böhmischen Gesel. d. Wiss.*, Prague. Also reprinted in Weber (1825).

GREEN, G. 1838 On the motion of waves in a variable canal of small depth and width. *Trans. Camb. Philos. Soc*, **6**, 457-462 (read May 15, 1837). Also in Ferrers (1871) pp. 223-230.

GREEN, G. 1839 Note on the motion of waves in canals. *Trans. Camb. Philos. Soc.*, **7**, 87-96 (read Feb. 18, 1839). Also in Ferrers (1871), pp. 271-280.

GREENHILL, A.G. 1887 Wave Motion in Hydrodynamics. *Amer. J. Math.*, **9**, 62-112.

HARMAN, P.M. (ed.) *Wranglers and Physicists; Studies on Cambridge Physics in the Nineteenth century.* Manchester: Manchester Univ. Press.

HICKS, W.M. 1881-2 Report on Recent Progress in Hydrodynamics. *Report of Brit. Assoc. for Advancement of Science* (1881) 57-88; (1882) 39-70.

HOWARTH, O.J.R. 1931 *The British Association for the Advancement of Science: a Retrospect 1831-1931.* London: The British Assoc.

KELLAND, P. 1840a On the Theory of Waves. *Report of Brit. Assoc. for Advancement of Science* (1840) pt. *ii*, pp. 50-52.

KELLAND, P. 1840b On the Theory of Waves, Part 1. *Trans. Roy. Soc. Edinburgh,* **14**, 497-545 (read April 1839).

KELLAND, P. 1844 On the Theory of Waves, Part 2. *Trans. Roy. Soc. Edinburgh,* **15**, 101-144 (read Jan. 1841).

KORTEWEG, D.J. & DE VRIES, G. 1895 On the change of form of long waves advancing in a rectangular canal, and on a new type of long stationary waves. *Philos. Mag.* (5) **39**, 422-443.

LAGRANGE, J.-L. 1781 Mémoire sur la Théorie du mouvement des Fluides. *Nouv. Mém. de l'Acad. de Berlin,* année 1781, pp.196-. Also in *Oeuvres de Lagrange,* **4**, 695-748, Paris: Gauthier-Villars 1889.

LAGRANGE, J.-L. 1786 Sur la manière de rectifier deux entroits des Principes de Newton relatifs à la propagation du son et au mouvement des ondes. *Nouv. Mém. de l'Acad. de Berlin,* année 1786. Also in *Oeuvres de Lagrange,* **5**, 591-609, Paris: Gauthier-Villars 1889.

LAGRANGE, J.-L. 1788 *Méchanique analitique.* Paris: la Veuve Desaint. Also (3rd ed.) in *Oeuvres de Lagrange,* **12** [section on waves is pp. 318-322].

LAMB, H. 1879 *A Treatise on the Mathematical Theory of the Motion of Fluids.* Cambridge: Cambridge Univ. Press.

LAMB, H. 1895 *Hydrodynamics.* 2nd ed., (4th ed. 1916, 6th ed. 1932). Cambridge: Cambridge Univ. Press.

LAPLACE, P.-S.M. DE. 1776 Suite des Récherches sur plusieurs points du Système du monde (XXV - XXVII). *Mémoires présentés par divers Savans à l'Académie Royale des Sciences de l'Institut de France* (1776) 525-552. [Sur les Ondes pp. 542-552.]

LARMOR, J. ed. 1907 *Memoir and Scientific Correspondence of the late Sir George Gabriel Stokes...* Cambridge: Cambridge Univ. Press [Selected correspondence only, excluding that with William Thomson, Lord Kelvin].

LOVE, A.E.H. 1912 Hydrodynamique (partie élémentaire) pp. 61-101 (Sect. 17); Développements concernant l'hydrodynamique pp. 102-208 (Sect. 18). Tom. IV, vol. 5 of *Encyclopédie des Sciences Mathématiques Pure et Appliquées*, ed. J. Molk & P. Appell, Paris: Gauthier-Villars, Leipzig: E.G. Teubner. (trs. of original German edition).

MCCOWAN, J. 1891 On the solitary wave. *Philos. Mag.*, **32**, 45-58; 553-555. Also **33** (1892) 236.

MICHELL, J.H. 1893 The highest wave in water. *Philos. Mag.*, **36**, 430-437.

NEWTON, I. 1687 *Philosophiae Naturalis Principia Mathematica.* London: Jussu Societatis Regiae ac Typis J. Streater. Also 1st English ed. trans. N. Motte 1729.

OSTROGRADSKY, M.A. 1832 Mémoire sur la propagation des ondes dans un bassin cylindrique. Mémoires présentés par divers Savans à l'Académie Royale des *Sciences de l'Institut de France*, **3**, 23-44 (read 1826).

PARIS, R. 1996 The mathematical work of G.G. Stokes. *Math. Today*, **32**, (3-4), 43-46, IMA, Southend-on-Sea.

PEACOCK, G. (ed.) 1855 *Miscellaneous Works of the late Thomas Young...*, 3v. London: John Murray.

POISSON, S.D. 1818 Mémoire sur la théorie des ondes, *Mémoires de l'Académie Royale des Sciences de l'Institut de France*, année 1816, 2nd ser., **1**, 70-186.

RANKINE, W.J.M. 1863 On the exact form of waves near the surface of deep water. *Philos Trans. Roy. Soc. London* (1863) 127-138 (read 27 Nov., 1862). Also in Rankine's *Miscellaneous Scientific Papers*, pp. 481-494, ed. W.J. Millar, London: Charles Griffin & Co. 1881.

RAYLEIGH, BARON (J.W. STRUTT) 1876 On Waves. *Philos. Mag.* (5) **1**, 257-279. Also in *Scientific Papers of John William Strutt, Baron Rayleigh 1899*, **1**, 251-271. Cambridge: Cambridge Univ. Press.

RAYLEIGH, BARON (J.W. STRUTT) 1878 Note on progressive waves. In *Theory of Sound*, **2** v., 2, 297-302. London: Macmillan & Co. Also *Proc. Lond. Math. Soc.*, **9**, No. 125.

REYNOLDS, O. 1877 On the rate of progression of groups of waves and the rate at which energy is transmitted by waves. *Nature*, **16**, (Aug. 23), pp. 343-344. Also in Reynolds' *Papers on Mechanical and Physical Subjects* (3 vols.) 1900, **1**, 198-203. Cambridge: Cambridge Univ. Press.

RUSSELL, J.S. & ROBISON, J. Report on Waves. *Rep. Brit. Assoc. for the Advancement of Science* (1837) pp. 417-496; *ibid.* (1840) pp. 441-443.

RUSSELL, J.S. 1842 Supplementary Report of a Committee on Waves. *Rep. Brit. Assoc. for the Advancement of Science* (1842) pt. *ii*, pp. 19-21.

RUSSELL, J.S. 1844 Report on Waves. *Rep. Brit. Assoc. for the Advancement of Science* (1844) pp. 311-390.

STOKES, G.G. *Mathematical & Physical Papers of Sir George Gabriel Stokes*, 5 vols. (1880-1905). Cambridge: Cambridge Univ. Press [here referred to as '*Papers*': these reprint most of Stokes papers, with some additions.].

STOKES, G.G. 1842a On the Analytical Condition of the Rectilinear Motion of Fluids, with Reference to a Paper of Professor Challis. *Philos. Mag.*, **21**, 297-300. [Not in *Papers.*]

STOKES, G.G. 1845 On the Theories of the Internal Friction of Fluids in Motion, and of the Equilibrium and Motion of Elastic Solids. *Trans. Camb. Philos. Soc.*, **8**, 287-319 (read April 1845).

STOKES, G.G. 1846 Report on Recent Researches in Hydrodynamics. Rep. Brit. Assoc. for the Advancement of Sci. (1846) 1-20.

STOKES, G.G. 1847 On the Theory of Oscillatory Waves. *Trans. Camb. Philos. Soc., viii*, (read March 1847). [Appendices and Supplement added in *Papers*, **1**, 1880.]

STOKES, G.G. 1848a, On the Constitution of the Lumeniferous Ether. *Philos. Mag.*, **32**, 343-349.

STOKES, G.G. 1848b, Notes on Hydrodynamics. III On the Dynamical Equations; IV Demonstration of a Fundamental Theorem. *Camb. & Dublin Math. Jour.*, **3**, 121-127, 209-219.

STOKES, G.G. 1849 Notes on Hydrodynamics. VI On Waves. *Camb. & Dublin Math. Jour.*, **4**, 219-240.

STOKES, G.G. 1851 (a) On the alleged Necessity for a new General Equation in Hydrodynamics. *Philos. Mag.*, **1** (ser. 4), 157-160. (b) On the Principles of Hydrodynamics, in Reply to Prof. Challis. *ibid.* 393-4. (c) On the Principles of Hydrodynamics. *ibid.* **2**, 60. [Not in *Papers.*]

STOKES, G.G. 1856 On the Effect of the Internal Friction of Fluids on the Motion of Pendulums. *Trans. Camb. Philos. Soc.*, **9**, 8-106 (read Dec. 9, 1850)

STOKES, G.G. 1880 Appendices and Supplement to a Paper on the Theory of Oscillatory Waves. *Papers* **1** (1880) 219-229, 314-326.

STOKES, G.G. 1883 On the Highest Wave of Uniform Propagation (Preliminary notice.) *Proc. Camb. Philos. Soc.*, **4**, 361-365 (read Nov. 1883).

STOKES, G.G. 1891 Note on the Theory of the Solitary Wave. *Philos. Mag.*, **31**, 314-316.

STOKES, G.G. 1895 The Outskirts of the Solitary Wave. from H. Lamb's *Hydrodynamics* (1895) p. 421.

STOKES, G.G. 1905a On the Maximum Wave of Uniform Propagation. Being a second Supplement of a Paper on the Theory of Oscillatory Waves. Typed ms., publ. posthumously in *Papers* **5**, 1905.

STOKES, G.G. 1905b Mathematical Tripos and Smith's Prize Examination Papers: reprinted in *Papers* **5**, 1905.

THOMSON, W. (BARON KELVIN) 1871 Ripples and Waves. *Nature*, **5**, 1-3.

WEBER, E.H. & WILHELM, E. 1825 *Wellenlehre auf Experimente gegründet....* Leipzig: Gerhardt Fleischer.

WILSON, D.B. 1987 *Kelvin & Stokes, a Comparative Study in Victorian Physics.* Adam Hilger, Bristol.

WILSON, D.B. ed. 1990 *The Correspondence between Sir George Gabriel Stokes and Sir William Thomson Baron Kelvin of Largs.* 2 vols. Cambridge, Cambridge Univ. Press.

WOOD, A. 1995 George Gabriel Stokes 1819-1903, an Irish mathematical physicist. Irish Math. Soc. Bull., **35**, 49-58.

YOUNG, T. 1807 *A Course of Lectures on Natural Philosophy* 2v. London: J. Johnson. 1845, 2nd ed., ed. Philip Kelland, London: Taylor & Walton.

Young, T. 1821 *Elementary Illustrations of the Celestial Mechanics of Laplace* [Book 1] London: John Murray. Also in Peacock (1855, pp. 141-158), 'Some Propositions on Waves and Sound'.

Is the Logarithmic Wind Law Valid Over the Sea?

A.-S. Smedman, X.G. Larsen and U. Högström

Department of Earth Sciences, Meteorology, Uppsala university, Uppsala, Sweden

ABSTRACT

The dependence of drag on the ocean of parameters representing wave state has been studied for near-neutral conditions with the aid of an extensive data set taken from the air-sea interaction station Östergarnsholm in the Baltic Sea. The measurements include turbulence flux, slow response, 'profile', data at several levels in the lowest 30 m above the sea surface and wave data from a Waverider Buoy anchored 4 km outside the island.

During conditions with developing sea, the drag is found to depend on inverse wave age, expressed with the parameter u_*/c_p (where u_* is friction velocity and c_p is the phase speed of the dominant waves) in agreement with recent findings over the world ocean. For such conditions, it is also demonstrated that the logarithmic wind law is indeed valid.

For the case of mixed sea/swell the situation is shown to be much more complicated. Firstly, in spite of the occurrence of near-neutral conditions, the data clearly show that the logarithmic wind law is not valid any more. Secondly, the drag coefficient, C_D is found to depend on *two* parameters representing the wave state: the wave age, u_*/c_p and a wave spectral ratio E_1/E_2, where E_1 is a measure of the energy of the relatively long waves (those having a phase velocity larger than the wind speed at 10 m) and E_2 a corresponding measure of the short wave energy. Thus, plotting C_D as a function of u_*/c_p gives a clear ordering of the data in parallel, sloping bands according to the value of E_1/E_2.

A tentative interpretation of the results suggests that, whereas very young and slow waves affect the atmospheric flow similar to rigid roughness elements, with the occurrence of longer waves, an entirely different mechanism gains successively more importance. For such waves dynamical coupling with the atmospheric turbulence is bound to occur. For those cases, it may be speculated that the often observed kink in the wind profile represents the upper bound of a wave-boundary-layer, which is thus, in the general case, an order of magnitude deeper than predicted and observed during growing sea conditions.

1. INTRODUCTION

As reviewed by Komen *et al.* (1998), correct parameterization of the drag over the ocean has considerable impact not only on a synoptic scale (the development of cyclones) but also on the climate. Although considerable efforts have been spent on the issue, the problem is far from settled, different experiments reported in the literature (see below for details) giving seemingly contradicting results. In particular the effect of swell on the resulting drag is still far from understood. In this paper several years worth of concurrent measurement of turbulent flux, mean atmospheric profiles and wave data are analysed in order to obtain further understanding of how the state of the sea influences the drag. The data have been gathered at the marine site Östergarnsholm in the Baltic Sea during the years 1998 and 1999. The general criteria for selection of data have been: wind from sector with long undisturbed fetch, near-neutral conditions and completeness of the data. In all, 431 60-minute data have been used in the analysis, see Section 2.

During neutral atmospheric conditions it is generally assumed that there is a logarithmic wind profile in the lowest ten meters or more. Over a *solid surface*, this is a well-established fact, supported by innumerable measurements in the laboratory as well as in the atmospheric surface layer. Thus, for flow over an aerodynamically rough surface we get

$$U = \frac{u_*}{\kappa} \ln(z/z_0) \tag{1.1}$$

where U is mean wind speed at height z, u_* is the friction velocity $= \sqrt{\tau/\rho}$, where τ is shearing stress, and $\tau/\rho = \sqrt{(\overline{u'w'})^2 + (\overline{v'w'})^2}$ and ρ air density, κ is von Karman's constant $= 0.40$ (Högström 1996) and z_0 the roughness length. Over a solid surface, z_0 is related to the size and geometry of the roughness elements at the surface. Eq. (1.1) is usually assumed to be valid over the *ocean* as well during neutral conditions. It is, however, not self-evident that this is the case over a moving and undulating surface in dynamic interplay with the overlaying atmosphere.

Measurements of wind profiles in neutral conditions over the ocean at reliable sites are rare. Most oceanic measurements report in fact only data from one level, usually 10 m and assume that Eq. (1.1) is valid. Introducing the drag coefficient

$$C_D = \left(\frac{u_*}{U_{10}}\right)^2 \tag{1.2}$$

assuming neutral conditions and that Eq. (1.1) is valid, gives

$$C_{DN} = \left(\frac{\kappa}{\ln(10/z_0)}\right)^2 \tag{1.3}$$

i.e. a unique relation between the neutral drag coefficient C_{DN} and the roughness length z_0. Note, however, that if the 'wave-boundary layer' should extend to an appreciable height - as suggested by the results of Hare *et al.* (1997) - the logarithmic law may be invalid, and then a drag coefficient determined from concurrent measurements of u_* and U_{10} are not likely to give physically meaningful values for z_0 with the aid of Eq. (1.3).

2. Site and measurements

The measurements analysed in this paper have been made at the Östergarnsholm station for air/sea interaction research, Figure 1. It consists of a an instrumented 30 m high tower situated at the southernmost peninsula of the low island Östergarnsholm and measurements with a 3D Waverider buoy moored

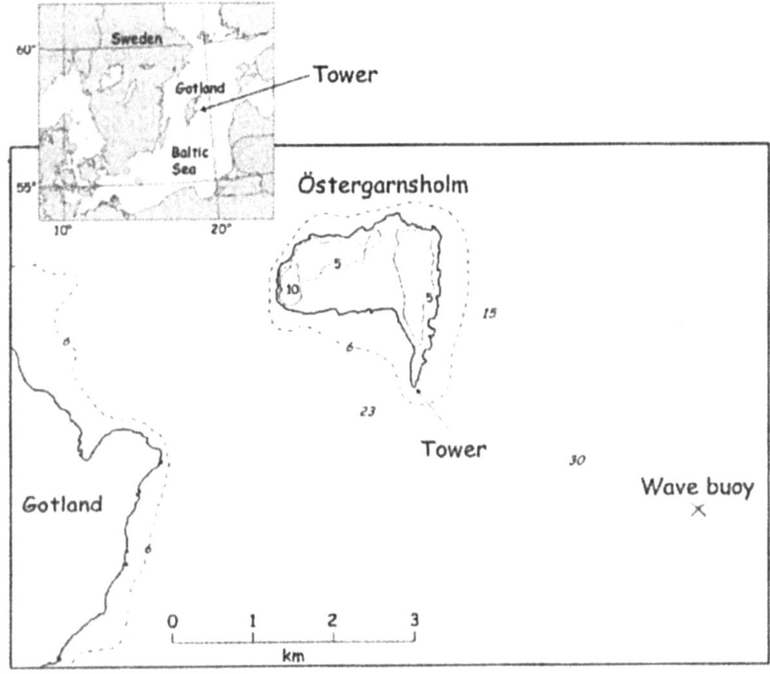

FIGURE 1. Map of the measuring site.

at 36 m depth about 4 km to the south-east of the tower. The instruments on the tower include eddy correlation measurements with Solent 1012R2 sonic anemometer at 8, 16 and 24 m above the ground and slow response, 'profile' sensors for wind speed and direction (light-weight cup anemometers and styrofoam wind vanes) and temperature at 5 levels. The base of the tower is situated 1 - 2 m above the sea level, the actual level changing over time as a result of the prevailing wind conditions over the Baltic Sea. Actual measuring heights above the sea level are derived from a sea level record at Visby, situated on the west coast of Gotland.

The sonic anemometers have been calibrated in a big wind tunnel, resulting in individual flow distortion correction matrices (cf. Grelle & Lindroth 1994). Also the cup anemometers have been individually calibrated in this big wind tunnel.

Wave data is recorded once an hour. The directional spectrum is calculated from 1600 s of data onboard the buoy. The spectrum has 64 frequency bands (0.025-0.58 Hz). The significant wave height is calculated by trapezoid method from frequency bands 0.05-0.58 Hz, and the peak frequency is determined by a parabolic fit.

For winds coming from the sector 80 - 220 degrees, there is undisturbed over water fetch for more than 150 km, and only data with this wind direction have been used here. About 10 km from the peninsula, the depth is 50 m, reaching below 100 m farther out. In Smedman *et al.* (1999) the possible influence of limited water depth on the tower measurements was studied in detail. Flux footprint calculations were done, showing that the turbulence instruments "see" areas far upstream of the island. This means that the measurements chosen for the present analysis are likely to be representative of open ocean conditions.

The 431 data of hourly means chosen for the analysis are from the years 1998 and 1999 and have been selected according to the following criteria:

- Wind coming from the sector with long fetch, 80 - 220 degrees
- Complete meteorological data and wave data
- Wave spectra with a single peak
- Near-neutral conditions. As it is earlier found that both u_* and the wind gradient (see below) may be influenced by swell (Smedman *et al.* 1994, Rutgersson *et al.* 2001) only the heat flux is used to describe near neutral stratification. Thus for unstable conditions: $0 < \overline{w'\theta'} < 0.01\text{ms}^{-1}\text{K}$, and for stable conditions: $0 > \overline{w'\theta} > -0.002\text{ms}^{-1}\text{K}$.
- The angle between the dominant waves and the wind $< 40°$
- Mean wind speed $> 2\text{ms}^{-1}$.

3. Criteria for characterising sea state

On dimensional grounds, Charnock (1955) derived the following expression for the roughness of the sea

$$z_0 = \alpha u_*^2/g \tag{3.1}$$

where α is the so-called Charnock parameter or dimensionless roughness and g is acceleration of gravity. Eq. (3.1) is often used in large-scale synoptic and climatic models with a constant value for α, typically in the range 0.01 - 0.03. For *pure wind seas*, several investigations have, however shown that α is a function of wave age, defined as

$$c_p/u_* \tag{3.2}$$

where c_p is the phase speed of the waves at the peak of the spectrum.

Drennan *et al.* (2000) analyse the directional wave spectra in order to make the decomposition into *E(wind sea)* and *E(swell)*. Without introducing a model, the Waverider data cannot be used to obtain directional wave spectra. Instead, we make a division of our one-dimensional spectra into two parts:

$$E_1 = \int_0^{n_1} S(n)dn, \tag{3.3}$$

$$E_2 = \int_{n_1}^{\infty} S(n)dn, \qquad n_1 = \frac{g}{2\pi U_{10}\cos\theta} \tag{3.4}$$

where n is frequency, $S(n)$ is the one-dimensional wave spectrum and θ is the angle between the waves at peak frequency and the wind at 10 m. Thus, n_1 is the frequency which corresponds to a phase speed c (derived with the deep-water linear dispersion relation) equal to the wind speed at 10 m, or more precisely, the component of the wind in the direction of the waves, $U_c =$

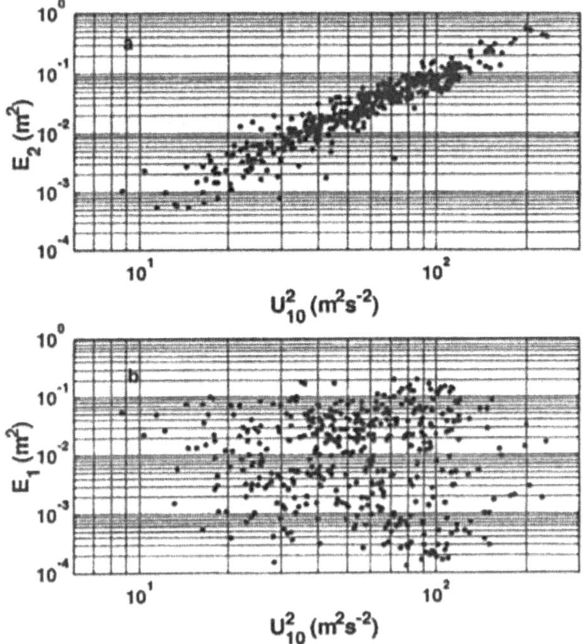

FIGURE 2. Wind sea part of the wave spectrum (E_2) plotted as a function of wind speed (a) and the low frequency part of the wave spectrum (E_1) plotted as a function of wind speed (b).

$U_{10}\cos\theta$. This separation thus results in a long-wave spectral part, E_1 and a short-wave spectral part E_2.

Figure 2 shows the relation between U_{10}^2 and E_2, Figure 2a, and U_{10}^2 and E_1, Figure 3b, respectively. The difference between the two plots is striking, the short wave part E_2, being a strong function of the square of the wind speed - in fact there is a linear relation for $U_{10}^2 > 10$ - whereas there is no relation at all between the long-wave part E_1 and wind speed. From this analysis it appears reasonable to refer to E_2 as *the wind sea part* of the spectrum. When swell is present it will, no doubt, appear in the low-frequency part E_1 of the spectrum, but this part may also contain relatively long waves resulting from a saturated wave spectrum. Figure 2 illustrates that the method used here (as a surrogate for an analysis of true directional spectra) for dividing the one-dimensional spectra into a 'wind-speed-independent part' E_1 and a 'wind sea part', E_2, is likely to give a reasonably correct result.

4. Analysis of the Neutral Wind Profile

During neutral conditions we expect the logarithmic wind law, Eq. (1.1) to be valid, implying that, in principle, we would get the same value for the roughness length z_0 from measurements of wind speed at *two* levels or more and from measurements of and wind speed at *one* level. This idea is tested in Figure 3,

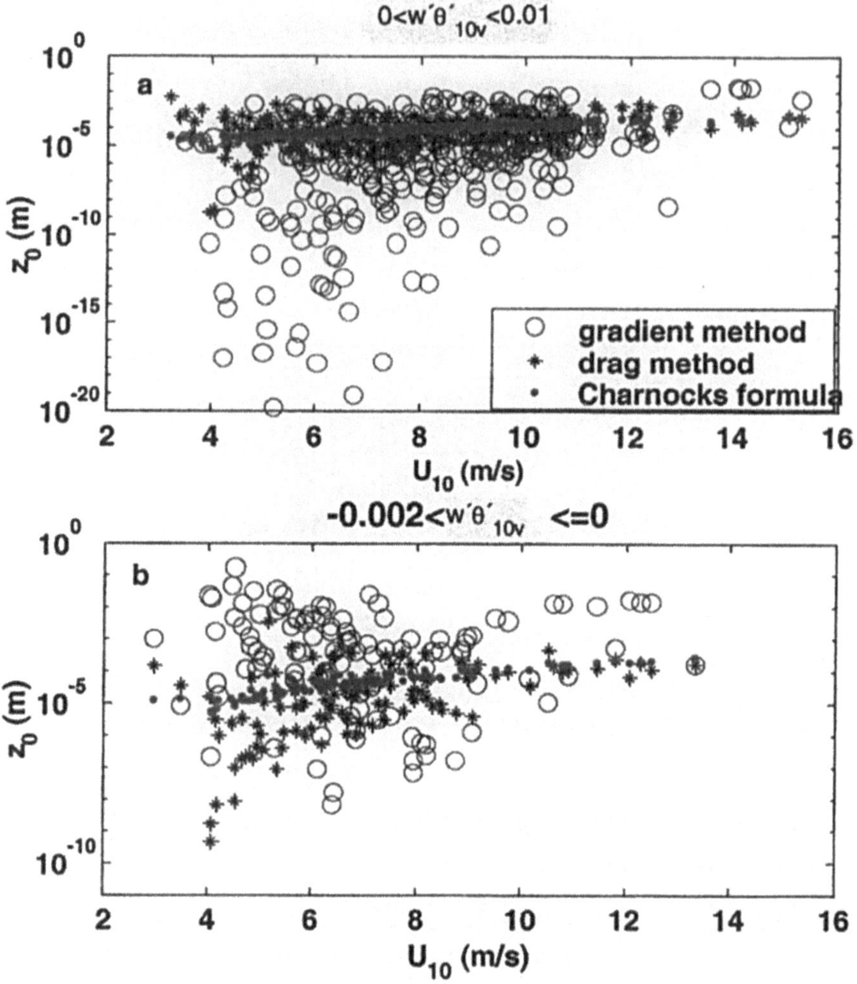

FIGURE 3. The roughness length z_o calculated with four different methods (see text) for slightly unstable stratification (a) and slightly stable stratification (b).

where various estimates of z_0, have been plotted against wind speed at 10m for slightly unstable data in Figure 3a and stable data in Figure 3b. The stars have been derived with Eq. (1.1) from measurements of the friction velocity u_* and *the wind speed at the same level,* U_{10}, this approach being denoted "the drag method" below. The open circles have been derived with the same equation but from measurements *of the wind speed at the two lowest levels,* 7 and 12 m above the ground, this approach being denoted "the gradient method" below. The filled circles have been derived with the Charnock expression, Eq. (3.1), the value of the Charnock parameter α being taken as 0.011 and u_* being obtained with the eddy correlation technique.

Figures 3a and b both show that there is generally a large difference between the z_0-values obtained with the gradient method (open circles) and corresponding values obtained with the drag method (stars). This systematic difference, however, decreases markedly for wind speeds above 10ms^{-1}. The gradient symbols (open circles) exhibit a systematically different pattern for the slightly unstable cases, Figure 3a, and for the slightly stable cases in Figure 3b. Thus, in Figure 3a most circles are much *below* the values derived with the drag method (stars), whereas most circles are *above* the stars in Figure 3b. This means that the gradient method appears to be extremely sensitive to stability variations very close to neutrality. Note that the general trend of the values derived with the drag method (stars) is not much different in figure 3a and 3b. The conclusion of Figure 3 is that the logarithmic wind law over the ocean appears to be valid only for wind speeds above 10ms^{-1} and for pure wind sea conditions. If z_0 is plotted as a function of E_1/E_2 the result is the same. It is clear from this graph (not shown here) that for pure wind sea, i.e. E_1/E_2 small, there is reasonable agreement between the profile data and the drag data, but already for as small values of E_1/E_2 as 0.05, the circles start to deviate. Note, however, that there is a correlation between high wind speed in general and growing sea. As discussed below, it is likely that the "growing-sea criterion" is the basic one, and that thus there may be cases when the wind speed is below 10ms^{-1} and still a logarithmic wind profile is obtained, and that there is likely to be cases with wind speed above this limit with a non-logarithmic wind profile.

Figure 4a shows examples of actual wind profiles in linear-log representation during a particular period of 30 hours with near-neutral conditions and E_1/E_2 varying widely with time. It is clear that most profiles are not straight lines, as would have been expected if they were logarithmic. Instead, most profiles appear to be a composite of two lines with different slope. The intersection of these lines moves first upwards with time, disappears above the highest measuring level during hours 13 -17, descending later again. Figure 4b shows the development of U_{10} (full line) and c_p (full line with circles) and Figure 4c E_1/E_2 for the same period of time. It is clear that E_1/E_2 increases from values originally below 0.1 to more than 10 during the first 17 hours or so, decreasing again to values around 0.1. There appears thus to be a variation of profile shape with wave age. During the period with large E_1/E_2 -values, wind speed is almost constant with height. As illustrated in Figure 4d, this gives extremely low apparent z_0-values.

The above discussion can be summarised accordingly. A logarithmic wind profile is obtained over the ocean only for growing seas. Then short waves, which move much slower than the wind, dominate. This is "seen" by the wind similarly to a surface with rigid roughness elements - hence the close similarity to results obtained for neutral atmospheric and laboratory conditions with flow over a rough surface. However, as soon as the wave spectrum starts to approach a saturated state, i.e. E_1/E_2 increases, waves which move with a speed close to or even larger than that of the wind at 10 m become of increasing importance. Then complex interactions between wave motions and atmospheric motions occur in the atmospheric surface layer at corresponding scales, as observed in the atmosphere by Rieder & Smith (1998), Hare *et al.* (1997) and others, and expressed in the following words by Donelan *et al.* (1993): "The young waves are

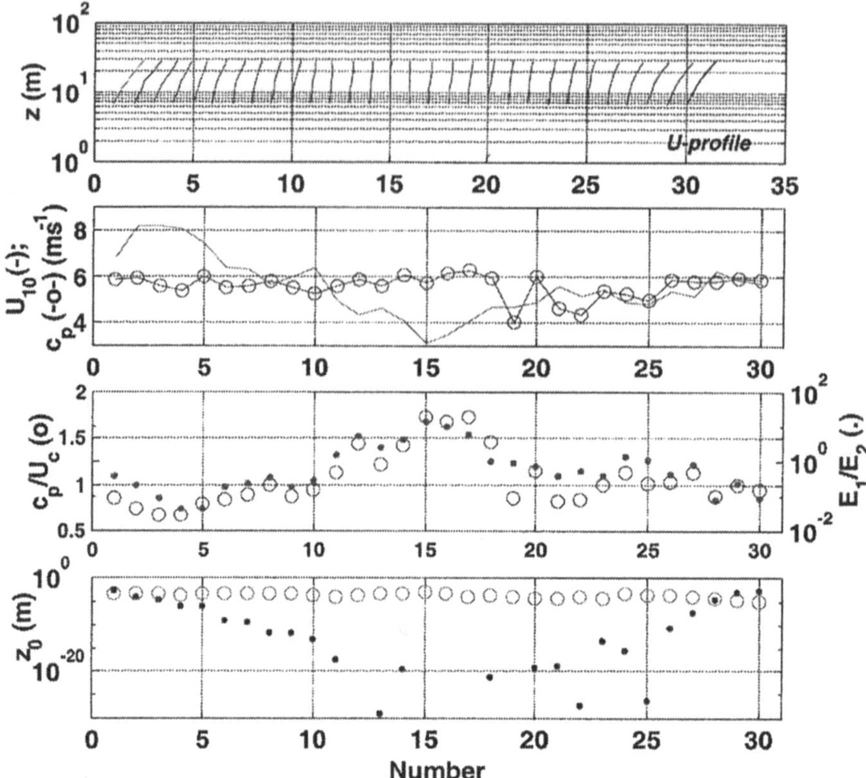

FIGURE 4. Example period of 30 hours showing (a) wind profiles, (b) wind speed (full line) and phase speed (∘), (c) wave age (∘) and E_1/E_2 (•) and (d) z_o calculated from the profiles (•) and from the drag method (∘).

believed to extract momentum from the wind by mechanisms - flow separation, viscous instability - different from those - instability of the turbulent shear flow in the boundary layer - that drive the longer, older wave component." The results of the present study show that this leads to a non-logarithmic wind profile, with a lower portion with less slope than in the upper portion. This might be interpreted as the occurrence of a wave-boundary-layer of the order 10m or more in depth. This view is partly supported by the pressure-velocity correlation measurements made by Hare *et al.* (1997) over the ocean. In the case of growing seas (E_1/E_2 small), the observed validity of the logarithmic law indicates that the wave-boundary-layer is indeed very shallow, as predicted by modelling studies (Belcher & Hunt 1996, Makin *et al.* 1995). For the case of strong swell, Smedman *et al.* (1999) showed that the long, dominating waves produce wave-induced momentum flux directed from the ocean surface into the atmosphere. It was concluded that the wave-boundary-layer in this case extended even beyond the highest measuring level, 30m.

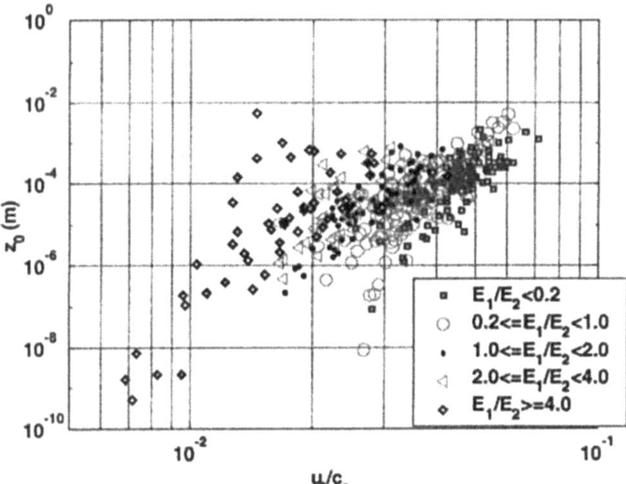

FIGURE 5. Apparent roughness z_{0a} as a function of inverse wave age stratified according to E_1/E_2.

5. VARIATION OF THE DRAG WITH SEA STATE

Figure 5 shows, on a logarithmic scale, z_0, derived with Eq. (1.1) from measurements of u_* (by the eddy correlation technique) and U_{10}, plotted against inverted wave age u_*/c_p, also on a logarithmic scale. Note, that z_0 in this general case is just an integration constant, which has the meaning of the roughness length that would match measured stress and wind speed at 10 m, *provided the profile from the surface up to 10 m would have been logarithmic. We may call it the apparent roughness length,* z_{0a}. The data in this graph has been stratified according to E_1/E_2. For increasing values of this parameter, the data are seen to line up in bands, roughly parallel to those of the crosses representing growing sea. The bands move gradually to the left in the figure with increasing E_1/E_2. The band farthest to the left is $E_1/E_2 > 4$ and is likely to represent swell conditions. Figure 5 vividly illustrates that the apparent roughness length is a function of two wave state variables, inverted wave age u_*/c_p, and the wave spectral ratio E_1/E_2. This means that for a certain value of inverted wave age, the apparent roughness length can take on a wide range of values, depending on the value of E_1/E_2, i.e. the ratio between the energy of the long and the short waves, according to Eqs. (3.3) and (3.4).

Figure 5 is based on the complete near-neutral data set, i.e. it contains both the slightly unstable and the slightly stable data separate plots (not shown here) of the unstable and the stable data show no systematic differences. This finding is in agreement with the conclusion from inspection of Figure 4a and b that the star symbols, which represent computations based on the drag method, have a similar trend both for unstable and stable conditions.

Figure 6 is a linear representation of C_D against u_*/c_p for the same data set. C_D has been obtained with the defining equation, Eq. (1.3), from measurements of u_* and U_{10}. Like in Figure 5, the data in this graph has been stratified

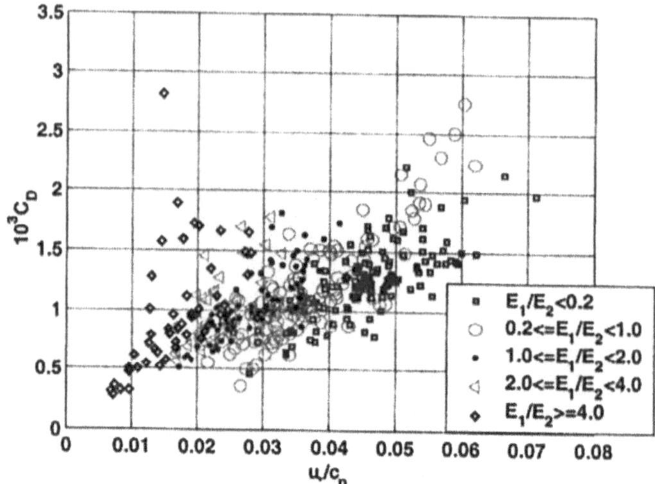

FIGURE 6. Drag coefficient C_D as a function of inverse wave age stratified according to E_1/E_2.

according to E_1/E_2. Also in this graph, the data line up in roughly parallel bands, and a wide range of C_D-values ensues for a given value of u_*/c_p according to the value of E_1/E_2.

A remarkable feature of both Figure 6 is the wide range of C_D-values encountered for the data with $E_1/E_2 > 4$, which are cases likely to represent swell. No systematic analysis has been made at this stage to clarify which factors are responsible for the group of surprisingly high C_D-values, but case studies indicate that the history of wave state development and concurrent development of wind direction and speed may be of importance. Thus the one case with C_D in excess of 2×10^{-3}, appeared in connection with pronounced shifts in these parameters.

6. DISCUSSION AND CONCLUSIONS

The (neutral) drag coefficient C_D, is found to be governed by two sea-state parameters in the general case: the (inverted) wave age parameter u_*/c_p (where c_p is the phase speed corresponding to the peak frequency of the wave spectrum) and the wave energy spectral ratio, E_1/E_2. Thus, for a given value of u_*/c_p, C_D can take on values within a wide range depending on E_1/E_2. The limit of $E_1/E_2 -> 0$ corresponds to developing sea, discussed above; $E_1/E_2 > 4$ is the swell-dominated case. In between these limits, there is a wide range of wave conditions representing seas in various degree of saturation.

The above findings imply that knowledge of the two wave-state parameters u_*/c_p and E_1/E_2 would be needed to give correct estimation of the stress in a numerical model of the flow over the ocean. This in turn, is only possible if a wave-model (such as WAM) is coupled to the atmospheric model. Even in the relatively simple case of developing seas, information from a wave-model is needed.

All results presented in this paper have been derived for near-neutral conditions and situations with one-peak wave-spectra. Situations with multi-peak spectra are common in the open ocean. The present analysis has been carried out for such cases as well, with similar result, although with increased scatter (not shown here).

REFERENCES

BELCHER, S.E. & HUNT, J.C.R. 1996 Turbulent shear flow over slowly moving waves. *J.Fluid Mech.*, **251**, 109-148.

ANCTIL, F. & DONELAN, M.A. 1996 Air-water momentum flux observations over shoaling waves. *J. Phys. Oceanogr.*, **26**, 1344 - 1353.

CHARNOCK, H. 1955 Wind stress on a water surface. *Q.J.Roy.Met.Soc.*, **81**, 639 - 640.

DONELAN, M.A., DOBSON, F.W., SMITH, S.D. & ANDERSON, R.J. 1993 On the dependence of sea surface roughness on wave development. *J. Phys. Ocean.*, **23**, 2143-

DRENNAN, W.M., GRABER, H.C., HAUSER, D. & QUENTIN, C. 2000 On the wave age dependence of wind stress over pure sea. *J. Geophys. Res.*, submitted.

GRELLE, A. & LINDROTH, A. 1994 Flow distortion by a Solent sonic anemometer: Wind tunnel calibration and its assessment for flux measurements over forest and field. *J. Atmos. Oceanic Technol.*, **11**, 1529-1542.

HARE, J.F., HARA, T., EDSON, J.B. & WILCZAK, J.M. 1997 A similarity analysis of the structure of airflow over surface waves. *J. Phys. Oceanography*, **27**, 1018-1037.

HÖGSTRÖM, U. 1996 Review of some basic characteristics of the atmospheric surface layer. *Boundary-Layer Meteorology*, **78**, 215 - 246.

KOMEN, G., JANSEN, P.A.E.M., MAKIN, V. & OOST, W. 1998 On the sea state dependence of the Charnock parameter. *The Global Atmosphere and Ocean System*, **5**, 367-388.

MAKIN, V.K., KUDRYATSEV, V.N. & MASTERBROEK, C. 1995 Drag of the sea surface. Boundary-Layer *Meteorolol.*, **73**, 159-182.

RIEDER, K.F. & SMITH, J.A. 1998 Removing wave effects from the wind stress vector. *J. Geophys. Res.*, **103**, No C1, 1363-1374.

RUTGERSSON, A., SMEDMAN, A. & HÖGSTRÖM, U. 2001 The use of conventional stability parameters during swell. *J. Geophys. Res.*, accepted.

SMEDMAN, A., TJERNSTRÖM, M. & HÖGSTRÖM, U. 1994 Near neutral marine atmospheric boundary layer with no surface stress: A case study, *J. Atmos. Sci.*, **23**, 3399-3411.

SMEDMAN, A., HÖGSTRÖM, U., BERGSTRÖM, H., RUTGERSON, A., KAHMA, K.K. & PETTERSON, H. 1999 A case study of air-sea interaction during swell conditions. *J. Geophys. Res.*, **104**, 25, 833-25, 851.

On the Accuracy of Ocean Winds and Wind Stress – An Emperical Assesment

P.K. Taylor and M.J. Yelland

James Rennell Division for Ocean Circulation and Climate, Southampton Oceanography Centre, UK

Abstract

In this paper we shall consider the accuracy of in situ measurements of wind and wind stress over the ocean, and also the contrasting characteristics of different wind stress parameterisations. There is much scatter in the drag coefficient or roughness length measurements for winds below 10 m/s. While this scatter may be caused by sampling limitations and other measurement errors, there is increasing evidence that swell waves may modify the effective surface roughness. However, at these lower wind speeds, the resulting uncertainty in the wind stress is very small, only a few percent of the magnitude of the wind stress at 20 m/s.

At wind speeds between 10 to 20 m/s the measurements from the open ocean are less scattered with both eddy correlation and inertial dissipation wind stress estimates giving similar values. At these higher wind speeds the instrumentation on meteorological buoys is relatively low compared to the height of the dominant waves. However we shall present data for wind velocity fluctuations and buoy motion which demonstrate that meteorological buoys can be used to adequately determine the wind velocity, and hence the wind stress, even in high wave conditions.

At wind speeds above 20 m/s, different parameterizations predict significantly different wind stress values. At these higher wind speeds we suggest that, compared to the behaviour at lower wind speeds, the sea surface roughness will increase less rapidly with increasing wind speed. Unfortunately the available wind stress data are few, particularly for winds above 25 m/s, and insufficient to test this prediction.

1. Introduction

In this paper we shall first compare parameterization formula for determining wind stress from wind speed and then consider the accuracy of our measurements of wind over the ocean. Wind stress is caused by the interaction between the wind and the surface roughness (represented by the drag coefficient, C_{D10n}, or the roughness length, z_0). Thus the standard "bulk aerodynamic" formula is:

$$\tau/\rho = u_*^2 = C_{D10n}(U_{10n} - U_0)^2 \tag{1.1}$$

where τ is the wind stress, ρ the air density, u_* the friction velocity, and U the wind speed. The subscript "$10n$" refers to the value at a height of 10m under neutral atmospheric stability, "0" refers to the sea surface.

Since the characteristics of the sea surface vary with the wind conditions, we need to know how the aerodynamic sea surface roughness varies with wind or sea state. How accurately must we know the value of C_{D10n}? Figure 1 shows a

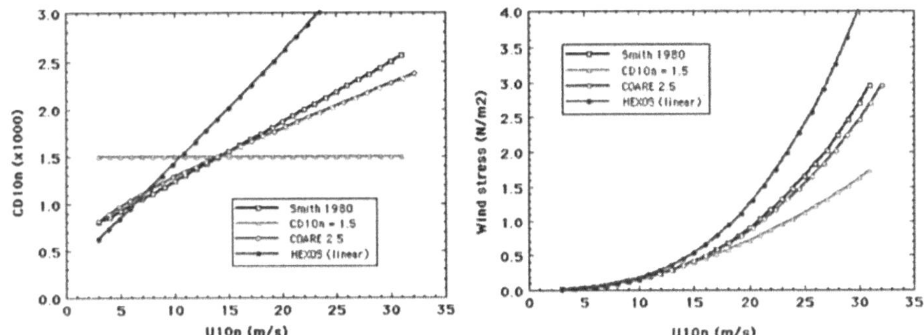

FIGURE 1. (left) three observed C_{D10n} to U_{10n} relationships: Smith (1980), TOGA COARE 2.5 (Fairall *et al.*, 1996), and a linear HEXOS fit (Smith 1992); for comparison $C_{D10n} = 1.5 \times 10^{-3}$ is also shown. (right) the corresponding variation of wind stress with U_{10n}.

range of relationships between C_{D10n} and U_{10n}, and the resulting wind stress which would be calculated. No matter which relationship we use, up to 12 m/s there is very little difference between the calculated wind stress value. Thus, although large in percentage terms, at wind speeds below 10 m/s variations in wind stress are small in absolute terms. This also holds for roughness variations ascribed to the effects of swell (Donelan *et al.*, 1997; Rieder & Smith 1998; Drennan*et al.*, 1999) which have been observed at these lower wind speeds.

At higher wind speeds Figure 1 shows that significant differences are obvious. In these conditions it is a good assumption that the stress is aligned in the wind direction, furthermore good agreement has been found between eddy correlation and inertial dissipation stress estimates (e.g. Drennan *et al.* 2002). The different roughness length parameterizations would be best tested using data at very high wind speeds ($U_{10n} > 25$ m/s for 10 minute mean), but there is an urgent need for more observations in those conditions. Our emphasis on higher wind speeds does not imply that knowledge of u_* is unimportant at lower wind speeds. Since stability is dependent on u_*^3, different drag parameterizations can significantly alter the calculated heat fluxes in regions such as the tropical warm pool (Taylor *et al.* 1998) which have an important role in the world climate.

Bonekamp *et al.* (2002) have recently compared various parameterization formula. They found that a simple, linear dependence of the drag coefficient on wind speed (Smith 1980; Yelland *et al.* 1998):

$$C_{D10n} = a + bU_{10n} \tag{1.2}$$

although dimensionally inconsistent, represented observed data sets surprisingly well. In contrast the dimensionally consistent formula of Charnock (1955):

$$z_0 = \alpha u_*^2/g \tag{1.3}$$

performed less well unless the Charnock parameter, α, was assumed to vary with the wave age (c_p/u_* where c_p is the phase speed of the dominant wave), for example:

$$z_0 = a(c_p/u_*)^{-b}(u_*^2/g) \tag{1.4}$$

Based on data from the main experiment in the Humidity Exchange Over the Sea (HEXOS) programme, Smith *et al.* (1992) suggested $a = 0.48$ and $b = 1$. This "HEXOS" formula performed slightly better than the linear formula (1.2) (Bonekamp *et al.* 2002), and was recommended by Komen *et al.* (1998) in their review of Charnock parameters. However Taylor & Yelland (2001a) noted that the coefficients a and b in (1.4) tended to vary from one data set to another, and suggested that z_0 is better parameterised in terms of the height and steepness of the dominant waves:

$$z_0/H_s = a(H_s/L_p)^b \tag{1.5}$$

where H_s is the significant wave height and L_p the wavelength of the waves at the spectral peak. Suggested values for the coefficients (which were poorly defined by the available data) were $a = 1200$ and $b = 4.5$. In contrast to wave age based formula, (1.5) predicts that all pure wind seas will have a similar wind speed to roughness relationship with no increase in the roughness for short fetch or short duration seas. However, compared to lakes, it does predict lower roughness for typical ocean sea states, and also implies that the Charnock parameter will appear to vary with wave age.

In the next section we will compare the performance of two roughness length parameterisations: the HEXOS formula (1.4) and the Taylor & Yelland (2001a) formula (1.5). Then, since the quality of a wind stress estimate depends on the quality of the wind data as well as the parameterisation formula, we will briefly consider the accuracy of wind data from ships and buoys (Section 3).

2. Comparison of parameterisation

2.1. *The HEXOS Experiment*

Taylor & Yelland (2001a) used the Janssen (1997) subset of the HEXOS data set to illustrate the limitations of wave age based scalings such as (1.4). As noted by Oost (1998), those HEXOS data which exhibited greater roughness were associated with waves for which L_p, H_s, and c_p were each greater than was generally observed at a similar wind speed. However the wave age was similar to, or older than, that for other data points for which the roughness was less (see Figure 1 in Taylor & Yelland, 2001a and Figure 2 below). First we should ask whether these, apparently anomalous data were reliable? There seems little doubt that the enhanced roughness for these cases was real; it was detected in the data from two different anemometers operated by two different research teams. We can also understand why the wave characteristics were different. In general the HEXOS sea states were fetch limited with a minimum fetch of about 180 km for winds from 280° T. Inspection of the data shows that the cases of longer wavelength waves corresponded to periods when the wind direction was from 300° T and higher, for which fetches of 300 to 500 km were possible. Application of standard wave formula for the appropriate fetches will predict waves of similar characteristics to those observed at the experiment site.

Given these wave characteristics, wave age based formula, such as that of Smith *et al.* (1992) despite being developed using the HEXOS data, can only

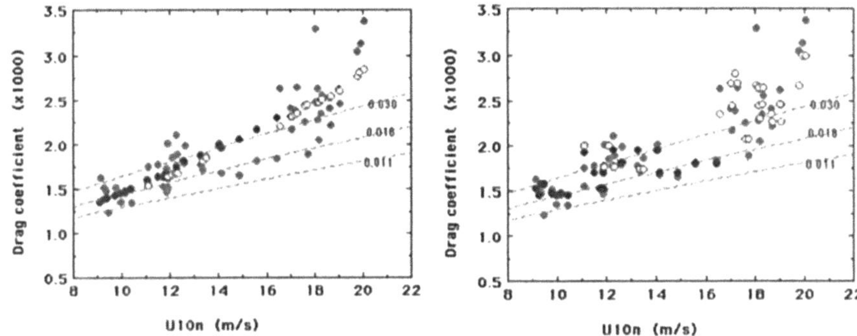

FIGURE 2. Observations of drag coefficient as a function of wind speed for HEXOS data selected for "pure wind sea" conditions (Janssen, 1997) are shown as grey points with values of the Charnock relationship (1.3) for three different Charnock parameters. The left plot shows (in black) values predicted by the HEXOS formula (Smith *et al.* 1992), while the right plot shows values predicted by Taylor & Yelland (2001a). Predictions for waves with $L_p > 72$ m (more than 4x the water depth) are shown by open symbols.

predict the overall trend of the observations (Figure 2). In contrast the Taylor and Yelland (2001a) formula represents the scatter of the observations to a much better degree.

2.2. *The RASEX experiment*

Taylor & Yelland (2001a) used the wave data of Johnson *et al.* (1998) to calculate roughness length values for comparison with the observations of Vickers & Mahrt (1997a) which had undergone quality control procedures as described by Vickers & Mahrt (1997b). As was noted by Johnson *et al.* (1998) the Smith *et al.* (1992) formula significantly over estimated the observed roughness (Figure 3). In contrast the Taylor & Yelland formula successfully predicted the magnitude of the roughness for the majority of the observed data.

2.3. *The SWS2 experiment*

The second Storm Wave Study experiment, SWS-2 (Dobson *et al.* 1999; Taylor *et al.*, 1999), took place over the Grand Banks off Newfoundland during October to November 1997. Wind stress data were obtained from a sonic anemometer mounted on a Nomad meteorological buoy (Figure 4a). Whereas the wind stress data for HEXOS and RASEX were obtained using the eddy correlation method, the SWS-2 data were obtained using the inertial dissipation method. The buoy was equipped with motion sensors from which the estimates of H_s and L_p (defined by the wavelength at the peak of the energy spectrum) were obtained using spectral analysis. Despite the mixed wind sea and swell conditions at this "open ocean" site, the mean values of roughness observed during SWS-2 experiment were well predicted by the Taylor & Yelland (2001a) formula (Figure 4b).

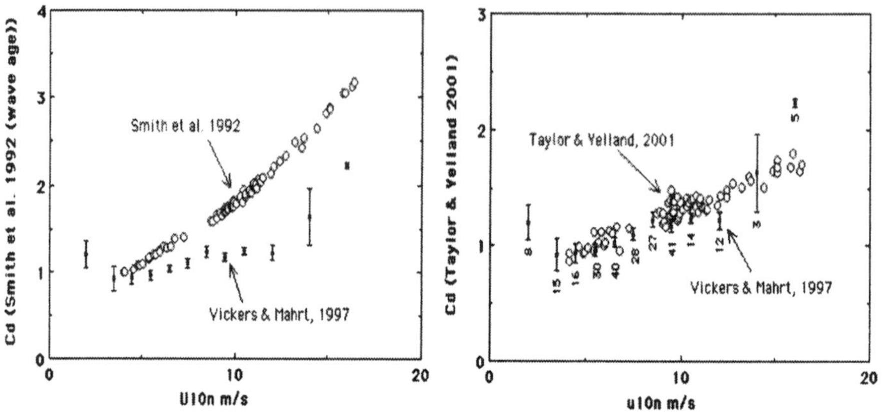

FIGURE 3. Mean observations of drag coefficient as a function of wind speed for the RASEX experiment (Vickers and Mahrt, 1997a): (left) as predicted by the Smith *et al.* (1992) HEXOS formula; (right) as predicted by Taylor & Yelland (2001a).

2.4. *Other data sets*

Taylor & Yelland (2001a) also showed that their formula predicted the observed roughness for other data sets: the Lake Washington data of Ataktürk & Katsaros (1999), and the wave tank data of Cheng & Mitsuyasu (1992) and Keller *et al.* (1992). They also compared with data from various studies on Lake Ontario (Donelan, 1982; Colton *et al.* 1995; Anctil & Donelan, 1996; Terray *et al.* 1996). For these latter data they found that, while on average their formula performed well for data with wave ages (c_p/u_*) above about 12, data at younger wave ages appear to be rougher than predicted. Such young waves were rare in the other data sets examined; however in the few cases which did occur, their formula also under-estimated the roughness. While this suggested the need for a further parameter in the formula, they did not include a wave-age dependence since it would have degraded the agreement obtained for the other, more extensive data sets.

2.5. *Roughness at high wind speeds*

As noted above, it is at the highest wind speeds that the largest differences occur between the roughness values predicted by the various parameterisations. Taylor & Yelland (2001a) suggested that, ignoring any enhanced roughness for very young waves (see previous section), at higher wind speeds the C_{D10n} to U_{10n} relationship should follow their "deep water, pure wind sea" curve. If true, this results in lower roughness values than would be obtained by extrapolating a typical linear relationship such as that of Yelland *et al.* (1998). There is little observational evidence; the high speed wind flume data of Kunishi & Imasato (1966), as reproduced by Kondo (1975), do lie about the predicted pure wind sea relationship but with much scatter (Figure 5). However the experimental problems are such that the reliability of these high wind speed flume data must be open to question (e.g. see Oost 1991).

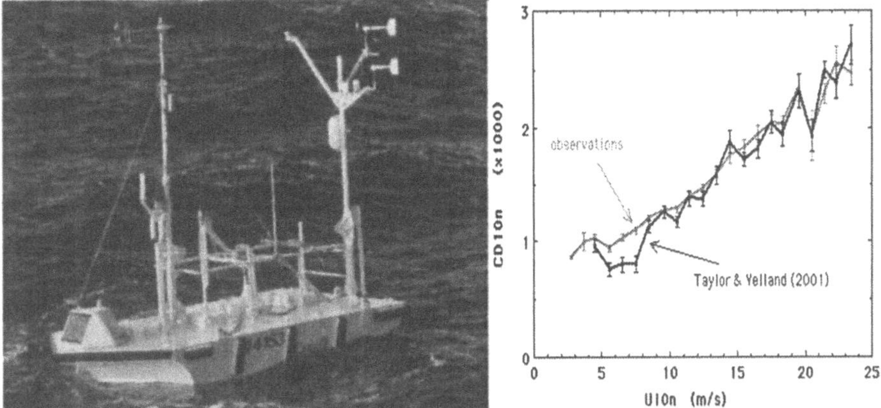

FIGURE 4. (a) The sonic anemometer on the forward mast of the SWS-2 Nomad buoy (b) Mean observations of drag coefficient as a function of wind speed for the SWS-2 experiment (from Taylor & Yelland, 2001a).

3. ACCURACY OF WIND DATA

3.1. *Ships*

All wind (and wind stress) measurements from ships must allow for the air flow distortion caused by the ship, otherwise large errors can occur. Computational Fluid Dynamics (CFD) studies such as those described by Yelland *et al.* (1998) show that it is not possible to site an anemometer in a position where the airflow has not been disturbed by the ship's presence. However, provided a well exposed anemometer position is chosen (as would be the case on a research ship), CFD can be used to estimate and adequately correct the wind velocity errors (Yelland *et al.* 2002). The CFD results also provide an estimate of the vertical displacement of the airflow which is needed for wind stress estimates determined using the inertial dissipation method. However, since the turbulence is parameterised, the effect of the ship on the turbulent structure, and hence the errors in wind stress estimated using the eddy correlation method, cannot be predicted.

Determining the biases in wind velocity measurements from a typical anemometer installation on a merchant ship is more difficult. The larger size of the ship will normally imply that the anemometer will be in a region of severe flow distortion for which CFD modelling may be less accurate. On the basis of wind tunnel studies, Yelland *et al.* (2001) suggest that the pattern of flow disturbance over the accommodation block will scale with the height difference from the main deck (or top of the deck cargo) to the wheelhouse top. The implication is that an anemometer on a tanker would be expected to over-estimate the wind significantly (up to 30%) whereas on a container ship an under-estimate is more likely.

3.2. *Buoy wind data*

Because of the potentially large flow distortion errors affecting data from ships, wind measurements from buoys might be expected to be a superior alternative.

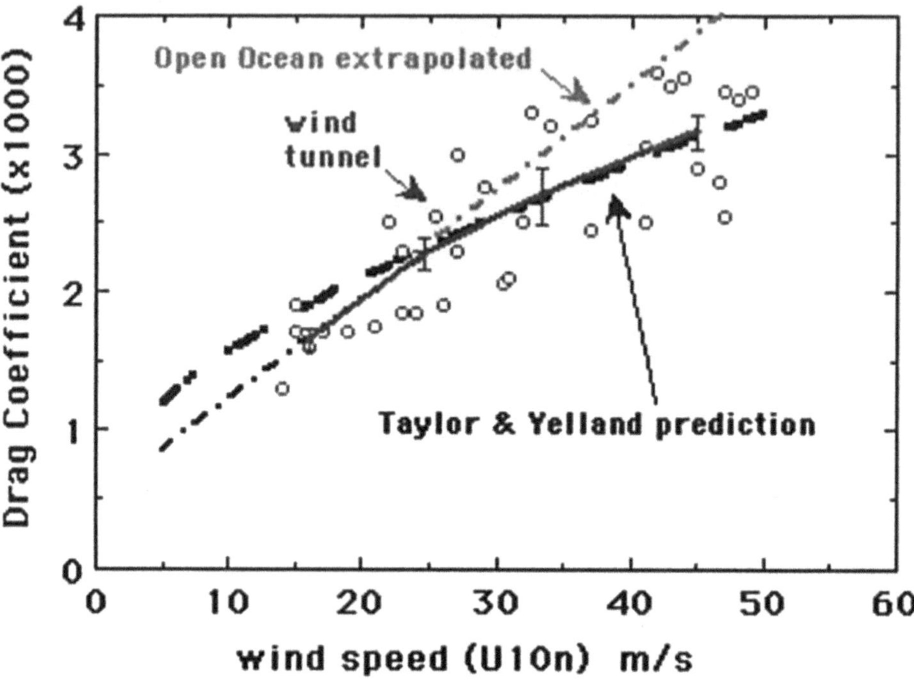

FIGURE 5. Drag coefficient as a function of wind speed for high wind sea conditions. The circles are the wind flume data of Kunishi & Imasato (1966) and the line with error bars are 10 m/s averaged values for these data. The chain line represents the Yelland *et al.* (1998) relationship while the heavy dashed line is the predicted relationship of Taylor & Yelland (2001a).

However a number of studies have questioned the accuracy of buoy data. For example: Gilhousen (1987) found vector averaged winds from buoys to be low by around 7% compared to platform data; Large *et al.* (1995) suggested that buoy mounted instruments under-estimate the wind speed due to the effect of waves on the wind profile (e.g. by 15% at 20m/s); Skey *et al.* (1995) suggested a typical under-estimate of around 20% due to the sheltering effects of large waves. During the SWS-2 experiment (see section 2.3) the instrumentation on the Nomad buoy and a nearby research ship, the RRS Hudson, allowed these problems to be investigated.

The low magnitude of vector averaged winds was found to be due to errors in the wind direction data. These were primarily caused by the standard practice for Canadian weather buoys of aligning the R. M. Young propeller vane anemometers with the 0° to 360° transition (where there is a 5° "deadband") toward the bow of the Nomad. Before quality control the vector - scalar wind speed difference was about 8%, similar in both magnitude and distribution to Gilhousen (1987; compare Figure 7a and his Figure 9). Using quality control procedures to remove the erroneous samples reduced the difference to around

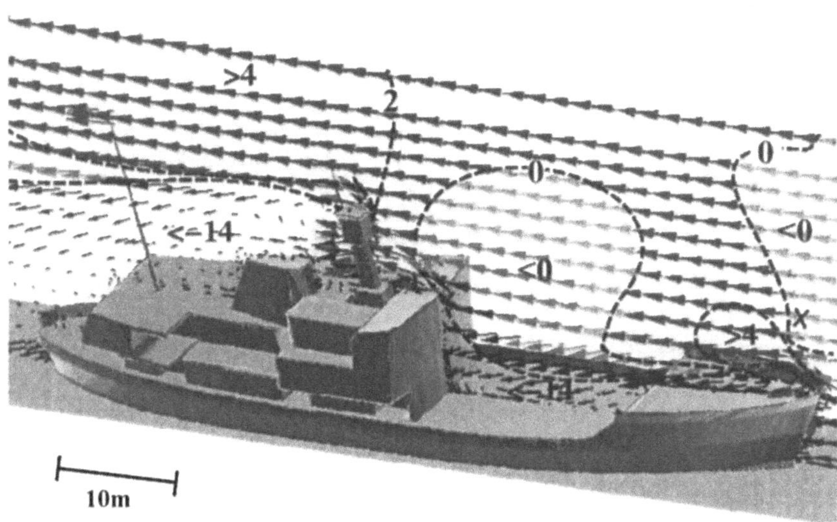

FIGURE 6. CFD results for bow-on flow over the research ship CSS Dawson. The numbers and shading indicate the wind speed error, as a percentage of the undisturbed value, on a vertical fore-aft plane through the bow-mast anemometer position which is shown by a cross (adapted from Taylor *et al.* 1999).

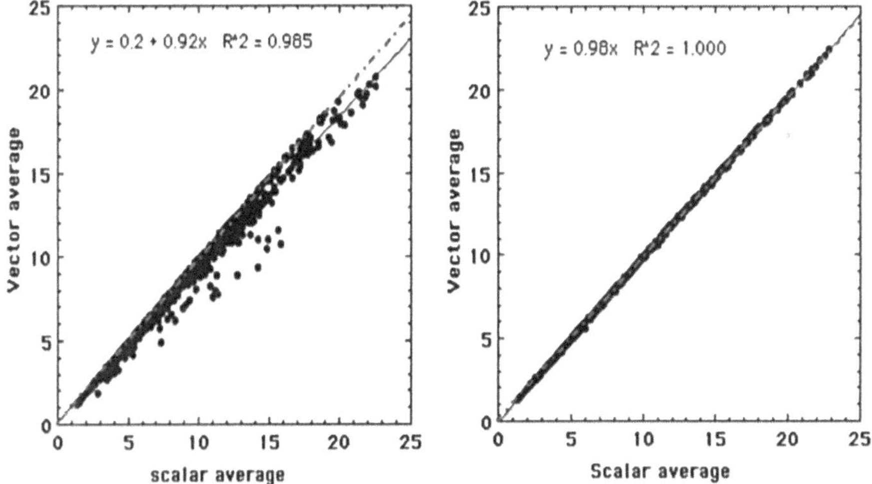

FIGURE 7. Scatter plots of vector averaged wind speeds against scalar averaged speeds for a R.M. Young propeller vane anemometer mounted on the Nomad buoy: (left) no quality control; (right) after quality control.

1% to 2% (Figure 7b). This residual difference is likely to have been due to wave induced crosswind motion of the buoy in which case the vector average would be the more accurate value.

The use of a motion package to record the 6 degrees of motion of the Nomad buoy allowed the "sheltering" effects of waves to be estimated. The method of

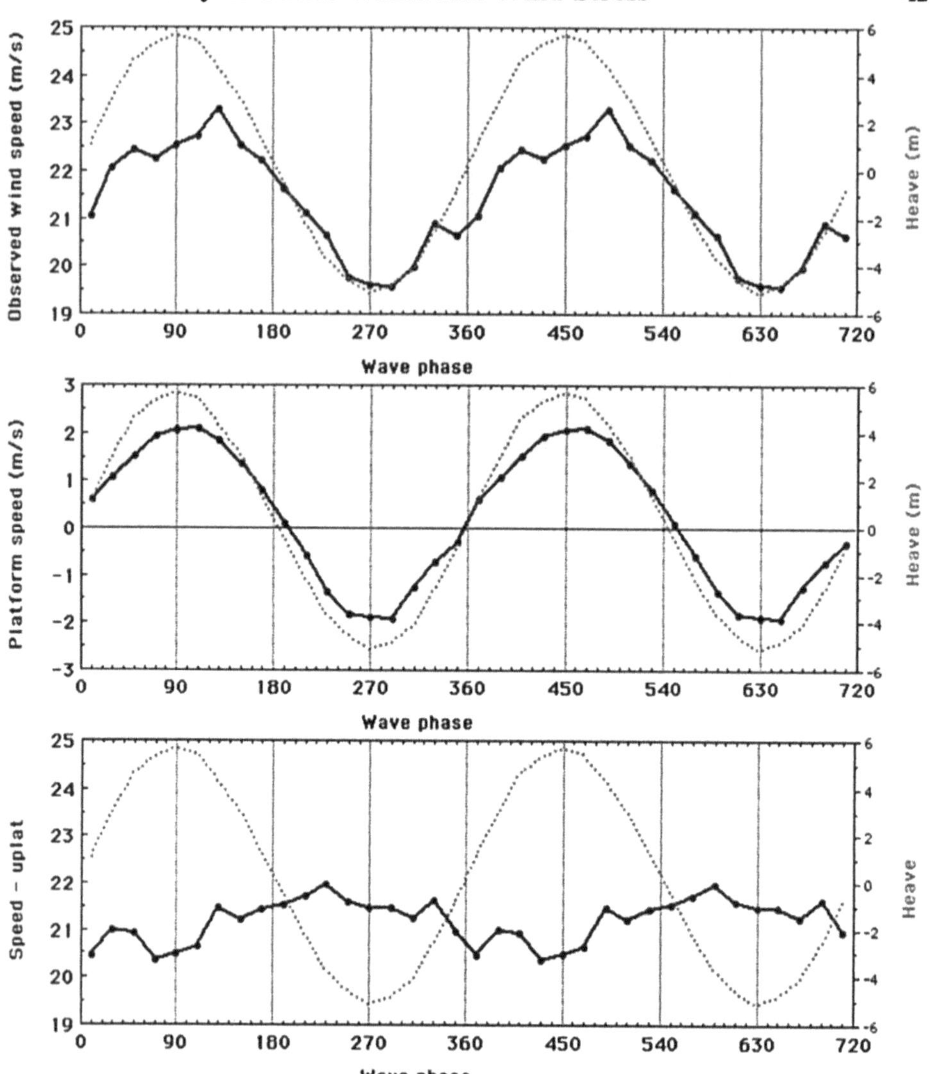

FIGURE 8. (a - top) The observed wind speed averaged by wave phase; (b - middle) the averaged platform speed where forward (into wind) is positive; (c - bottom) the wind speed corrected for platform motion. The grey line on each plot is the buoy heave (righthand scale).

analysis was based on "zero up-crossing analysis". Thus a wave was defined as the period between successive upward crossings of the mean sea level by the buoy, and represented as a phase angle from 0° to 360°. The variation of other parameters (such as the wind speed) were then averaged in terms of the wave phase. Some preliminary results are shown in Figure 8 where, for clarity, two cycles of the averaged wave phase are shown. In this example, the mean wind speed was around 20 m/s and the significant wave height about 8 m implying peak to trough wave heights significantly greater than the anemometer height (5.5 m). The observed, phase averaged wind speed was highest just after the wave peak and lowest in the wave trough (Figure 8a), the apparent fluctuations being similar to the 20% reported by Skey *et al.* (1995). However the motion package showed that the buoy was moving bodily into the wind at the wave peak and away from the wind in the troughs (Figure 8b). When the wind data were corrected for that motion the residual variation in the mean wind was small, and probably not significant. If the largest waves were considered separately, the wind did appear to be highest over the wave peak, however for the data set as a whole the preliminary analysis showed no evidence for wave "sheltering".

Comparison of the buoy and ship winds also suggested that any "sheltering" effects were small. While the wind stress estimates from the two platforms were in good agreement, particularly at the higher wind speeds (e.g. see Taylor & Yelland 2001b), the wind speed data suggested that the Nomad instrument was reading low, but only by around 3%. Given that both anemometers were found to have changes between pre- and post- cruise calibrations of around 2%, it is not clear whether the observed wind speed difference was real.

4. Summary

The correct parameterisation for roughness length is still controversial. The differences between parameterisations are most important at higher wind speeds (say greater than 10 m/s, and definitely when over 15 m/s). There is an urgent need for more data at very high wind speeds (> 25 m/s). With regard to the accuracy of the wind observations, ships must have well exposed anemometers with air-flow corrections estimated and applied. Careful measurements from buoys can achieve acceptable accuracy (better than a few %) however there is evidence that not all operational buoys do not achieve that potential.

Acknowledgements: The SWS-2 experiment received funding from the Canadian federal Program of Energy Research and Development (PERD) and the UK Ministry of Defence/Natural Environment Research Council Joint Grant Scheme project "Coastal and Open Ocean Wind Stress. The motion correction algorithms for the Nomad data were supplied by Jeff Hare & Chris Fairall (NOAA, Boulder).

REFERENCES

Anctil, F. & Donelan, M.A. 1996 Air-water momentum flux observations over shoaling waves. *J. Phys. Oceanogr.*, **26**, 1344-1353.

Ataktürk, S.S. & Katsaros, K.B. 1999 Wind stress and surface waves observed on Lake Washington. *J. Phys. Oceanogr.*, **29**, 633-650.

Bonekamp, H., Sterl, A., Komen, G., Janssen, P.A.E.M., Taylor, P.K. & Yelland, M.J. 2002 Statistical comparisons of observed and ECMWF modeled open ocean surface drag (in press *J.Phys.Oceanogr.*,).

Charnock, H. 1955 Wind stress on a water surface. *Q. J. Roy. Met. Soc.*, **81**, 639-640.

Cheng, Z. & Mitsuyasu, H. 1992 Laboratory studies on the surface drift current induced by wind and swell. *J. Fluid Mech.*, **243**, 247-259.

COLTON, M.C., PLANT, W.J., KELLER, W.C. & GEERNAERT, G.L. 1995 Tower-based measurements of normalized radar cross-section from lake-Ontario - evidence of wind stress dependence. *J. Geophys. Res.*, **100**, 8791-8813.

DOBSON, F.W., ANDERSON, R.J., TAYLOR, P.K. & YELLAND, M.J. 1999 Storm wind study II: open ocean wind and sea state measurements. in *Proc Symp. on The Wind-driven air-sea interface: electromagnetic and ccoustic sensing, wave dynamics and turbulent fluxes* (Ed. M.L. Banner), Univ. of New South Wales, Sydney, Australia, 11-15 January 1999, 295-296.

DONELAN, M. A. 1982 The dependence of the aerodynamic drag coefficient on wave parameters. *First international conference on meteorology and air-sea interaction of the coastal zone*, American Meteorological Society, Boston, Mass., 381-387.

DONELAN, M.A., DRENNAN, W.M. & KATSAROS, K.B. 1997 The air-sea momentum flux in conditions of wind sea and swell. *J. Phys. Oceanogr.*, **27**, 2087-2099.

DRENNAN, W.M., GRABER, H.C. & DONELAN, M.A. 1999 Evidence for the effects of swell and unsteady winds on marine wind stress. *J. Phys. Oceanogr.*, **29**, 1853-1864.

DRENNAN, W.M., GRABER, H.C., HAUSER, D. & QUENTIN, C. 2002 On the wave dependence of wind stress over pure wind seas, . *J. Geophys. Res.*, (accepted).

FAIRALL, C.W. *et al.* 1996 Bulk parametrisation of air-sea fluxes for TOGA COARE. *Journal of Geophysical Research*, **101**, 1295-1308.

GILHOUSEN, D.B. 1987 A field evaluation of NDBC moored buoy winds. *Journal of Atmospheric and Oceanic Technology*, **4**, 94-104.

JANSSEN, P.J.A.M. 1997 Does wind stress depend on sea-state or not? - a statistical error analysis of HEXMAX data. *Boundary-Layer Meteorol.*, **83**, 479-503.

JOHNSON, H.K., HOEJSTRUP, J., VESTED, H.J. & LARSEN, S.E., 1998 Dependence of sea surface roughness on wind waves. *J. Phys. Oceanogr.*, **28**, 1702-1716.

KELLER, M.R., KELLER, W.C. & PLANT, W.J. 1992 A wave tank study of the dependence of X band cross sections on wind speed and water temperature. *J. Geophys. Res.*, **97**, 5771-5792.

KOMEN, G., JANSSEN, P.A.E.M, MAKIN, V. & OOST, W. 1998 On the sea state dependence of the Charnock parameter. em The Global Atmosphere and Ocean System, **5**, 367-388.

KONDO, J. 1975 Air-sea bulk transfer coefficients in diabatic conditions. *Boundary Layer Meteorology*, **9**, 91-112.

KUNISHI, H. & IMASATO, N. 1966 On the growth of wind waves by high-speed wind flume. *Ann. Disaster Prevention. Res. Inst.*, Kyoto Univ. (in Japanese), **9**, 667-676.

LARGE, W.G., MORZEL, J. & CRAWFORD, G.B. 1995 Accounting for surface wave distortion of the marine wind profile in low-level ocean storms wind measurements. *J. Phys. Oceanogr.*, **25**, 2959-2971.

OOST, W.A. 1991 The wind profile in a wave flume. *Journal of Wind Engineering and Industrial Aerodynamics*, **37**, 113-121.

OOST, W.A. 1998 The KNMI HEXMAX stress data - a revisit. *Boundary-Layer Meteorol.*, **86**, 447-468.

RIEDER, K.F. & SMITH, J.A. 1998 Removing wave effects from the wind stress vector. *J. Geophys. Res.*, **103**, 1363-1374.

SKEY, S.G.P., BERGER-NORTH, K. & SWAIL, V.R. 1995 Detailed measurements of winds and waves in high sea states from a moored Nomad weather buoy, preprints 4th Internat. Workshop on Wave hindcasting and forecasting, Banff, 16-20 Oct. 1995, 213-223.

SMITH, S.D. 1980 Wind stress and heat flux over the ocean in gale force winds. *J. Phys. Oceanogr.*, **10**, 709-726.

SMITH, S.D., ANDERSON, R.J., OOST, W.A., KRAAN, C., MAAT, N., DECOSMO, J., KATSAROS, K.B., DAVIDSON, K.L., BUMKE, K., HASSE, L. & CHADWICK,

H.M. 1992 Sea surface wind stress and drag coefficients: the HEXOS results. *Boundary-Layer Meteorol.*, **60**, 109-142.

TAYLOR, P.K., YELLAND, M.J., DOBSON, F.W. & ANDERSON, R.J. 1999 Storm wind study II: wind stress estimates from buoy and hhip. in *Proc Symp. on The wind-driven air-sea interface: electromagnetic and acoustic sensing, wave dynamics and turbulent fluxes* (Ed. M.L. Banner), Univ. of New South Wales, Sydney, Australia, 11-15 January 1999, 353-354.

TAYLOR, P.K., JOSEY, S.A. & KENT, E.C. 1998 A comparison of climatological, model derived and observed air-sea flux values for the COARE area. *CLIVAR/GEWEX conf. COARE 98*, Boulder, Co., USA, 7-14 July 1998, WCRP-107, WMO/TD 940, WMO, Geneva, 249-250.

TAYLOR, P.K., KENT, E.C., YELLAND, M.J. & MOAT, B.I. 1999 The accuracy of marine surface winds from ships and buoys. *CLIMAR 99, WMO Workshop on Advances in Marine Climatology*, Vancouver, 8-15 Sept. 1999.

TAYLOR, P.K. & YELLAND, M.J. 2001a The dependence of sea surface roughness on the height and steepness of the waves, *J. Phys. Oceanog.*, **31**, (2), 572-590.

TAYLOR, P. K. & YELLAND, M.J. 2001b "Comment on 'On the effect of ocean waves on the kinetic energy balance and consequences for the inertial dissipation technique'." *J. Phys. Oceanogr.*, **31**, 2532-2536.

TERRAY, E.A., DONELAN, M.A., AGRAWAL, Y.C., DRENNAN, W.M. KAHMA, K.K., WILLIAMS III, A.J., HWANG, P.A. & KITAIGORODSKII, S.A. 1996 Estimates of kinetic energy dissipation under breaking waves. *J. Phys. Oceanogr.*, **26**, 792-807.

VICKERS, D. & MAHRT, L. 1997a Fetch limited drag coefficients. *Boundary-Layer Meteorol.*, **85**, 53-79.

VICKERS, D. & MAHRT, L. 1997b Quality control and flux sampling problems for tower and aircraft data. *J. Atmos. & Oceanic Tech.*, **14**, 512-526.

YELLAND, M.J., MOAT, B.I., TAYLOR, P.K., PASCAL, R.W., HUTCHINGS, J. & CORNELL, V.C. 1998 Measurements of the open ocean drag coefficient corrected for air flow disturbance by the ship. *J. Phys. Oceanogr.*, **28**, 1511-1526.

YELLAND, M. J., MOAT, B.I. & TAYLOR, P.K. 2001 Airflow distortion over merchant ships. Southampton Oceanography Centre, Southampton, UK. *Contract Report for AES/BIO. SOC Internal Report No. 74.*, 32 pp.

YELLAND, M.J., MOAT, B.I., PASCAL, R.W. & BERRY, D.I. 2002 CFD model estimates of the airflow distortion over research ships and the impact on momentum flux measurements. *(submitted to J. Atmos. Oceanic Tech.)*

Wind-Over-Waves Coupling

V.K. Makin[1] and V.N. Kudryavtsev[2]

[1] Netherlands Meteorological Institute (KNMI), De Bilt, The Netherlands,
[2] Marine Hydrophysical Institute, Sebastopol, Ukraine

Abstract

Wind-over-waves coupling is a modern theory of microscale air-sea interaction, which allows to relate the sea drag directly to the properties of wind waves and peculiarities of their interaction with the wind. Interaction of waves with ocean surface phenomena explains variability of fluxes. Role of short and dominant waves in supporting the sea drag is discussed.

1. Introduction

Interaction between Earth's atmosphere and the oceans occurs at the air-sea interface, via surface fluxes of momentum, heat, moisture and gases. These surface fluxes serve as lower boundary conditions for the general circulation models of the atmosphere and provide upper boundary conditions for general circulation models of the ocean. Any natural or anthropogenic impact on the air-sea interface will therefore change the surface fluxes and influence the global and mesoscale atmosphere and ocean circulations, that is, climate and weather. Correct description of surface fluxes is thus of importance to predict long and short term climate changes. That requires the understanding of the physics of the microscale air-sea interaction, which is to a great extent determined by the wind-wave interaction.

The last decade has seen tremendous progress in modelling air-sea fluxes, with the emergence of new ideas of, inter alia, how the wind generates waves, how a spectrum of waves mediates momentum and other transfer, and how breaking waves impact exchanges between the atmosphere and oceans. The common thread running through these new ideas is the strong dependence of the fluxes on the wave properties. Hence current understanding shows how any changes to the waves will have impacts on the surface fluxes. These developments mean that we are now in a position to study systematically the impacts of currents, swell, slicks and other complicating factors on surface fluxes. A theory which allows the assessment is called the wind-over-waves coupling (WOWC) theory. A modern WOWC theory was recently developed by Makin *et al.* (1995), Makin & Kudryavtsev (1999), Kudryavtsev *et al.* (1999), Kudryavtsev & Makin (2002), and Makin & Kudryavtsev (2002). Here a concise description of this theory is given.

2. CONCEPT OF WIND-OVER-WAVES COUPLING

Wind-over-waves coupling is a modern theory of microscale air-sea interaction, which allows to relate the sea drag directly to the properties of wind waves and peculiarities of their interaction with the wind and ocean surface phenomena, and to explain the formation of fluxes and their variability. The approach is based on the conservation equation for integral momentum:

$$u_*^2 = \tau^\nu + \overline{p\frac{\partial\eta}{\partial x}}, \tag{2.1}$$

where u_* is the friction velocity, τ^ν is the viscous surface stress, $\tau^f = \overline{p\partial\eta/\partial x}$ is the form drag at the sea surface, and a bar denotes statistical averaging. Equation (2.1) reflects a fundamental fact that the stress $\tau = u_*^2$ at the surface is formed by viscous stress and the form drag τ^f. The form drag of the sea surface η is a correlation of the wave-induced surface pressure field p with the wave slope $\partial\eta/\partial x$. As a matter of fact the second term on the right-hand side of equation (2.1) becomes dominant for moderate and high winds. It becomes immediately clear that that are waves that are responsible for formation of the stress and its variation. We keep in mind the following scheme. The atmosphere provides the energy input to the wave field. Waves grow and adjust themselves to the atmosphere. As waves support the stress the atmosphere in turn is adjusted to waves. So, the atmosphere and waves form a self-consistent system, which is in equilibrium. If any ocean surface phenomenon such as currents of any origin, swell, slicks or even rain changes the property of the wave field the balance in the system atmosphere-waves is broken, and the system adjusts itself to a new equilibrium. This explains the variability of fluxes as a result of interaction of waves with the ocean surface phenomena.

3. THE MODEL

Equation (2.1) preassumes stationary and spatial homogeneous conditions. The sea surface is described statistically in terms of the directional wave variance spectrum $F(\mathbf{k})$,where $\mathbf{k}$ is the wavenumber vector. The wind direction coincides with the mean direction of waves propagation and the wave spectrum is symmetrical relative to that direction. Relating the form drag in (2.1) to geometrical properties of the surface (described in terms of the wave spectrum) and to the properties of the energy exchange between waves and the wind, the stress at the surface is related or coupled directly to the sea state.

3.1. *The form drag*

Two main mechanisms of the wind-wave interaction that support the form drag are distinguished. When the wavy surface is regular (in a sense that there are no wave breaking events) the wind flows over the wave smoothly, i.e. the surface is streamlined. This regime of wind-wave interaction is described in terms of the non-separeted sheltering mechanism (Belcher & Hunt 1993), which provides the energy flux to waves from the wind. A part of the form drag supported by the non-separated sheltering mechanism: the wave-induced stress τ_w^f can be written

$$\tau_w^f = \int_k \int_\theta \beta c^2 B(k,\theta) \cos\theta d \ln k d\theta, \tag{3.1}$$

where $B = k^4 F$ is the saturation wave spectrum, c is the phase speed, θ is the angle, and β is the dimensionless energy flux to waves or the growth rate parameter. The growth rate parameter is taken in the form

$$\beta = C_\beta \left(\frac{u_*}{c}\right)^2, \tag{3.2}$$

where the proportionality coefficient is dependent on wave parameters

$$C_\beta = c_\beta \kappa^{-1} \ln \frac{\pi}{k z_c}, \tag{3.3}$$

$z_c = \exp[\kappa c/(u_* \cos\theta)]$, and c_β is a constant close to 2. Notice, that $C_\beta \to 0$ both for very long waves (c/u_* is large) and for very short waves(k is large).

It is a common knowledge that waves intensively break on the sea surface. There is increasing experimental evidence that breaking waves play a significant role in the dynamics of the lower atmosphere (e.g. Melville 1996). A significant augmentation of the surface local stress above breaking waves is reported in laboratory experiments (e.g. Banner 1990; Giovanangeli *et al.* 1999). In these studies it has been established that the air flow separation (AFS) from the crest of breaking waves is responsible for this augmentation. The impact of the air flow separation from breaking waves on the sea drag was accounted for in Kudryavtsev & Makin (2001). They assumed that the sea surface can be presented as a streamlined surface covered by areas, where the air flow separation takes place. The air flow separation occurs intermittently on the sea surface, where wave breaking fronts arise. It was further assumed that the stress due to separation is proportional to the pressure drop Δp_s on the forward side of the breaking front and to the total length of wave breaking fronts $\sum l_i$ (details see in Kudryavtsev & Makin 2001). The quantity $1/S \sum l_i$ is the average total length of breaking fronts per unit surface introduced originally by Phillips (1985)

$$\frac{1}{S} \sum l_i = \Lambda(\mathbf{c}) d\mathbf{c}, \tag{3.4}$$

where the distribution $\Lambda(\mathbf{c})$ represents the surface density of the total length of wave breaking fronts that have velocities in the range $\mathbf{c}$ to $\mathbf{c} + d\mathbf{c}$. The drop of pressure induced by the separation acts on the wave breaking front during a short period of time and then disappears. The pressure drop can be estimated by using the analogy between the AFS from breaking waves and separated flows typical of the backward facing step. It can be thus parameterized as

$$\Delta p_s = \frac{1}{2} \gamma {u_s}^2, \tag{3.5}$$

where γ is an empirical constant close to 1, u_s is the reference speed defined as a positive difference between the mean wind speed at a reference level specified here at $z = 1/k$ and the phase speed of the wave

$$u_s = \frac{u_*}{\kappa} \cos\theta \ln\frac{1}{kz_0} - c, \tag{3.6}$$

where z_0 is the roughness parameter defined through the logarithmic wind profile

$$U(z) = \frac{u_*}{\kappa} \ln\frac{z}{z_0} \tag{3.7}$$

extending to the surface from a height where the wind velocity is not influenced by wave motions. Introducing these assumptions it was shown that the separation stress supported by the AFS from all waves has a general form

$$\tau_s^f = \varepsilon_b \gamma \int_{\mathbf{c}} u_s^2 \cos\theta k^{-1} \Lambda(\mathbf{c}) d\mathbf{c}, \tag{3.8}$$

where $\varepsilon_b = 0.5$ is the characteristic slope of the breaking wave. Notice, that even if a fast wave propagating with the phase velocity close to or faster than the mean wind speed breaks it will not support the stress because the separation cannot take place under these conditions.

3.1.1. *Stress supported by the AFS from equilibrium range of short gravity waves*

The separation stress supported by short gravity waves in the equilibrium range of the spectrum $\tau_{s_{eq}}^f$ was obtained by Kudryavtsev & Makin (2001). Following the approach by Phillips (1985) the distribution function $\Lambda(\mathbf{c})$ is directly related to the average rate of the energy dissipation per unit area by breakers with velocities between $\mathbf{c}$ and $\mathbf{c}+d\mathbf{c}$. It was further assumed that under the steady condition the energy dissipation due to wave breaking is equal or proportional to the energy input from the wind in the equilibrium range of the wind wave spectrum. The total length of wave breaking fronts can be then expressed in terms of the saturation spectrum as

$$\Lambda(\mathbf{c}) \sim \frac{\beta}{b} B(\mathbf{c}) k^{-1} \tag{3.9}$$

with $b = 0.01$ being an empirical constant. With (3.9) the separation stress (3.8) can be written as

$$\tau_{s_{eq}}^f = \varepsilon_b \gamma b^{-1} \int_\theta \int_{k<k_m} u_s^2 \beta(k,\theta) B(k,\theta) \cos\theta d\theta d\ln k. \tag{3.10}$$

The integration over the wavenumber k in (3.10) is done in the wavenumber range satisfying the condition $k < k_m$, where $k_m = 2\pi/\lambda_m$ rad m^{-1} and the wavelength $\lambda_m = 0.1$ m. This condition reflects the fact that waves shorter than λ_m rather generate parasitic capillaries than break as discussed by Kudryavtsev *et al.* (1999). The generation of parasitic capillaries prevents the formation of the sharp surface slope and hence prevents the separation of the air flow from these short waves.

3.1.2. *Stress supported by the AFS from dominant waves*

When seas are young dominant waves (waves at the spectral peak) break intensively (Babanin *et al.* 2001). They contribute to the separation stress and impact the sea drag. The range of dominant waves is defined here by the condition $k \leq 2k_p$, where k_p is the spectral peak wavenumber. The assumption that dissipation is balanced by wind input is not valid in the range of the spectral peak and Equation (3.9) does not hold there. The statistics of dominant waves breaking will be described by a breaking wave model based on a concept of a threshold level. The model is based on the description of statistical properties of a Gaussian wave surface, where it is assumed that the wave breaking event takes place when the sea surface exceeds some threshold level. The detailed analysis of statistical properties for a random, moving, Gaussian surface is given by Longuet-Higgins (1957). He derived a general expression for the mean length of a contour for the crossection of the wavy surface by a plane of a constant height ζ_0 per unit area $\overline{s}$ (his Equation 2.3.16). ζ_0 is defined as the height of a counter above which the onset of dominant wave breaking occurs. It is assumed that when the surface level exceeds the 'threshold' level ζ_0, a strong and sudden (explosive) instability erupts and causes the onset of breaking. Assuming further that dominant waves can be presented as a superposition of narrow band random surface waves it can be shown (Makin & Kudryavtsev 2002) that the length of the contours of the breaking zone is

$$\overline{s} = \frac{1}{\pi} k \exp\left(-\frac{\varepsilon_T^2}{\varepsilon_d^2}\right), \tag{3.11}$$

where $\varepsilon_d = H_d k_p/2$ is the dominant wave steepness (H_d is their significant wave height) and $\varepsilon_T = \sqrt{2}\zeta_0 k_p$ is a tuning constant. Taking into account that the length of breaking fronts is approximately twice less than $\overline{s}$, the average total length per unit surface area of breaking fronts of dominant waves is

$$\Lambda(\mathbf{c})d\mathbf{c} = \frac{1}{2}\overline{s} \tag{3.12}$$

and the separation stress (3.8) supported by dominant waves is

$$\tau_{s_d}^f = \frac{\varepsilon_b \gamma}{2\pi} u_{sd}^2 \exp\left(-\frac{\varepsilon_T^2}{\varepsilon_d^2}\right). \tag{3.13}$$

Here the reference wind speed for dominant waves u_{sd}

$$u_{sd} = \frac{u_*}{\kappa} \ln \frac{\varepsilon_b}{k_p z_0} - c_p \tag{3.14}$$

is specified at the level just above breaking dominant waves, i.e. at $z = \varepsilon_b/k_p$, and c_p is the phase speed at the spectral peak.

3.2. *Viscous stress*

Patching the linear wind profile inside the viscous layer with the logarithmic wind profile above it, the viscous stress can be written

$$\tau^\nu = (\kappa d)^{-1} \ln\left(\frac{\delta}{z_0}\right) u_*^2, \tag{3.15}$$

where

$$\delta = d\frac{\nu}{u_*} \tag{3.16}$$

is the thickness of the viscous sublayer, ν is the molecular viscosity, $d = 12$ is a constant.

3.3. *Resistance law of the sea surface*

Equation (2.1), where viscous stress is calculated according to (3.15) and the form drag τ^f can be evaluated through (3.1), (3.10) and (3.13), describes the resistance law of the sea surface relating the stress to properties of the wave field. Given the wind speed at a specified height and a wave spectrum Equation (2.1) is solved by iterations to provide the sea surface stress.

3.4. *Specification of the wave spectrum*

To obtain the stress the wave spectrum should be known. It can be described by an empirical model or by a physical model of the wave spectrum. The former will provide a 'one-way' coupling: adjustment of the atmosphere to a given wave field. To study the self-consistent adjustment of the atmosphere and waves, and potentially the variability of fluxes a physical model of the wave spectrum is necessary. To calculate the stress due to the AFS supported by short gravity waves $\tau^f_{s_{eq}}$, the equilibrium part of the wave spectrum defined at $k < k_m$ has to be known. The calculation of the separation stress supported by dominant waves $\tau^f_{s_d}$ requires the shape of the spectrum at the spectral peak, while the calculation of the wave-induced stress τ^f_w requires the shape of the spectrum in the wavenumber range from capillary waves to the spectral peak. A composite model of the wave spectrum (Kudryavtsev *et al.* 1999) describes the saturation spectrum $B(k, \theta)$ in the full wavenumber range from few millimeters up to the spectral peak. It consists of two parts: the low and the high wavenumber spectrum

$$B(k, \theta) = B_l(k, \theta) + B_s(k, \theta). \tag{3.17}$$

The shape of the low wavenumber (at the spectral peak) spectrum B_l defined by the inverse wave age parameter U_{10}/c_p is given by the empirical model by Donelan *et al.* (1985). The shape of the high wavenumber spectrum B_s results from the physical model developed by Kudryavtsev *et al.* (1999). The model is based on the energy balance equation and accounts for wind input, viscous dissipation, dissipation due to wave breaking (including energy losses due to generation of parasitic capillaries by short gravity waves), and nonlinear three-wave interaction. As it will be shown the short waves support most of the sea drag, that is why their description through a balance of physical mechanism is crucial. The stress is defined by the saturation spectrum B. The saturation equilibrium spectrum B_s in turn depends on the stress via the wind input. Thus the wind waves and the atmospheric boundary layer are strongly coupled forming a self-consistent dynamical system.

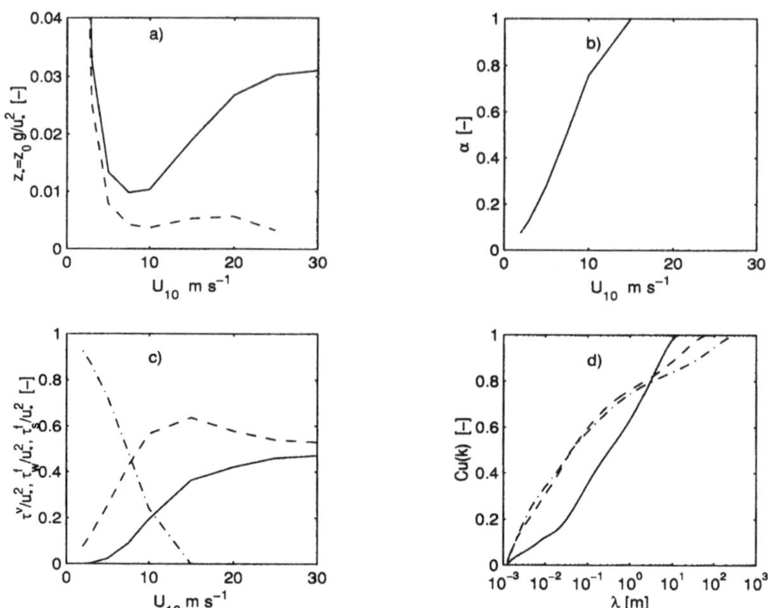

FIGURE 1. (a) Charnock parameter $z_0 g/u_*^2$ versus U_{10}. Model results: solid line, separation stress is accounted for; dashed line, separation stress is not accounted for. (b) The same as in (a) but for the coupling parameter α. (c) Stress contributions. Solid line, stress due to separation τ_s^f/u_*^2; dashed line, wave-induced stress τ_w^f/u_*^2; dashed-dotted line, viscous stress τ^ν/u_*^2. (d) Cumulative spectrum $Cu(k)/\tau^f$ of the form drag τ^f versus the wavelength λ. Solid line, wind speed $U_{10} = 5\text{ms}^{-1}$; dashed line, $U_{10} = 10\text{ms}^{-1}$; dashed-dotted line, $U_{10} = 20\text{ms}^{-1}$.

4. ROLE OF SHORT WAVES IN SUPPORTING THE SEA DRAG

We first analyze a case of a fully developed sea specified by the inverse wave age parameter $U_{10}/c_p = 0.83$. In Figure 1a the dimensionless roughness length or the Charnock parameter $z_0 g/u_*^2$ is shown as a function of the wind speed. The solid line represents the solution of the full model (the stress due to the AFS is accounted for), while the dashed - the model solution when the stress due to the AFS is not accounted for. The additional stress supported by the AFS is responsible for a well pronounced wind speed dependence of the Charnock parameter shown in Figure 1a. In the range of the wind speed from 10 ms^{-1} to 20 ms^{-1} the Charnock parameter increases twice in correspondence with field measurements (e.g. Yelland & Taylor 1996). A strong increase of the Charnock parameter at low winds reflects the transition of the sea surface from the aerodynamically rough to the smooth condition.The coupling parameter α defined as the ratio of the form drag to the total drag $\alpha = \tau^f/u_*^2$ is shown in Figure 1b. At high wind speeds most of the drag is due to the form drag, while at low wind speeds the viscous stress dominates. To show the role of the air flow separation in the momentum transfer the contribution to the total stress u_*^2 of viscous stress τ^ν/u_*^2, the wave-induced stress τ_w^f/u_*^2, and the stress due to the AFS τ_s^f/u_*^2 as a function of the wind speed is shown in Figure 1c. Notice, that $\tau^\nu/u_*^2 + \tau_w^f/u_*^2 + \tau_s^f/u_*^2 = 1$. For low wind speeds $U < 5$ ms^{-1}

viscous stress dominates the sea surface drag while the role of the form drag is negligible. With the increase of the wind speed the role of the form drag becomes pronounced. At the wind speed $U > 10$ ms^{-1} the surface drag is mainly supported by the wave-induced and the AFS stresses. The relative role of the stress due to the AFS increases with increasing the wind speed and for high wind speeds it supports about 50% of the total stress.In Figure 1d the cumulative spectrum of the form drag τ^f as a function of the wavelength λ is shown to illustrate the role of waves from different ranges of the wavenumber in supporting the stress. For all wind speeds shown about 80% of the form drag is supported by waves shorter than few meters. Short gravity-capillary and capillary waves support a significant part of the stress especially for higher winds. If these waves are damped by e.g. oil, so does the stress which they support. That will results in lower values of the drag coefficient as compared to a case of a clean sea surface (Grodskii *et al.* 1999) This is an example how an ocean surface phenomenon changing the properties of waves impacts the stress.

For a fully developed sea dominant waves do not contribute to the stress. Though their breaking occurs (Banner *et al.* 2000) they propagate with the phase speed exceeding the mean wind speed. The separation cannot occur under such conditions. All the separation stress due to the AFS comes from waves in the equilibrium range. The wave-induced flux to dominant waves is also very small because the growth rate parameter is small ($C_\beta \to 0$).

5. Role of dominant waves in supporting the sea drag

For young seas characterized by increased inverse wave age parameter, the role of dominant waves in forming the sea drag becomes more and more important (Makin & Kudryavtsev 2002). Supporting about 10% of the total stress at $U_{10}/c_p = 2$, and about 20% at $U_{10}/c_p = 3$, their contribution becomes dominant for very young seas $U_{10}/c_p = 5$.

5.1. *Wave age dependence of the sea drag*

The Charnock parameter $z_0 g/u_*^2$ as a function of the inverse wave age parameter based on the friction velocity u_*/c_p is shown in Figure 2. Calculations are done for the wind speed $U_{10} = 7.5$ m s^{-1} and $U_{10} = 20$ m s^{-1} and the inverse wave age parameter $0.83 < U_{10}/c_p < 25$. Data are compiled from Donelan *et al.* (1993), their Figure 2. Model results (as well as data) show a clear increase of the Charnock parameter with increasing the inverse wave age. This increase is explained by the model by the increased steepness of young waves and thus increased separation stress due to the AFS from dominant waves. The Charnock parameter has a maximum around $u_*/c_p = 0.3 \div 0.4$ ($U_{10}/c_p = 7$). For higher values of the inverse wave age parameter corresponding to very young waves typical for small water bodies and laboratory conditions, the Charnock parameter decreases. This is explained by the fact that the wind wave spectrum is very narrow in the wavenumber space, and the separation stress from the equilibrium range becomes smaller. The dominant waves are very peaked but small in height, and the reference speed (3.14) is rapidly dropping reducing the separation stress from dominant waves. Both effects lead to decreasing of the

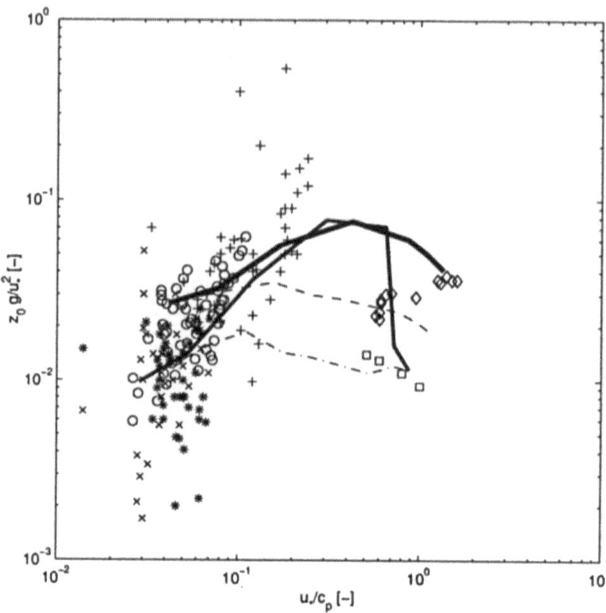

FIGURE 2. Charnock parameter z_0g/u_*^2 versus inverse wave age u_*/c_p. Model results: thin solid line - $U_{10} = 7.5$ ms^{-1}; thick solid line - $U_{10} = 20$ ms^{-1}; dashed-dotted line - $U_{10} = 7.5$ ms^{-1} for $\tau_d^s = 0$; dashed line - $U_{10} = 20$ ms^{-1} for $\tau_d^s = 0$. Symbols indicate data: circles - North Sea; pluses - Lake Ontario; stars - Atlantic Ocean, long fetch; x-marks - Atlantic Ocean, limited fetch; diamonds and squares - wave tanks.

Charnock parameter. Despite the scatter of the data is huge, the model results in general agree well with measurements.

To outline the role of separation from dominant waves we switched off this stress in the model and plot results in the same Figure. There is only a marginal increase of the Charnock parameter with increasing inverse wave age. This suggests that the separation from dominant waves is responsible for the observed behavior of the Charnock parameter with inverse wave age. It is also clear that for waves close to a fully developed $u_*/c_p < 0.1$ ($U_{10}/c_p < 1.5$) the dominant waves do not support the separation stress as already explained in the previous section.

5.2. *Depth dependence of the sea drag*

The fact that the AFS from dominant waves contribute a noticeable part to the total stress (sea drag) suggests a mechanism, which could explain a known experimental fact that the drag coefficient is higher in the shallow waters as compared to the open ocean data (e.g. Geernaert 1990). When waves propagate into the shallow water long waves begin to feel the bottom and become steeper. Hence, the depth limited spectra is expected to be more peaked compared to spectra in the deep water (Young & Verhagen 1996). That leads to the enhanced breaking of dominant waves. The enhanced breaking leads to the enhanced separation of the air flow from dominant waves, which gives rise to the total stress (Makin & Kudryavtsev 2002).

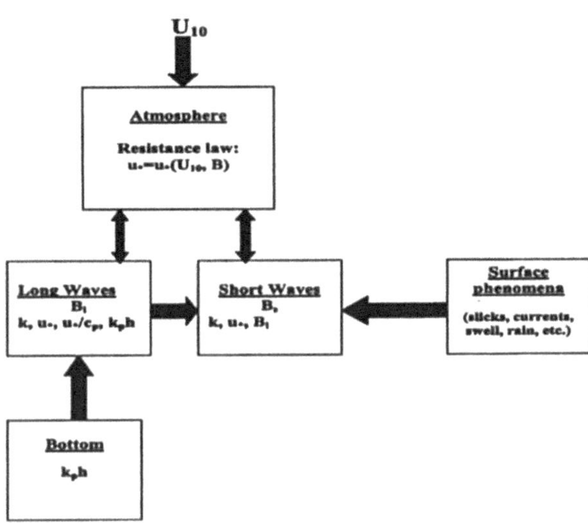

FIGURE 3. Schematic description of wind-over-waves coupling. Arrows indicate impacts.

6. SUMMARY

Wind-over-waves coupling - a modern theory of microscale air-sea interaction is presented. The theory allows to relate the sea drag directly to the properties of wind waves and peculiarities of their interaction with the wind and to explain the formation of fluxes. Waves interact with ocean surface phenomena, change their properties - and thus the stress they support. This explains the variability of fluxes. Schematic description of wind-over-waves coupling is presented in Figure 3. Waves play a crucial role in supporting the sea drag. In a fully developed sea almost all the stress is supported by short waves at moderate and high winds. The air flow separation from short waves plays a significant role supporting to about 50% of the stress at high wind speeds. The wind speed dependence of the Charnock parameter is explained by the separation from short waves, while the role of dominant waves is negligible. However, the separation of the air flow from dominant waves begin to play a significant role in supporting stress for young seas and explains the wave age and the finite bottom depth dependencies of the sea drag. WOWC theory provides a framework to study sea surface fluxes and their variability.

REFERENCES

BABANIN, A.V., YOUNG, I.R. & BANNER, M.L. 2001 Breaking probability for dominant waves on water of finite constant depth. *J. Geophys. Res.* **106**, 11659-11676.

BANNER, M.L. 1990 The influence of wave breaking on the surface pressure distribution in wind wave interaction. *J. Fluid Mech.* **211**, 463-495.

BANNER, M.L., BABANIN, A.V. & YOUNG, I.R. 2000 Breaking probability for dominant waves on the sea surface. *J. Phys. Oceanogr.,* **30,** 3145-3160.

BELCHER, S.E. & HUNT, J.C.R. 1993 Turbulent shear flow over slowly moving waves. *J. Fluid Mech.* **251**, 109-148.

DONELAN, M.A., HAMILTON, J. & HUI, W.H. 1985 Directional spectra of wind generated waves. *Phil. Trans. R. Soc. London, Ser. A* **315**, 509-562.

DONELAN, M.A., DOBSON, F.W., SMITH, S.D. & ANDERSON, R.J. 1993 On the dependence of sea surface roughness on wave development. *J. Phys. Oceanogr.,* **23**, 2143-2149.

GEERNAERT, G.L. 1990 Bulk parameterizations for the wind stress and heat fluxes. In *Surface Waves and Fluxes, Vol. 1*, G.L. Geernaert & W.J. Plant (eds.), Kluwer Academic Publishers, Dordrecht, pp. 91-172.

GIOVANANGELI, J.P., REUL, N., GARAT, M.H. & BRANGER, H. 1999 Some aspects of wind-wave coupling at high winds: an experimental study. In *Wind-Over-Wave Couplings*, S.G. Sajjadi *et al.* (eds.), Oxford University Press, pp. 81-90.

GRODSKII, S.A., KUDRYAVTSEV, V.N. & MAKIN, V.K. 1999 An estimate of the influence of the sea surface pollution on short wind waves and parameters of the atmospheric boundary layer. *Morskoy Gidrophizicheskiy Zhurnal*, **6**, 3-14, (in Russian). (Engl. transl.: GRODSKII, S.A., KUDRYAVTSEV, V.N. & MAKIN, V.K. 2001 Evaluation of the influemce of surface films on short wind waves and the characteristics of the boundary layer of the atmosphere. *Physical Oceanography.* **11**, No. 6, 495-508.

KUDRYAVTSEV, V.N. & MAKIN, V.K. 2001 The impact of air-flow separation on the drag of the sea surface. *Boundary-Layer Meteorol.* **98**, 155-171.

KUDRYAVTSEV, V.N., MAKIN, V.K. & CHAPRON, B. 1999 Coupled sea surface-atmosphere model 2. Spectrum of short wind waves. *J. Geophys. Res.* **104**, 7625-7639.

LONGUET-HIGGINS, M.S. 1957 The statistical analysis of a random moving surface. *Phil. Trans. R. Soc. London, Ser. A* **249**, 321-387.

MAKIN, V.K. & KUDRYAVTSEV, V.N. 1999 Coupled sea surface-atmosphere model 1. Wind over waves coupling. *J. Geophys. Res.* **104**, 7613-7623.

MAKIN V.K. & KUDRYAVTSEV, V.N. 2001 Impact of dominant waves on sea drag. *Boundary-Layer Meteorol.*, **103)**, 83-99.

MAKIN V.K., KUDRYAVTSEV, V.N. & MASTENBROEK, C 1995 Drag of the sea surface. *Boundary-Layer Meteorol.* **79**, 159-182.

MELVILLE, W.K. 1996 The role of surface-wave breaking in air-sea interaction. *Ann. Rev. Fluid Mech.* **28**, 279-321.

PHILLIPS, O.M. 1985 Spectral and statistical properties of the equilibrium range in wind generated gravity waves. *J. Fluid Mech.* **156**, 505-531.

YELLAND, M. & TAYLOR, P.K. 1996 Wind stress measurements from the open ocean. *J. Phys. Oceanogr.* **26**, 541-558.

YOUNG, I.R. & VERHAGEN, L.A. 1996 The growth of fetch limited waves in water of finite depth. Part 2. Spectral evolution. *Coastal Eng.* **29**, 79-99.

Sea Surface Roughness Parameterization

I. Papadimitrakis and A.I. Papaioannou
National Technical University of Athens Department of Civil Engineering
Hydraulics, Water Resources and Maritime Engineering

Abstract

The contribution of wind-generated waves to the deep-sea surface roughness, in the absence of a swell, is examined under neutral stability conditions. Based on a modified Banner and Melville wave breaking mechanism, a critical frequency is determined for partitioning the spectrum of surface waves among 'short' and 'long' wave components, where the latter do not (always) contribute to the surface roughness. This partitioning allows estimation of an effective sea surface roughness as a fraction of the sea surface variance (i.e. a fraction of the zero order moment of the spectrum). When extensive, large-scale, wave breaking occurs near the spectral peak frequency the entire wave spectrum contributes to the surface roughness.

The proposed model for estimating the critical-partitioning frequency, and thereafter the effective sea surface roughness and drag coefficient, utilizes both observed spectra collected from North Aegean and other Greek seas and theoretical formulations of the latter.

The critical frequency, as well as the surface roughness (z_0) and the drag coefficient (C_d) appear to be functions of two essential parameters of the wave field, namely the wave age and the significant slope. The critical frequency is found to decrease with (increasing) the inverse wave age and the significant slope approaching asymptotically the spectral peak frequency and causing a greater part of the spectrum to contribute to the sea surface roughness (as expected).

Results are compared with other field observations of both z_0 and C_d. Similarities and discrepancies among the various data sets, as well as among our results and theoretical considerations are discussed. The importance of the significant wave slope and of the stage of wave development in determining the drag coefficient is also examined.

1. Introduction

The influence of surface wave condition on the wind-stress (or drag) coefficient above an air-water interface, in the absence (and in fewer cases in the presence) of a swell and under neutral stability conditions, has been long recognized by a number of investigators, as for example: Kitaikorodskii & Volkov (1965), Melville (1977), Donelan (1982), Huang *et al.* (1986) and more recently by Donelan *et al.* (1993), Makin *et al.* (1995), Anctil & Donelan (1996), Makin (1999), Makin & Kudryavtchev (1999), Kudryavtchev & Makin (2001) and Taylor & Yelland (2001). The associated research efforts span over a period of almost four decades and have analyzed the subject matter from different perspectives.

Such interfacial effects are usually incorporated in the roughness length (z_0)

of a water surface ruffled by wind-generated waves in the absence and/or presence of a swell, a quantity which characterizes the roughness of an air-water interface and appears in expressions of the mean wind velocity profile, assuming that the latter follows the logarithmic 'law of the wall'. The existence of a logarithmic wind velocity profile above water waves, and under neutral stability conditions, has been well established for both oceanic and laboratory conditions. Thus, the wind-stress (or drag) coefficient, defined as the square of the ratio of the wind friction velocity, u_*, and the time mean wind velocity, U_z, measured (usually) at the height, z, of 10 m from the mean water level, is expressed as: $C_d = [\kappa/\ln(z/z_0)]^2$, where $\kappa(\approx 0.40 - 0.41)$ is the Von Karman constant. Preferably, the difference between the mean wind velocity U_{10} and a suitable surface drift current velocity U_s (i.e. $U_{10} - U_s$) is used in the denominator of the above definition ratio.

Certainly, z_0 depends on both the local wind conditions and the characteristics of the traveling surface waves. However, due to the complexities of the latter, the lower limit of the log-linear profile is not well established and most attempts to assess the effects of wave characteristics on the wind-stress coefficient have not been quite conclusive thus far.

For intermediate values of the parameter $\tilde{\omega}_p(= \omega_p u_*/g)$, Kitaigorodskii (1968) has shown that the effective height of the roughness elements (h_o) of a moving wave surface is related to the 1-D spectrum of the waves, $S(\omega)$, by the following relationship:

$$h_0 \approx \left[\int_0^\infty S(\omega)\exp\{-2\kappa g/(u_*\omega)\}\,d\omega\right]^{1/2} \tag{1.1}$$

where $S(\omega), \omega$ and g represent wave spectral density, angular frequency and gravitational acceleration, respectively, and the subscript p refers to spectral peak values; z_0 is considered to be proportional to h_0. Equation (1.1) is based on various assumptions that are frequently violated in the field.

Melville (1977) has proposed a model for z_0, based on the premise that the roughness elements are the small scale breaking waves for which $c \cong u_*$, where c represents the celerity of high frequency waves. His approach, however, maybe oversimplified due to the assumptions made.

C_d is also expressed as: $[\kappa/\ln\{(\alpha C_d)^{-1}Fr^{-2}]^2$, where $Fr = U_{10}/(gz)^{1/2}$ or $(U_{10} - U_{10})/(gz)^{1/2}$; α is a non-dimensional constant (Charnock's constant $= z_0 g/u_*^2$) which, however, depends on the roughness Reynolds number ($Re = u_* z_0/\nu$) and the stage of surface wave development; ν is the air kinematic viscosity. Under steady state, thermally neutral stability conditions and in the absence of a swell, the stage of wave development maybe expressed by the wave age, c_p/u_* (or c_p/U_{10}), and/or some characteristic slope of the wave field. Donelan *et al.* (1993), for example, has suggested that under such conditions $\alpha \approx 0.025(c_p/u_*)$, but various other expressions for a are available in the literature, covering different dynamical regimes. An interesting expression for α is, in the authors opinion, that proposed by Melville (1977), despite the limitations of his theory. Melville's expression incorporates the influence of the presence of a swell on a via the swell slope, B, and a wave age which is based on the swell phase speed (assuming that the swell frequency corresponds to the peak frequency of the wave spectrum). With that expression of α, the trend in the variation of C_d with the swell slope is captured satisfactorily, when the air flow regime changes from smooth to fully rough, in general agreement with field data.

It has also been suggested by Kitaigorodskii & Donelan (1984) that for large and small values of the parameter $\tilde{\omega}_p[= \omega_p u_*/g = u_*/c_p = (c_p/u_*)^{-1}]$, respectively, h_0 maybe approximated by either of the relationships:

$$h_0 \approx \left[\int_0^\infty S(\omega)\,d\omega\right]^{1/2}, \qquad h_0 \approx a' u_*^2/g \tag{1.2}$$

where a' is another non-dimensional constant similar to Charnock's α. It must be emphasized that the field and laboratory data of α, when plotted as a function of $(c_p/u_*)^{-1}$, present a lot of scatter. Similarly, plots of C_d against U_{10} show a lot of scatter. Although some of the data disparity maybe attributed to experimental uncertainties, it is the lack of incorporation of important physical processes in past descriptions of C_d, which is responsible for most of this scatter. As Donelan *et al.* (1993) have stated, a complete formula for wave roughness, z_0, ought to be able to describe both field and laboratory data equally well, under a variety of conditions.

The dependence of z_0 on u_* varies in form, as u_* changes from a value of about 10^{-2} to 10 m/s, and this reflects the difference in the dynamics of the various flow regimes (from smooth to fully rough) encountered in the field, as the wind speed increases from low to extreme values (Krauss & Businger 1994).

This work provides an alternative parameterization of the roughness length, z_0, that corresponds to a spectrum of wind-waves formed at an air-water interface, in the absence of a swell and under neutral thermal stability conditions. It utilizes the older ideas of Banner & Melville (1976) and Melville (1977), but avoids the weakness of Melville's assumption. The proposed parameterization is based on kinematic wave breaking criteria, other theoretical formulations and/or field observations, and is given in terms of parameters expressing the stage of wave development. The layout of this paper has, briefly, as follows. Some theoretical arguments are presented in section 2 for determining a critical frequency, ω_{cr}, to partition the spectrum of waves and estimate an effective surface wave variance, a_e, that is considered to be equal to z_0. The important parameters that describe the characteristics of this frequency are also identified. Some forms of the wave spectrum and the available observations from Greek seas, used to verify our z_0 model, are presented briefly in sections 3 and 4, respectively. Section 5 describes the derived results and compares them with similar data available from the literature. Finally, section 6 comments shortly on the behavior of the derived results and makes further suggestions for additional work.

2. THEORETICAL CONSIDERATIONS

The proposed model for the surface wave roughness, z_0, exploits, modifies and extends the ideas proposed by Melville (1977) about (certain) breaking waves that contribute to the formation of z_0, utilizing the Banner & Melville (1976) air flow separation and breaking mechanism.

Experimental evidence in the field and theoretical calculations suggests that, mainly the high frequency waves contribute to the roughness and drag of the surface, as the low frequency components (which can also be regarded as swell) modulate simply the flow field on both sides of the interface. This is certainly true, except when extensive, large scale breaking occurs near the spectral peak frequency. Therefore, it is safe to argue that all waves, in the spectrum, with frequency $\omega \leqslant \omega_{cr}$ will not contribute to the formation of surface roughness length. Thus, ω_{cr} is an appropriate frequency that partitions the spectrum of waves into two regions, one of low frequency and another of high frequency components where only the latter region contributes to the formation of surface roughness. The effective variance of the surface wave spectrum which, then, contributes to the formation of z_0 is given as:

$$a_e = \left[a_{\text{rms}} - 2\int_0^{\omega_{cr}} S(\omega)\, d\omega\right]^{1/2}, \quad a_{\text{rms}} = \left[2\int_0^{\infty} S(\omega)\, d\omega\right]^{1/2} \tag{2.1a, b}$$

Banner & Melville (1976) demonstrated that flow separation occurs in the vicinity of a breaking wave, which has a stagnation point near its crest, and this finding has been also substantiated by theoretical computations. Now, using kinematic criteria, it can be argued that a (single) gravity wave breaks when the orbital velocity at the wave crest, augmented by the local wind-induced drift current, q_c, exceeds the phase speed, c, of the wave, viz.:

$$c \leqslant u^c_{orb} + q_c \tag{2.2}$$

Here, u^c_{orb} is the orbital velocity induced by the wave at its crest (neglecting the effects of a viscous sub-layer there), and q_c is the corresponding tangential net surface drift (the difference between the total drift and mass transport). According to Phillips (1977), q_c may be calculated from the following relationship:

$$q_c = c - u^c_{orb} - [(c - u^c_{orb})^2 - q_0(2c - q_0)]^{1/2} \tag{2.3}$$

where q_0 represents the value of q at the intersection of the waveform with the mean sea level (MSL) and maybe taken as either $0.03U_{10}$ or $0.53u_*$, although there is experimental evidence that it monotonically decreases with the wind speed. Since for a monochromatic wave of frequency ω and phace speed c, in the gravity range, $\omega = g/c$, it is safe to argue that for a spectrum of waves where all waves with phase speeds $c \leqslant u^c_{orb} + q_c$ break and contribute to the formation of surface roughness, the critical frequency, ω_{cr}, can be determined as:

$$\omega_{cr} = g/(u^c_{orb} + q_c) \tag{2.4}$$

Furthermore, since for a monochromatic wave of amplitude $\boldsymbol{a}$ and frequency $\boldsymbol{\omega}, u^c_{orb} = \boldsymbol{\omega}.\boldsymbol{a}$, it might be argued that for a spectrum of waves a characteristic orbital velocity (corresponding, say, to an appropriate peak value) maybe determined as:

$$u^c_{orb} = \left[2\int_0^\infty \omega^2 S(\omega)\,d\omega\right]^{1/2} \quad \text{or} \quad u^c_{orb} = \left[2\int_0^\infty gk^2 S(k)\,dk\right]^{1/2} \qquad (2.5a,b)$$

where k is the wave number corresponding to ω and related to it through the dispersion relationship.

It has been suggested that now (i.e., in the presence of a spectrum of waves) q_c can be calculated from an expression similar to (2.3) with c replaced by c_p, the phase velocity of the dominant wave. Equations (2.1)–(2.5), along with an appropriate form of the wave spectrum, can yield z_0.

Combining now Eqs. (2.3) and (2.4) yields:

$$\omega_{cr}/\omega_p = \left\{1 - [(1-B_{ef})^2 - \gamma_1(2-\gamma_1)]^{1/2}\right\}^{-1} \qquad (2.6)$$

where $B_{ef} = u^c_{orb}/c_p$ and $\gamma_1 = q_0/c_p \approx 0.03(U_{10}/c_p)$. Equation (2.6) also yields $(\omega_{cr})_{\min} = \omega_p$ when $(B_{ef})_{\max} = 1 - [\gamma_1(2-\gamma_1)]^{1/2}$, assuming that $U_{10}/c_p \leqslant 66.67$, a relationship usually satisfied in the field. For the range of wave ages encountered in the field, $B_{ef} < (B_{ef})_{\max}$ and $\omega_{cr} > \omega_p$. B_{ef} expresses a characteristic slope of the wave field, and for a monochromatic wave of amplitude $\boldsymbol{a}$ and wave number $\boldsymbol{k}$ it reduces to $\boldsymbol{a.k}$ (as $\omega/c = k$). Equation (2.6) suggests that ω_{cr} depends essentially on two parameters, namely the wave age c_p/U_{10} and a characteristic slope of the wave field. It is noted that the ratio c_p/u_* is a more appropriate wave age parameter than c_p/U_{10}, since u_* is a representative velocity of the wind velocity profile, better than U_{10}.

A statistical measure of the steepness of a wave field is the significant slope, § (Huang *et al.* 1981), although other measures of this quantity have appeared in the literature, as for example the rms slope, a quantity that is however more difficult to obtain. The significant slope appears to be an important parameter of a wind-generated wave field and is defined as: $\S = a_{\text{rms}}\sqrt{2}\lambda_p$ with $\lambda_p = 2\pi/k_p$ and $k_p = \omega_p/g$. B_{ef} is related to § through a simple relation of the form: $B_{ef} = a_1.\pi.\S$, where a_1 is a numerical constant whose absolute value depends on the form of the spectrum used to evaluate u^c_{orb}. Figures 1(a,b) show the variation of ω_{cr}/ω_p as a function of either the inverse wave age (with the characteristic slope as a parameter) or as a function of § (and the inverse wave age as a parameter). As expected, ω_{cr}/ω_p decreases with increasing the inverse wave age (and/or characteristic slope) when the latter (or the former) remains constant.

With the critical frequency, ω_{cr}, known the effective variance a_e, contributing to the formation of surface roughness length z_0, can be easily calculated, since:

$$a_e = \left[2\int_{\omega_{cr}}^\infty S(\omega)\,d\omega\right]^{1/2} \qquad (2.7)$$

Expressions for a_e, in terms of the wave age and the significant slope, can be obtained with the aid of the analytical form of the wave spectrum and by specifying the frequency interval where ω_{cr} falls in. When extensive, large scale breaking occurs near the spectral peak, it is reasonable to assume that the whole wave spectrum contributes to the dynamic roughness. In that case it is

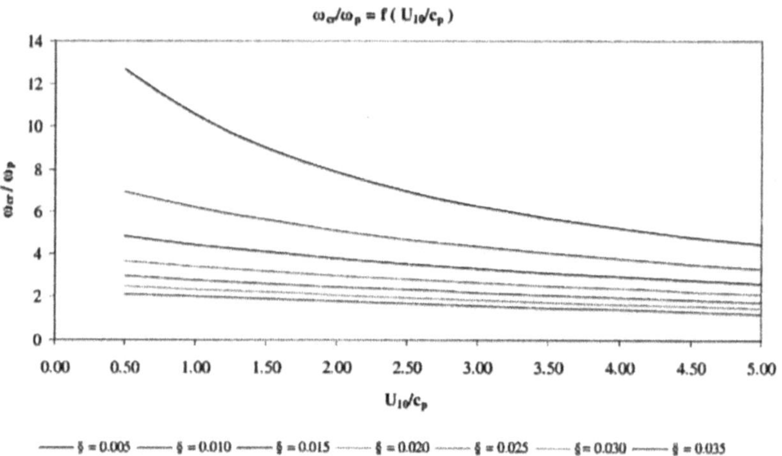

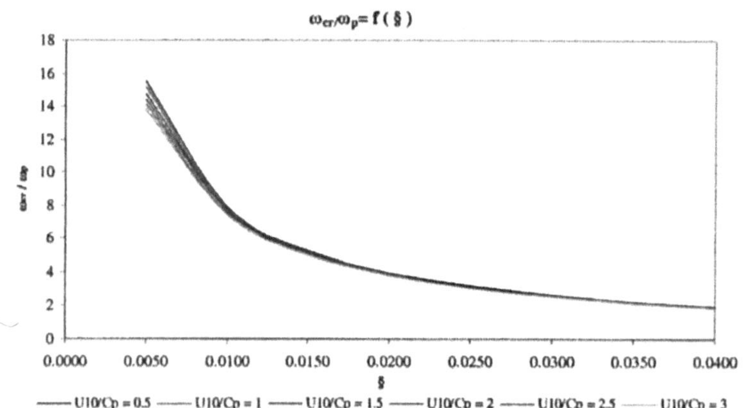

FIGURE 1. (a): Variation of ω_{cr}/ω_p with U_{10}/c_p and § as a parameter; (b): Variation of ω_{cr}/ω_p with § and U_{10}/c_p as a parameter.

safe to write that: $a_e \approx a_{\rm rms}$, an assumption that has been proven valid when $a^+_{\rm rms} \geqslant 300$, where $a^+_{\rm rms} = a_{\rm rms} u * /\nu$. In this study, it is assumed that $z_0 \approx a_e$.

Since, z_0 depends on ω_{cr}, it follows that C_d may depend on both parameters that characterize the wave field, i.e. U_{10}/c_p and §. Yet calculations of a_e (for predictive purposes) with independent values of the significant slope and of the inverse wave age should be done with caution as field and laboratory measurements have shown that, for wind-generated waves, there is a dependence between the two parameters. Indeed, various investigators at the past have provided expressions that relate independently the non-dimensional quantities $\hat{\omega}_p (= \omega_p U_{10}/g = U_{10}/c_p)$ and $\hat{\alpha}_{\rm rms} (= a_{\rm rms} g/\sqrt{2} U^2_{10})$ to the non-dimensional fetch $\hat{x} (= xg/U^2_{10})$, under both laboratory and field conditions, thus allowing various relationships of the form $U_{10}/c_p = Fn(\S)$ to be readily obtained which cover different dynamical regimes. Recently, Elfouhaily et al. (1997), based on field data, provided general expressions for variables similar to $\hat{\omega}_p$ and $\hat{\alpha}_{\rm rms}$ as a function of another non-dimensional fetch. Such expressions cover both limited and unlimited fetch areas and, when combined properly, yield expressions that relate, again, the wave age and the significant slope of the wave field.

The idea of partitioning the wave spectrum into two regions is not new, as Donelan (1982) had proposed for ω_{cr}, empirically, the frequency $2\omega_p$ but his subsequent approach and calculations to obtain C_d are completely different from ours. Furthermore, the idea that wave breaking occurs at the forward face of the wave and close to its crest, at a place where the airflow separates from the wave forming a stagnation point, has been recently exploited by Kudryavtsev & Makin (2001) in an attempt to calculate the roughness length, z_0, of the sea surface and subsequently the drag coefficient, C_d. Their findings substantiate the idea that airflow separation, and the associated wave breaking, play a key role in the formation of the sea surface drag especially at moderate to high winds. At moderate wind speeds, the high frequency wavelets (shortest gravity waves) dominate in the formation of roughness length, z_0, due to the high surface density of their breaking crests, but at very high wind speeds the longer waves contribute as well to the formation of the separation stress, and consequently to C_d. These findings are in accord with the ideas described previously, regarding the contribution of the various waves in the spectrum to the formation of z_0, as ω_{cr} shifts towards ω_p with increasing wind speed and $a_e(\approx z_0)$ approaches $a_{\rm rms}$, or at least the part of the surface under the spectrum from ω_p and beyond.

Since ω_{cr} depends on the form of the wave spectrum, as both u^c_{orb} and q_c depend on the latter, it is necessary to describe the form of the spectrum used in the calculations of ω_{cr}. It is possible, however, to obtain ω_{cr} from observed spectra of a wave field, that is from continuous time series observations of the sea surface displacement.

3. Spectral considerations

To obtain the roughness length, z_0, as a function of wind speed, an empirical wave spectrum has to be used for calculating the required quantities u^c_{orb} and q_c from the wave characteristics. Here we use the empirical spectrum proposed by Donelan *et al.* (1985), although other forms of the spectrum are now being considered in further calculations, in order to obtain an estimate of the sensitivity of the derived results with regard to the spectral form used, as the theoretical description of a directional wave spectrum still remains the subject of intensive research efforts.

The directional wave spectrum proposed by Donelan et al. (1985), and modified slightly by Banner (1990), has the form $S(\omega, \phi) = S_1(\omega)D(\omega, \phi)$, where:

$$S_1(\omega) = a_p g^2 \omega^{-5} \left(\frac{\omega}{\omega_p}\right) \exp\left\{-\left(\frac{\omega_p}{\omega}\right)^4\right\} \gamma_d^{\Gamma} \tag{3.1a}$$

and

$$D(\omega, \phi) = \frac{1}{2}\beta \frac{1}{\cosh^2\{\beta[\phi - \overline{\phi}(\omega)]\}} \tag{3.1b}$$

$$a_p = 0.006 \left(\frac{U_{10}}{c_p}\right)^{0.55}, \qquad 0.83 < \frac{U_{10}}{c_p} < 5 \tag{3.2}$$

$$\beta = 2.61 \left(\frac{\omega}{\omega_p}\right)^{1.3}, \qquad 0.56 < \frac{\omega}{\omega_p} \leqslant 0.95 \tag{3.3}$$

$$\beta = 2.28 \left(\frac{\omega}{\omega_p}\right)^{-1.3}, \qquad 0.95 < \frac{\omega}{\omega_p} \leqslant 1.6 \tag{3.4}$$

$$\beta = 10^{\{-0.4+0.8393\exp[-0.567\ln(\omega/\omega_p)]\}}, \qquad \frac{\omega}{\omega_p} > 1.6 \tag{3.5}$$

$$\gamma_d = 1.7, \quad 0.83 \leqslant \frac{U_{10}}{c_p} < 1, \quad \gamma_d = 1.7 + 6\log\left(\frac{U_{10}}{c_p}\right), \quad 1 \leqslant \frac{U_{10}}{c_p} < 5 \tag{3.6}$$

$$\Gamma = \exp\left(-\frac{(\omega/\omega_p - 1)^2}{2\sigma_d^2}\right), \quad \sigma_d = 0.08\left[1 + 4\left(\frac{U_{10}}{c_p}\right)^3\right], \quad 1 \leqslant \frac{U_{10}}{c_p} < 5 \tag{3.7a, b}$$

where ϕ is the (absolute) mean wind direction, and $\overline{\phi}(\omega)$ is the mean propagation direction of waves at frequency ω with respect to the same orientation system.

Upon integration of the expression $S(\omega, \phi)$ in the direction space, from 0-2π, and assuming that $\phi - \overline{\phi}(\omega) = 0$, the 1-D wave spectrum, $S(\omega)$, is obtained, viz.:

$$S(\omega) = \frac{1}{2} a_p g^2 \omega^{-5} \left(\frac{\omega}{\omega_p}\right) \exp\left[-\left(\frac{\omega_p}{\omega}\right)^4\right] \gamma_d^{\Gamma} \left[\frac{1 - e^{-4\pi\beta}}{1 + e^{-4\pi\beta}}\right] \tag{3.8}$$

Expressions (2.5) and (2.7), with the above form of the 1-D spectrum, are now integrated up to the cut-off frequency ω_{co}, which corresponds to a wave length of about 6 mm (Makin *et al.* 1995), to yield u_{orb}^c and ω_{cr}.

The following alternative expressions for $S(\omega)$, which are based on laboratory and field measurement, as well as on theoretical views expressed during the last several years, have also been used to test the sensitivity of the derived results to the particular form of the spectrum used. In the gravity range, this *composite* spectrum has the form:

$$\frac{S(\omega)\omega_p^5}{\beta g^2} = \begin{cases} \left(\dfrac{\omega}{\omega_p}\right)^9, & 0 \leqslant \omega \leqslant \omega_p \\ \left(\dfrac{\omega}{\omega_p}\right)^{-9}, & \omega_p < \omega \leqslant 1.5\omega_p \\ \dfrac{1}{2}\left(\dfrac{\omega}{\omega_p}\right)^{-4}, & 1.5\omega_p < \omega \leqslant 2\omega_p \\ \left(\dfrac{\omega}{\omega_p}\right)^{-5}, & 2\omega_p < \omega \leqslant \omega_g \end{cases} \tag{3.9}$$

In the gravity-capillary and pure capillary frequency ranges:

$$S(\omega) = \begin{cases} \beta_1 u_* U_{10} g \omega^{-4}, & \omega_g < \omega \leqslant \omega_c \\ \beta_2 u_* \gamma^{2/3} \omega^{-7/3}, & \omega_c < \omega \leqslant \omega_\nu \end{cases} \tag{3.10}$$

where: $g_* = g + \gamma k^2$, $\omega_g = \min(3\omega_p, 5 \times 2\pi = 31.4)$, $\omega_c = \left(\frac{4g}{\gamma}\right)^{1/4} \cong 85.1$ rad/sec (or 13.5 Hz). Here γ is the ratio of (air-water) surface tension and water density, and β represents Phillips equilibrium constant. The above spectrum has a number of interesting properties and has been found to also fit well laboratory wave spectra. However, due to space limitation we shall refrain from commenting on these properties here.

Upon proper integration and taking into account the dispersion relationships for the gravity, gravity-capillary and pure capillary ranges, the following relationships are obtained:

$$\frac{1}{2}a_{\rm rms}^2 = 0.2624\beta g^2 \omega_p^{-4}, \quad \beta \cong 1.905\left(\alpha_{\rm rms}^2 \frac{\omega_p^4}{g^2}\right), \quad u_{orb}^c \cong 1.416 a_{rms}\omega_p \tag{3.11a, b, c}$$

With the aid of significant slope, §, and a proper dispersion equation the above relationships yield:

$$\beta = 15.24\pi^2 §^2, \quad u_{orb}^c \cong 4\pi § c_p = B_{ef} c_p, \quad B_{ef} = 4\pi § \tag{3.12a, b, c}$$

The particular finding $\beta = 15.24\pi^2 §^2$ is in excellent agreement with the results of Huang *et al.* (1981), who found that $\beta = 16.04\pi^2 §^2$, a value which was said to describe best the JONSWAP observations.

4. Data acquisition and processing

The time series data, used to test the proposed concepts regarding ω_{cr} and z_0, were obtained with the aid of oceanographic buoy MEDOUSA designed and constructed by E.A.N.T., a Greek company for Development and Marine Technology, in co-operation with the *Department of Ocean Engineering* and

Naval Architecture of N.T.U.A. This particularly buoy has provided records of the water surface displacement along with directional wave information, as well records of the wind velocity and its direction measured at the height of 5 m above MSL, plus other meteorological and water quality information.

The data were collected in intervals of 20 minutes, every 3 hours, with a rate of 3 samples per sec, providing 8 sets of data daily. They were taken from various places of the Central and North Aegean sea (i.e. from Kavala, Porto Lagos, Alexandroupolis and from Saronikos Gulf) during the period of 1996-1998, although there are more data available from Cretan sea which, however, have not been analyzed yet.

Since the frequency response of the buoy is limited to a few Hz, as its size prohibits it to respond to the very small scale wavelets (ripples, etc.) corresponding to frequencies of about 5 Hz and above, it was decided to analyze only a part of the available records, i.e. those with significant wave height, H_s, exceeding (say) 30 cm, as in these cases the high frequency noise (accompanying the measurements) remains limited. The selected records were properly filtered digitally and subjected thereafter to FFT to produce the spectral characteristics necessary for the subsequent calculations of u^c_{orb} and z_0. Such characteristics include ω_p, H_s, various moments of the spectra (m_2, m_4), the spectral bandwidth and u^c_{orb}. Smoothed spectra from the above sea locations have a form very similar to that published in the literature (e.g., to those described by the JONSWAP experiment). Only single peaked spectra were considered in this investigation, as the presence of a second well formed peak at a frequency (usually) lower than that of the spectral peak indicates the presence of a swell whose interaction with the wind-generated waves may alter the dynamics of the wave field and our analysis may not be valid under such circumstances.

Because the upper frequency encountered in the observed spectra (constructed from the available time series) is usually $< \omega_{co}$, the integration of second moment of that spectrum was limited to the observed frequencies. This fact is reflected in a graph of the ratio $(\omega_{cr}/\omega_p)_{theo}$ against $(\omega_{cr}/\omega_p)_{obs}$ shown in Fig. 2, where the 1st and 2nd subscripts outside parenthesis denote that the corresponding ω ratios have been obtained using the theoretical spectra description and the observed spectra information, respectively. In such a graph the majority of data points falls slightly above the 45° line, with a best fit of the form: $y = 1.24x$. It is interesting to note that this fit is even better (meaning that the straight-line slope of the data has a value less than the above) when the *composite* form of the spectrum is used.

To calculate the wave age, c_p/U_{10}, the available wind velocity, U_5, was converted to U_{10} by simply exploiting the log-linear relationship of the wind velocity profile, whereas c_p was obtained from ω_p via the dispersion relationship. B_{ef} was then obtained as u^c_{orb}/c_p and the ratio ω_{cr}/ω_p from the corresponding expression (2.6) as a function of B_{ef} and $\gamma_1 = 0.03(U_{10}/c_p)$.

With a given U_{10}/c_p it is also possible to construct the corresponding theoretical spectrum, say according to Donelan *et al.* (1985, Eqs. 3.1-3.8), as all of the spectrum parameters are functions of U_{10}/c_p. Then, B_{ef} and ω_{cr} are readily obtained, which in turn allows estimation of $z_0(= a_e), u_*$ (from the log-

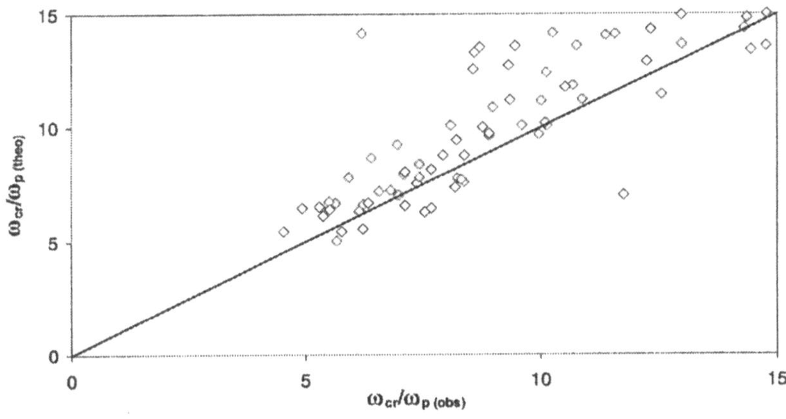

FIGURE 2. Plot of $(\omega_{cr}/\omega_p)_{theo}$ against $(\omega_{cr}/\omega_p)_{obs}$.

law expression of the wind velocity profile) and finally C_d. As an alternative (and/or additional) way of calculating z_0 and C_d, and further cross-checking the derived results, the *composite* spectrum form, described previously, may be used with the values of u_* known from the previous calculations (as a first estimate) to calculate again a_e, etc. and to obtain an idea of the sensitivity of the results to the spectral form used. In this case, § is obtained from the known $a_{\rm rms}$ and ω_p values, then B_{ef} is calculated as 4π§, and finally the ratio ω_{cr}/ω_p is obtained from the known values of B_{ef} and the wave age.

5. RESULTS

Although the observed $a_{\rm rms}$ values plotted as a function of U_{10}/c_p show considerable scatter, when correlated in the form $a_{\rm rms} = (U_{10}^2/g)(U_{10}/c_p)^{a_2}$, proposed by Donelan *et al.* (1993), yield a_1=0.065 and $a_2 = -1.683$, values very closed to 0.055 and -1.7 quoted by them. Hence it may be concluded that the surface wave variance in the Greek Sea locations where the data were collected from, behaves in a fashion similar to its counterpart quantity obtained from other seas. Normalized values of z_0 with either of $a_{\rm rms}$ and H_s have been correlated with the inverse wave age and/or significant slope or with the ratio $H_s/L_p(L_p = \lambda_p)$ which expresses some sort of wave slope. C_d values have also been correlated with either $U_{10}, c_p/u_*$ and/or § in various forms. The corresponding expressions found are given as follows:

$$a_{rms} = 0.065\left(\frac{U_{10}^2}{g}\right)\left(\frac{U_{10}}{c_p}\right)^{-1.683}, \qquad \frac{z_0}{H_s} \approx 0.03\left(\frac{H_s}{L_p}\right)^{0.1845} \qquad (5.1a, b)$$

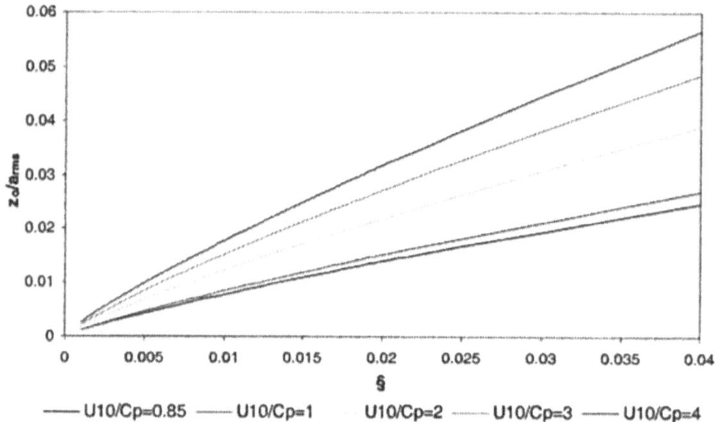

FIGURE 3. Variation of $z_0/a_{\rm rms}$ with § and U_{10}/c_p as a parameter according to equation (5.2c).

$$\frac{z_0}{a_{\rm rms}} = 3.458\S^{1.239}, \quad \frac{z_0}{a_{\rm rms}} = 0.034\left(\frac{U_{10}}{c_p}\right)^{0.631}, \quad \frac{z_0}{a_{\rm rms}} = 0.403\left(\frac{U_{10}}{c_p}\right)^{0.537}\S^{0.841} \qquad (5.2a,b,c)$$

$$10^{-3}C_D \cong 0.1769U_{10} + 2.008, \qquad 10^{-3}C_D \cong 4.008\left(\frac{c_p}{u_*}\right) \qquad (5.3a,b)$$

$$C_D = 0.015\S^{0.295}\left(\frac{U_{10}}{c_p}\right)^{-0.126} \qquad (5.3c)$$

Figure 3 shows the variation of $z_0/a_{\rm rms}$ against §, with U_{10}/c_p as a parameter (the latter ranging from 0.85 to 4.95) according to Eq. (5.2c). Figure 4 shows the variation of C_d against § with U_{10}/c_p as a parameter, according to equation (5.3c), including the variation of C_d with the same parameters, according to the assumptions made by Nikuradse (1932) i.e: $z_0 = a_{\rm rms}/30$, and Businger (1971) i.e: $z_0 = a_{\rm rms}/7.5$.

Finally, Fig. 5 shows the variation of $z_0/a_{\rm rms}$ against U_{10}/c_p, according to Eq. (5.2b), including the respective regression curves from Donelan *et al.* (1993) which correspond to their equations (6), (7) and (8).

6. Discussion and conclusions

The trend of the derived correlation curves (and/or the data) appearing in Figs. 3, 4 and 5, as both the normalized roughness length and drag coefficient vary either with the inverse of wave age, a kind of wave slope or both, appears similar to that published in the (recent) literature quoted in the introduction. More specifically, similar plots appear in Donelan *et al.* (1993), Anctil & Donelan (1996), Taylor & Yelland (2001) and Kudryavtsev & Makin (2001). Exact, quantitative comparison of our plots and expressions with the corresponding expressions provided in the literature is, perhaps, not appropriate as

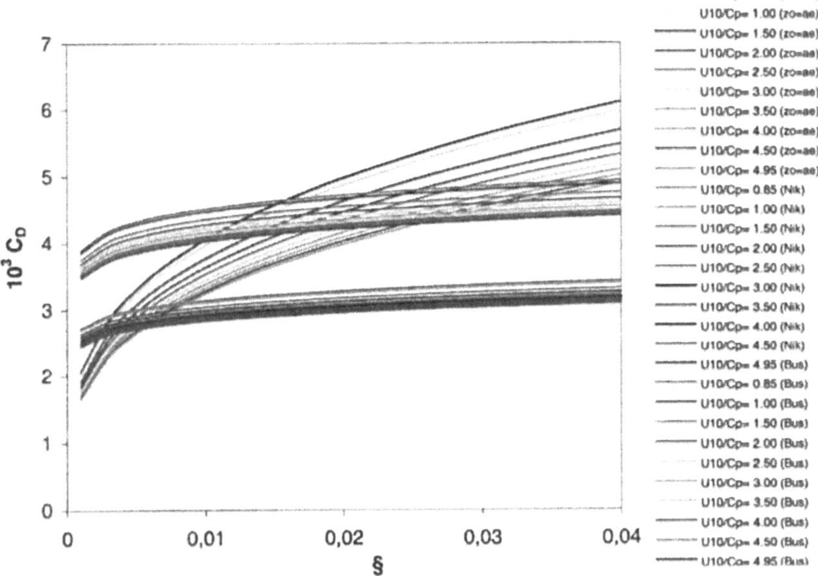

FIGURE 4. Variation of C_D with § and U_{10}/c_p as a parameter according to Eq.(5.3c).

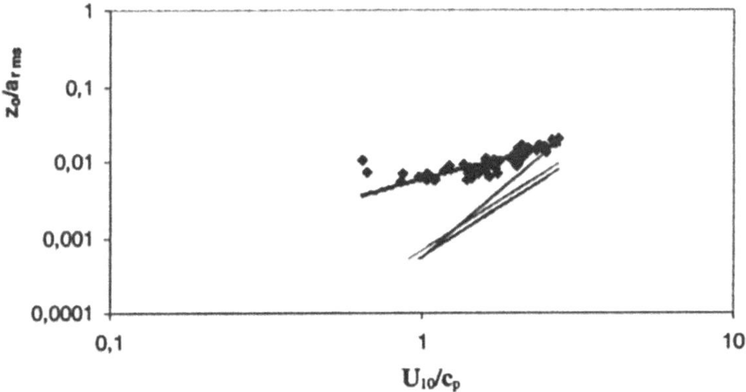

FIGURE 5. Variation of $z_0/a_{\rm rms}$ with U_{10}/c_p: ◇ Observed data; — Regression line according to Eq. (5.2b); Donelan's *et al.* (1993) regression lines :—, —, —, their Eqs. 6, 7, 8, respectively.

the latter expressions do not use the same slope definition (usually the rms surface slope appears there) and in some of them the arms values refer to the short waves alone, not to mention that some of the literature data have been obtained from shoaling areas where other effects may have influenced the exact behavior of the derived results.

It is worth commenting now on the behavior of both the drag coefficient, C_d, against § and U_{10}/c_p as a parameter, shown in Fig. 4, and of the ratio $z_0/a_{\rm rms}$ against U_{10}/c_p, shown in Fig. 5. It is seen, in Fig. 4, that the curves of C_d, produced with the aid of Donelan *et al.* (1985) spectrum lie in between the curves produced with the aid of Nikuradse (1932) and Businger *et al.* (1971) assumptions for values of the significant slope in the range 0.05-0.025, whereas below and above these § values C_d remains smaller and larger, respectively. This is, however, expected as at very low wind speeds (and thus low U_{10}/C_p values) the flow becomes super smooth whereas at very high wind speeds (and high U_{10}/C_p values) the flow becomes super rough, the density of high frequency wavelets becomes high and breaking may also occur. For very small values of §, C_d does not show the rising trend expected, as in these very low wind speeds (corresponding to the smooth flow regime) the surface roughness is mainly attributed to viscous effects which have been neglected in our formulation. The trend in the variation of C_d values with §, shown in Fig. 4, is also similar to that of other field data analyzed with the aid of Melville's theory, as described in the introduction.

It is also interesting to note that the variation of $z_0/a_{\rm rms}$ against U_{10}/c_p (in Fig. 5) shows some discrepancy when compared with the respective regression curves of Donelan et al. (1993), corresponding to their equations (6), (7) and (8). Our normalized z_0 values appear to be larger than their literature counterparts, at least in the specific range of U_{10}/c_p values between 0.65-2. This, in conjunction with the fact that the sea surface variance (arms) behaves as expected (in the literature), leads to the likely conclusion that our $z_0/a_{\rm rms}$ values may be somehow overestimated when plotted in a log-linear scale against U_{10}/c_p, in the range of $0.65 < U_{10}/c_p < 2$. Assuming that the literature results are correct and correspond to conditions similar to those covered by the parameters used in the preceding analysis, the above discrepancy may be partly attributed to the fact that Donelan's regression curves represent a fit to a cloud of corresponding data points with a lot of scatter, as their Fig. 1 shows, in contrast with our data which cluster around the regression line given by Eq. (5.2b). It is, therefore, possible that some of Donelan *et al.* (1993) actual data points cover the same area as ours, and the discrepancy among the regression lines, in Fig. 5, may indicate that in the presence of large (data) scatter linear regression is not appropriate for representing either the data variation or their trend.

The above discrepancy could also, possibly, be explained by the fact that the breaking process is stochastic in nature and as such it would appear necessary to include the intermittent (in time) and sporadic (in space) nature of the breaking wave fronts (near their crests) in the formulation of a more appropriate expression for the effective variance of the water surface, a_e. Thus, z_0 could be expressed as:

$$z_0 \approx a_e = \int_{\omega_{cr}}^{\omega_{co}} P_B S(\omega)\, d\omega \tag{6.1}$$

where $P_B(\omega)$ is the probability of breaking of a wave front with frequency ω. Expressions for $P_B(\omega)$ may be derived in terms of the basic parameters used in this work (or may be found in the literature), but this is beyond the scope of present work.

The discrepancy among our results and the curves of Donelan *et al.* (1985) appears to be greater at low U_{10}/c_p values and diminishes gradually as U_{10}/c_p approaches the value of 2. Since the values of U_{10}/c_p=0.83, 1.5, 2-3 and close to 5 correspond to fully developed, mature, young and very young seas, respectively, the discrepancy shown in Fig. 5 might be consistent with the fact that for a fully developed sea wave steepnesses are very mild, wave breaking is limited or non existing (perhaps, only micro-breaking is present in the field) and $P_B(\omega)$ is expected to be quite small. Thus, the inclusion of $P_B(\omega)$ in Eq. (6.1) would force the value of a_e to become smaller and bring our results closer to the field observations. For mature seas and, particularly, for young seas the wave steepnesses are greater as the young waves are choppy and peaky, breaking occurs more frequently and, thus, $P_B(\omega)$ is expected to get values much larger than for fully developed seas. This, in turn, would cause a_e to get larger values and bring our data, again, closer to the literature data.

In concluding, it is mentioned that this research effort is continued in the frame of further graduate thesis work conducted at the NTUA, and will include further tests of our theory with other spectral forms (as, for example, that proposed by Elfouhaily *et al.* 1997, Kudryavtsev & Makin, 2001), including the influence of stochastic nature of wave breaking and the presence of a swell in the formulation of the surface wave roughness.

REFERENCES

ANCTIL, F. & DONELAN, M.A. 1966 Air-water flux observations over shoaling waves, *J. Phys. Oceanogr.*, **26**, 1344-1353.

BANNER, M.L. & MELVILLE, W.K., 1976 On the separation of air flow over water waves, *J. Fluid Mech.*, **77**(4), 825-842.

BANNER, M.L. 1990 Equilibrium spectra of wind waves, *J. Phys. Oceanogr.*, **20**, 966-984.

BUSINGER, J.A., WYNGAARD, C.J., IZUMI, I. & BRADLEY, E.F. 1971 Flux-profile relationships in the atmospheric surface layer, *J. Atmos. Sciences*, **28**, 181-189.

DONELAN, M.A. 1982 The dependence of aerodynamic drag coefficient on wave parameters, Proc. Int. Air-Sea Interactions in Coastal Zone, Am. Meteorol. Soc., May 10-14, The Hague, Netherlands, 381-387.

DONELAN, M.A., HAMILTON, J. & HUI, W.H. 1985 Directional spectra of wind-generated waves, Philos. Trans. R. Soc. London, Ser. A., 315, 509-562.

DONELAN, M.A., DOBSON, F.W., SMITH, S.D. & ANDERSON, R.J. 1991 On the dependence of the sea surface roughness on wave development, *J. Phys. Oceanogr.*, **23**, 2143-2149.

ELFOUHAILY, T., CHAPRON, B., KATSAROS, K. & VANDEMARK, D. 1997 A unified directional spectrum for long and short wind-driven waves, *J. Geophys. Res.*, **102**, No. C7, 7733-7742.

HUANG, N.E., LONG, S.R. & BLIVEN, L.F. 1981 On the importance of the significant slope in empirical wind-wave studies, *J. Phys. Oceanogr.*, **11**, 569-573.

HUANG, N.E., LONG, S.R., BLIVEN, L.F. & DELEONIBUS, P.S. 1986 A study of the relationship among wind speed, sea state and the drag coefficient for a developing wave field, *J. Geophys. Res.*, **91**, No C6, 7733-7742.

KITAIGORODSKII, S.A. & VOLKOV, V.A. 1965 Calculation of turbulent heat and humidity fluxes in an atmospheric layer near water surface, Bull. (IZV.), Acad. Sci. USSR, Atmos. Oceanic Phys., 1(12).

KITAIGORODSKII, S.A. 1968 On the calculation of the aerodynamic roughness of the sea surface, Bull. (IZV.), Acad. Sci. USSR, *Atmos. Oceanic Phys.*, 4(8).

KITAIGORODSKII, S.A. & DONELAN, M.A., 1984 Wind-wave effects on gas transfer, Gas Transfer at Air-Water Surfaces, W. Brutsaert & G.H. Jirka (Eds.), Reidel, 147-170.

KRAUS, E.B. & BUSINGER, J.A. 1994 Atmosphere-Ocean Interactions, Oxford Monographs on Geology and Geophysics, 27, Oxford University Press, new York.

KUDRYAVTSEV, V.N. & MAKIN, V.K. 2001 The impact of air-flow separation on the drag of the sea surface, Boundary-Layer Meteor., 98, 155-201.

MAKIN, V.K., KUDRYAVSTEV, V.N. & MASTENBROEK, C. 1995 Drag of the sea surface, Boundary Layer Meteor.,159-182.

MAKIN, V.K., 1999 Coupling of waves with the atmosphere, Wind-Over-Wave Couplings, S.G. Sajjadi, N.H. Thomas & J.C.R. Hunt (Eds.), 69-79, Oxford University Press, New York.

MAKIN, V.K. & KUNDRYAVSTEV, V.N. 1999 Coupled sea-surface atmospheric model. 1. Wind over waves coupling, *J. Geophys. Res.*, **104**, No. C4, 7613-7623.

MELVILLE, W.K. 1977 Wind stress and roughness length over braking waves, *J. Phys. Ocean.*, **7**, 702-710.

NIKURADSE, J. 1932 Gesetzmessigkeiten der turbulenten stremung in glatten Rohren, Forschungsheft 356, Berlin. Phillips, O.M., 1977, The Dynamics of the Upper Ocean, 2nd Ed., Cambridge University Press, 336 pp.

TAYLOR, P.K. & YELLAND, M.J. 2001 The dependence of sea surface roughness on the height and steepness of the waves, *J. Phys. Oceanogr.*, **21**, 572-590.

Direction of Wind Stress Vector Over Waves

A.A. Grachev[1], C.W. Fairall[2], J.E. Hare[3] and J.B. Edson[4]

[1] University of Colorado CIRES/NOAA ETL, Boulder, Colorado, USA
[2] NOAA Environmental Technology Laboratory, Boulder, Colorado, USA
[3] Univeristy of Colorado CIRES/NOAA ETL, Boulder, Colorado, USA
[4] Woods Hole Oceanographic Institution, Woods Hole, Massachussetts, USA

Abstract

In this study, directional characteristics of the wind stress in different wind-wave regimes are considered. We discuss results based on measurements made during three field campaigns onboard the R/P *FLIP* in the Pacific in 1993 and 1995 (SCOPE, MBL II, and COPE). We find a strong influence of the surface waves on the wind stress direction especially for light winds. In general, the stress is a vector sum of the (i) pure shear stress (turbulent and viscous) aligned with the mean wind shear, (ii) wind wave-induced stress aligned with the direction of the pure wind sea waves, and (iii) swell-induced stress aligned with the swell direction. The direction of the wind wave-induced stress and the swell-induced stress components may coincide with, or be opposite to, the direction of wave propagation (pure wind waves and swell respectively). As a result, the stress vector may deviate widely from the mean wind flow including cases when stress is directed across or even opposite to the wind.

1. Introduction

Determination of the wind stress, $\vec{\tau}$, (or momentum flux, τ) over oceans is important in modelling atmospheric and oceanic dynamics, radar remote sensing, and other applications. In practice the stress is usually measured by a sonic anemometer at level of order 10 m above the sea surface. The stress vector at some level well above the viscous sublayer may be represented directly by the following relation (eddy-correlation method):

$$\vec{\tau} = \vec{\tau}_x + \vec{\tau}_y = -\rho < u'w' > \vec{i} - \rho < v'w' > \vec{j} \tag{1.1}$$

where $\vec{\tau}_x = -\rho < u'w' > \vec{i}$ and $\vec{\tau}_y = -\rho < v'w' > \vec{j}$ is the longitudinal and lateral stress respectively, $< >$ is a time or/and spatial averaging operator, and a prime denotes fluctuations about a mean value. It is common practice to align the X-axis with wind direction. The friction velocity squared, u_*^2, is given by the magnitude of the stress, $|\vec{\tau}|$, divided by air density ρ. Stress magnitude is numerically equal to the magnitude of the momentum flux τ. For this reason, it is a common practice not to differentiate between stress and momentum flux. The stress direction a, relative to the wind direction, is given by

$$\alpha = \arctan\left(< v'w' > / < u'w' >\right), \tag{1.2}$$

where α is positive (negative) when $\vec{\tau}$ is to the right (left) of the wind vector.

Previous investigations of the wind stress in the marine surface layer have primarily focused on determination of scalar variables, the stress magnitude (or u_*), drag coefficient, and roughness length. However, the stress vector is often aligned with a direction different from the mean wind flow. With the exception of several papers, there has been a general lack of investigation concerning the stress vector direction relative to the mean wind and surface waves' direction.

Based on field measurements, Geernaert *et al.* (1993), Rieder *et al.* (1994, 1996) reported high values of the crosswind component τ_y and the angle α in (1.2). They observed that when swell propagates at an oblique direction with respect to the local wind, the stress vector has a direction, which is, in general, a blend between the wind direction and the swell direction. Rieder *et al.* (1996) find that the direction of the wind stress and the whitecap motion are observed to be generally co-linear, with both lying between the mean wind and the swell. Hwang & Shemdin (1988) reported that short waves (the ocean surface roughness) are not aligned in the wind direction in the presence of swell. The angle of the principal axis of the joint probability distribution of wave slopes is found to point in a direction between wind and swell. It is important to know the principal direction of short waves, since the returned radar signals are more closely associated with Bragg backscattering from centimetre-scale roughness surface elements. Observations on Lake Ontario made by Colton *et al.* (1995), and spaceborne scatterometer measurements by Rufenach *et al.* (1998) in the equatorial Pacific, and Cornillon & Park (2001) over the Gulf Stream showed that the mean wind direction derived from the scatterometer cross section is probably more closely related to the surface stress than the wind velocity.

In light wind speed regimes, the influence of the surface waves on wind stress direction is more dramatic. It is common that in calm weather conditions, the wind and stress vectors are not aligned, and often, the wind and stress directions are nearly opposite (Drennan *et al.* 1999). However during light winds, deviations of the stress vector direction from the wind vector are not random. Grachev & Fairall (2001) showed that in the case of light winds and background swell, the stress vector lies in the obtuse angle between the wind direction and the opposite wave direction; i.e., it is facing the opposite direction from the wave propagation (see details in the next section). For moderate and strong winds ($U \geqslant 5-6$ ms^{-1}), $\vec{\tau}$ basically lies in the acute angle between the wind direction and the wave direction as found previously by Geernaert *et al.* (1993) and Rieder *et al.* (1994, 1996).

In this study, data collected by the National Oceanic and Atmospheric Administration (NOAA) Environmental Technology Laboratory (ETL) and Woods Hole Oceanographic Institution (WHOI) during three R/P *FLIP* expeditions in 1993 and 1995 are used to extend the prior analysis by including additional experimental evidence of the influence of surface wave direction on the stress vector direction.

2. Direction of the stress vector over waves

The above discussion shows that the direction of the wind stress is governed by both wind and swell directions. For this reason it makes sense to split the

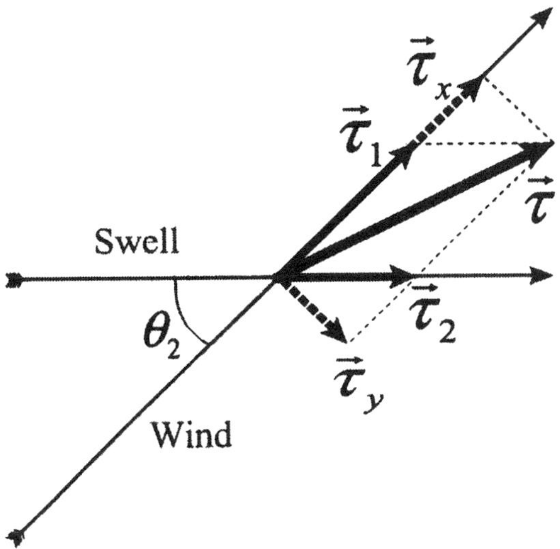

FIGURE 1. Decomposition of the stress vector, $\vec{\tau}$, into $\vec{\tau}_x$ and $\vec{\tau}_y$ and in a wind-associated coordinate system, and into $\vec{\tau}_1$ and $\vec{\tau}_2$ in a wind-swell coordinate system.

stress vector in two vectors aligned with wind, $\vec{\tau}_1$, and swell, $\vec{\tau}_2$, respectively:

$$\vec{\tau} = \vec{\tau}_1 + \vec{\tau}_2, \tag{2.1}$$

Figure 1 shows a decomposition of the stress vector in the rectangular coordinate system, $\vec{\tau}_x$ and $\vec{\tau}_y$, and in the wind-swell coordinate system, $\vec{\tau}_1$ and $\vec{\tau}_2$ (θ_2 is a relative angle between the wind and swell). In fact, we replace a fixed rectangular Cartesian reference frame associated with the wind alone and used in (1.1) by a fixed non-rectangular reference frame associated with the wind and swell directions, Equation (2.1). The relationships between stress components in these two coordinate systems are given by (see Figure 1):

$$\tau_x = \tau_1 + \tau_2 \cos\theta_2, \qquad \tau_y = \tau_2 \sin\theta_2 \tag{2.2}$$

$$\tau_1 = \tau_x - \tau_y \arctan\theta_2, \qquad \tau_2 = \tau_y / \sin\theta_2 \tag{2.3}$$

Note that (2.1) – (2.3) are not useful in the case $\sin\theta_2 = 0$, when the wind is aligned with the swell (in the same or opposite direction).

Let us consider some reasons which can cause a deviation of the stress vector from a wind direction. Over the sea, the total stress, $\vec{\tau}$, can be expressed as the vector sum of shear stress, $\vec{\tau}_{shear}$, and wave-induced stress, $\vec{\tau}_{wave}$:

$$\vec{\tau} = \vec{\tau}_{shear} + \vec{\tau}_{wave} \equiv \vec{\tau}_{visc} + \vec{\tau}_{turb} + \vec{\tau}_{wave}, \tag{2.4}$$

where $\vec{\tau}_{shear}$ is the sum of viscous, $\vec{\tau}_{visc}$, and turbulent shear stress, $\vec{\tau}_{turb}$. In general, all constituents in (2.4) depend on height above the sea surface, z. The amplitude of wave-induced pressure perturbations falls off exponentially with z and, therefore, well away from the surface $\vec{\tau}_{wave}$ tends to zero (e.g. Makin &

Kudryavtsev 1999). However, $\vec{\tau}_{wave}$ is not described by a simple exponentially decaying profile but has a more complicated nonmonotonic structure (Hare et al. 1997). The layer where the influence of $\vec{\tau}_{wave}$ cannot be neglected is known as the wave boundary layer (WBL).

The viscous stress and the turbulent stress in (2.4) are aligned with the direction of the mean wind shear at the reference level, z, since both $\vec{\tau}_{visc}$ and $\vec{\tau}_{turb}$ are associated with the local wind shear, $d\overline{U}/dz$. Note that, in many cases the shear vector coincides with the wind vector. The wave-induced stress on the sea surface $z = \eta$, $\vec{\tau}_{wave}(\eta)$, is associated with the atmospheric pressure distribution across the front and rear faces of the waves, and in the case of a pure unimodal 1-dimensional wave field (e.g. Belcher & Hunt 1998; Sullivan *et al.* 2000), $\vec{\tau}_{wave}(\eta)$ is aligned with the direction of wave propagation. Wave-induced momentum flux, $\vec{\tau}_{wave}(\eta)$, shows strong dependence on wave age and ranges from positive to negative values (Sullivan *et al.* 2000). For young seas the longitudinal component of wave-induced stress is positive, i.e. $(\tau_{wave})_x > 0$ (respective to wind direction, the X-axis). With increasing wave age, $(\tau_{wave})_x$ decreases, reaches zero, and reverses sign in the case of old seas, $(\tau_{wave})_x < 0$ (Belcher & Hunt 1998; Sullivan *et al.* 2000). The fact that the wave-induced stress can be positive as well as negative is a key point to understanding the stress vector orientation over ocean waves.

Oceanic waves occur over a broad frequency band and extend from capillary to long gravity waves. It is generally assumed that surface gravity waves can be separated into pure wind sea and swell waves (two peaks in the wave spectra are expected). These waves are different in origin and differ in properties. Wind surface waves are short waves and travel much more slowly than the wind, while swell are long and fast traveling ocean waves. Wind sea waves are generated by a local wind, while swells are generated in other areas at other times. Swells have a period and wavelength that is not associated with the local wind. For example, the equatorial Pacific wave climate is generally associated with a systematic swell generated in the Gulf of Alaska and in the southern Pacific Ocean during the local winter seasons. This swell travels thousands of kilometres, propagating into the equatorial region where the light winds are usually prevailing.

Thus, it makes sense to split the surface wave-induced stress into two parts, $\vec{\tau}_{wave}(\eta) = \vec{\tau}_{wave1}(\eta) + \vec{\tau}_{wave2}(\eta)$, where $\vec{\tau}_{wave1}(\eta)$ and $\vec{\tau}_{wave2}(\eta)$, are due to pure wind waves and swell, respectively. Combining this assumption and Eq. (2.4) for $z = \eta$ yields

$$\vec{\tau}(\eta) = \vec{\tau}_{shear}(\eta) + \vec{\tau}_{wave1}(\eta) + \vec{\tau}_{wave2}(\eta). \tag{2.5}$$

Generally, wind waves and swell propagate in different directions (i.e., θ_1 and θ_2 are relative angles between the wind and wind waves and swell, respectively). However, the direction of wind waves is frequently close to the wind direction (e.g. Geernaert *et al.* 1993, and Rieder *et al.* 1996). That is, $|\theta_1| \approx 1$, and therefore the vector $\vec{\tau}_{wave1}(\eta)$ is aligned with the wind direction.

In the case of complex sea (mixed wind sea and swell), $\vec{\tau}_{wave1}(\eta)$ and $\vec{\tau}_{wave2}(\eta)$ are governed by their own wave age. Since the swell usually travels faster and short waves more slowly than the wind, in the majority of cases $\tau_{wave1}(\eta) > 0$, and $\tau_{wave2} < 0$ (respective to wind direction, the X axis). However, reverse signs are also possible in transient conditions. The case $\tau_{wave1} < 0$ is associated with decaying wind conditions; e.g. after the passage of a storm or gale, when the total stress may reverse sign to negative (Smedman *et al.* 1994). The case $\tau_{wave2}(\eta) > 0$ is associated with strong winds travelling in the same direction as the swell or the counter-swell. The last leads to enhancement of the total stress (e.g., Drennan *et al.* 1999). Unlike the wave-induced stress, the shear stress is always positive, i.e. $\tau_{shear}(\eta) > 0$. The case $\tau_{shear}(\eta) < 0$ is an exotic situation, and it could be associated with the nonzero sea surface velocity with the same direction as the wind (due to surface currents). Determination of the $\vec{\tau}_{shear}$ direction is not so obvious. As mentioned above, the shear stress vector usually coincides with the mean wind direction, therefore $\vec{\tau}_{shear}$ and $\vec{\tau}_{wave1}$ are approximately co-directional. This results in

$$\vec{\tau}_1 = \vec{\tau}_{shear} + \vec{\tau}_{wave} \text{ and } \vec{\tau}_2 \equiv \vec{\tau}_{wave2}. \tag{2.6}$$

However, the situation may be more intricate, and $\vec{\tau}_{shear}$ may be aligned with the total stress vector $\vec{\tau}$ rather than with the wind itself. Once airflow blows over region where swell propagates at an oblique direction with respect to the local wind, the stress vector at the sea surface deviates from the wind direction. The stress exerted on the sea surface generates the capillarity and small scale surface waves (Hwang & Shemdin 1988), and the whitecap streaks (Rieder *et al.* 1996) aligned in the stress direction. Bragg backscattering from these centimeter-scale surface roughness elements associated with the surface stress determines the microwave signature of the surface. Small-scale roughness elements produced by the stress on the surface are associated with the local wind shear above the sea, which controls the momentum transfer between the air and sea. Note that the wind stress is associated with the local wind shear, $d < u > /dz$ rather than with the mean wind itself. This explains a deviation of the stress vector from the wind direction on the surface and beyond the WBL. In the case when $\vec{\tau}_{shear}$ is aligned with $\vec{\tau}$ rather than with the wind, relationships (2.6) should be replaced by

$$\vec{\tau} - \vec{\tau}_{shear} = \vec{\tau}_{wave2} + \vec{\tau}_{wave1} \text{ and } \vec{\tau} \parallel \vec{\tau}_{shear}. \tag{2.7}$$

One may consider that during the initial stage of the wind blowing at an oblique angle to the swell direction, $\vec{\tau}_{shear}$ coincides with the mean wind. Then, $\vec{\tau}_{shear}$ turns to the swell direction, and in the equilibrium state, it is co-directional with the total surface stress. However, it is clear that a final turn of $\vec{\tau}_{shear}$ from the wind direction to the $\vec{\tau}$ direction will not qualitatively change the orientation of the $\vec{\tau}$ vector relative to wind and swell directions.

According to the above discussion, the vector $\vec{\tau}_1$ may face into the wind direction ($\tau_1 > 0$) or in the opposite direction ($\tau_1 < 0$). Similarly, $\vec{\tau}_2$ may face into the swell direction ($\tau_2 > 0$), or into the counter-swell direction ($\tau_2 < 0$). Combining these cases gives all possible situations associated with the stress direction.

When the wind has a component in the direction of swell propagation ($\cos\theta_2 > 0$), four situations are possible. The typical case is associated with moderate and strong winds or weak swells. Both vector components are positive, $\tau_1 > 0$ and $\tau_2 > 0$ (this case is shown in Figure 1), and the total stress vector lies in the acute angle between the wind direction and the swell direction and is facing in the wind/wave direction. This case was described by Geernaert *et al.* (1993) and Rieder *et al.* (1994, 1996) in detail. As the wind speed decreases, the swell-induced stress, τ_2, decreases, reaches zero, and reverses sign, but τ_1 is still positive. This case is associated with light winds and strong background swell (swell regime), $\tau_1 > 0$ and $\tau_2 < 0$. The stress vector in this case is within the obtuse angle created by the wind direction and the direction opposite to the swell propagation. This regime was reported by Grachev & Fairall (2001). In a particular case, the stress vector may be directed perpendicular to the wind ($\tau_2 = 0$ but $\tau_y \neq 0$). The two next cases are associated with decaying wind conditions ($\tau_1 < 0$), e.g. after the passage of a storm or gale. In these situations, the wind speed drops in a short time period, but the local wind waves still travel with high phase velocities that lead to upward momentum transfer from ocean to atmosphere. The case when $\tau_1 < 0$ and $\tau_2 < 0$ is associated with light winds. The stress vector in this case lies at an acute angle between directions opposite to both wind and swell. One may expect that this regime rarely occurs. It seems likely that the case where $\tau_1 < 0$ and $\tau_2 > 0$ is more improbable than even the previous case because it requires that wind speed be high enough to satisfy the condition $\tau_2 > 0$, but at the same time have a decaying wind regime.

When the wind blows against the swell ($\cos\theta_2 < 0$), waves always drag the airflow and extract energy. Therefore, a stress component is directed against the swell direction regardless of the wind speed magnitude, i.e. in the counter swell cases $\tau_2 < 0$. The situation $\tau_1 > 0$ and $\tau_2 < 0$ is a regular case associated with steady state wind conditions, whereas the case in $\tau_1 < 0$ and $\tau_2 < 0$ is a decaying wind regime. A comparison of the approach with the field data is given in the next section.

3. Data analysis

We use data collected by the NOAA/ETL and WHOI during three campaigns aboard the R/P *FLIP* to examine the behaviour of the stress direction. Data were taken in the Pacific in September 1993 during the San Clemente Ocean Probing Experiment (SCOPE), in April - May 1995 during the Marine Boundary Layer II (MBL II) experiment, and in September 1995 during the Coastal Ocean Probing Experiment (COPE). Descriptions of the ETL/WHOI seagoing flux system and some other relevant information about the data can be found in Fairall *et al.* (1997), Edson *et al.* (1998), and Grachev & Fairall (2001). The data cover a wide range of the wind/wave conditions being mutually complementary. The range of the stress directions varies from cases when stress direction coincides with wind direction to cases when stress is directed across or even opposite the wind.

To investigate the stress direction, we divide the data into several groups based on the wind speed and direction of the wind relative to the dominant wave direction. Consider a case of moderate and strong winds, $U \geqslant 5-6$ ms^{-1}, blowing in the direction of the swell propagation, $\cos\theta_2 > 0$. This case is best illustrated with the COPE data. During COPE, the wind often blew

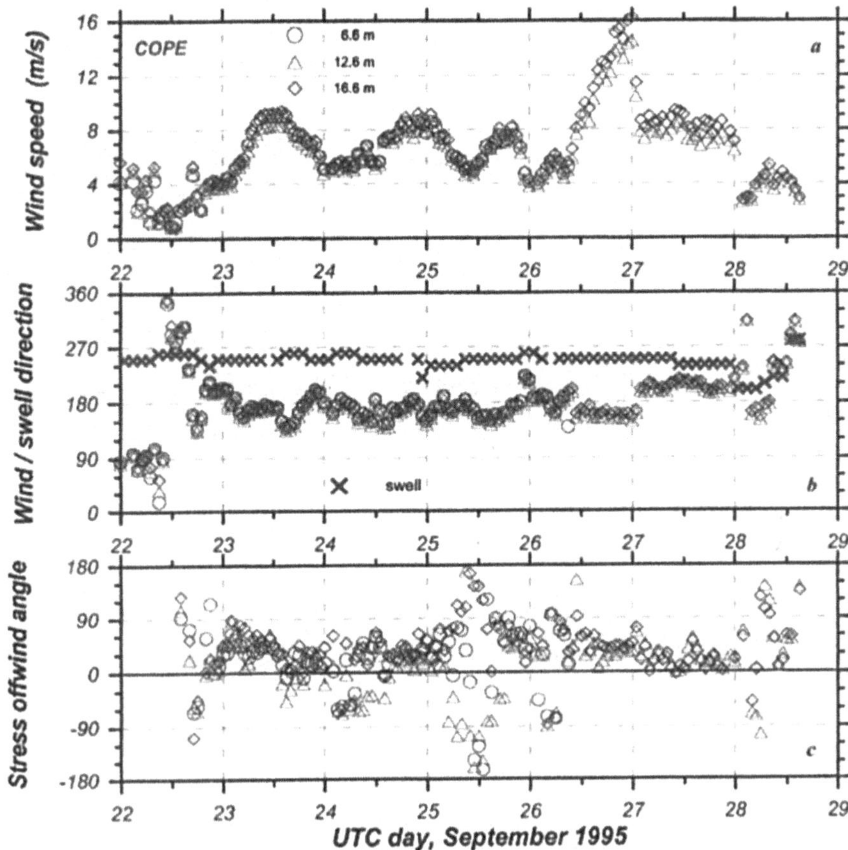

FIGURE 2. Time series of (a) wind speed, (b) wind and swell direction, and (c) stress offwind angle, , [see Equation (1.2)], during COPE. All angles are calculated using the meteorological convention ("from").

obliquely to the swell direction (Figure 2b). Large angles between the wind and swell (swells travel approximately cross-wind) cause high values of the cross-wind component with the same sign. As a result, the stress off-wind angle (1.2) deviates substantially from zero, and $\alpha \approx 30°$ (even for high winds (Figure 2a,c). Figure 2 shows that with the exception of light wind and counter swell events, the stress vector over September 23 – 28 generally lies between the wind direction (about 180°) and the swell direction (about 260°). Note, that the mean stress direction is generally in line with the wind and dominant wave direction, provided the wind and waves travel in about the same direction. This situation was observed during the moderate and strong wind events in SCOPE and for most of the time during MBL II.

Figure 3 shows the decomposition of the wind stress in the reference frame associated with the wind and swell direction (i.e., $\vec{\tau}_1$ and $\vec{\tau}_2$ coordinates) for moderate and strong winds. All three *FLIP* data sets presented in Figure 3

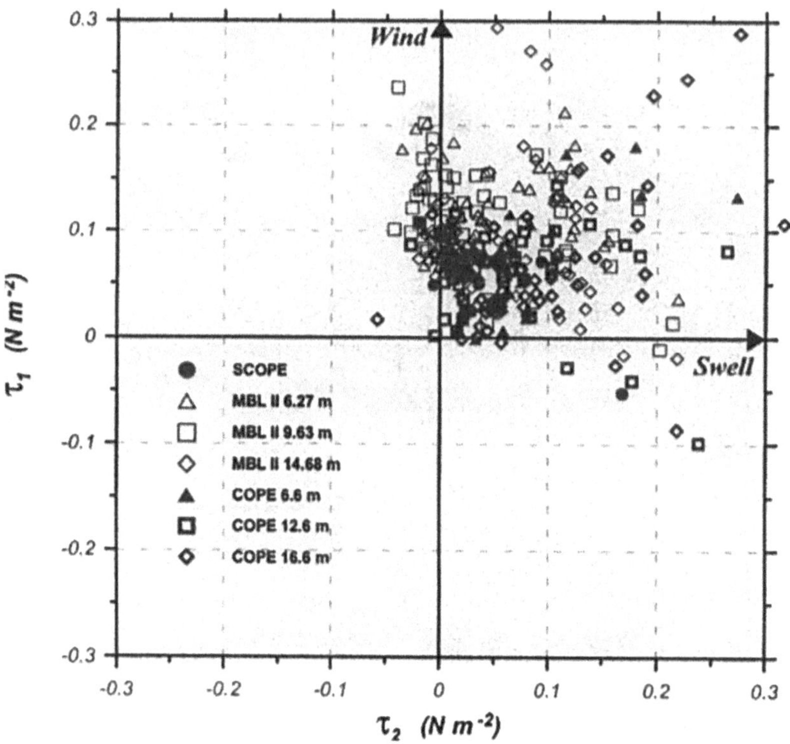

FIGURE 3. Decomposition of the stress vector in the wind-swell coordinate system for the case where swell propagates in the wind direction with moderate and strong winds. Different symbols correspond to data obtained during different cruises.

satisfy the following criteria: $\cos\theta_2 > 0$, $|\sin\theta_2| > 0.2$ and $U > 6$ ms^{-1}. Points in Figure 3 are clustered more in the upper right quadrant, $\tau_1 > 0$ and $\tau_2 > 0$ The stress vector lies in an acute angle between the wind direction and the wave direction and is facing the wind/wave direction. This result agrees with previous studies, such as Geernaert *et al.* (1993), Rieder *et al.* (1994, 1996).

In the case when $\tau_2 < 0$ with τ_1 still positive, the stress vector lies at an obtuse angle between the wind direction and the opposite wave direction, it is facing the direction which is opposite to the direction of wave propagation. This situation is usually associated with winds $2 \leqslant U \leqslant 4-6$ ms^{-1} in the presence of background swell. This case was described by Grachev & Fairall (2001), where they showed that a deviation of the direction of the stress vector from the wind vector during light winds is not random, and it is governed by both the swell direction and the wind direction. Figure 4 presents the directional characteristics of the wind and surface stress vectors as a function of the wind speed during the SCOPE. In this case, the wind direction is approximately between 210-330°. A northwest swell was moderate but almost always present, and the direction of the waves was very constant (about 300°). As wind speed decreases the stress vector deviates significantly from the wind (Figure 4b) and swell (Figure

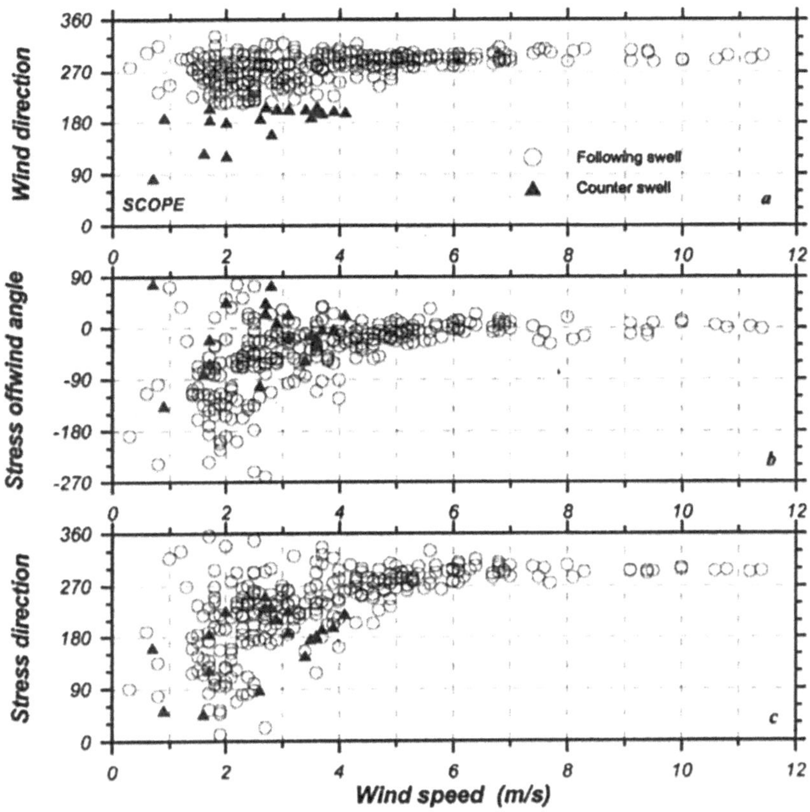

FIGURE 4. Wind and stress directions during SCOPE as function of wind speed: (a) the true wind direction, (b) stress offwind angle, α, [see Equation (1.2)], and (c) the true stress direction.

4c) direction. The stress vector turns in a counter-clockwise direction through about 180°, and finally it is nearly opposite to the wind and swell directions (Figure 4b,c). The regime where the surface stress is aligned opposite to the wind direction corresponds to upward momentum transfer (Grachev & Fairall 2001). Figure 5 shows a decomposition of the wind stress for this regime at $\vec{\tau}_1$ vs. $\vec{\tau}_2$ coordinates. Most of the points are grouped in the upper left quadrant, as predicted in Section 2. The cluster of points is stretched along the bisector. This is associated with self-correlation in the $\boldsymbol{\tau}_1$ and $\boldsymbol{\tau}_2$ in Equation (2.3), when $\boldsymbol{\tau}_y$ is significant compared to $\boldsymbol{\tau}_x$, and $|\theta_2|$ is small. However, this fact does not impact our approach, since self-correlation can change the relative positions of the points inside the quadrant but cannot change the quadrant itself, i.e. signs of $\boldsymbol{\tau}_1$ and $\boldsymbol{\tau}_2$.

The cases shown in Figures 3 and 5 are most common situations and are statistically more representative. However, the COPE data at $t = 23.6$ in Figure 2 gives a good example of a counter swell case, $\cos\theta_2 < 0$. In this situation,

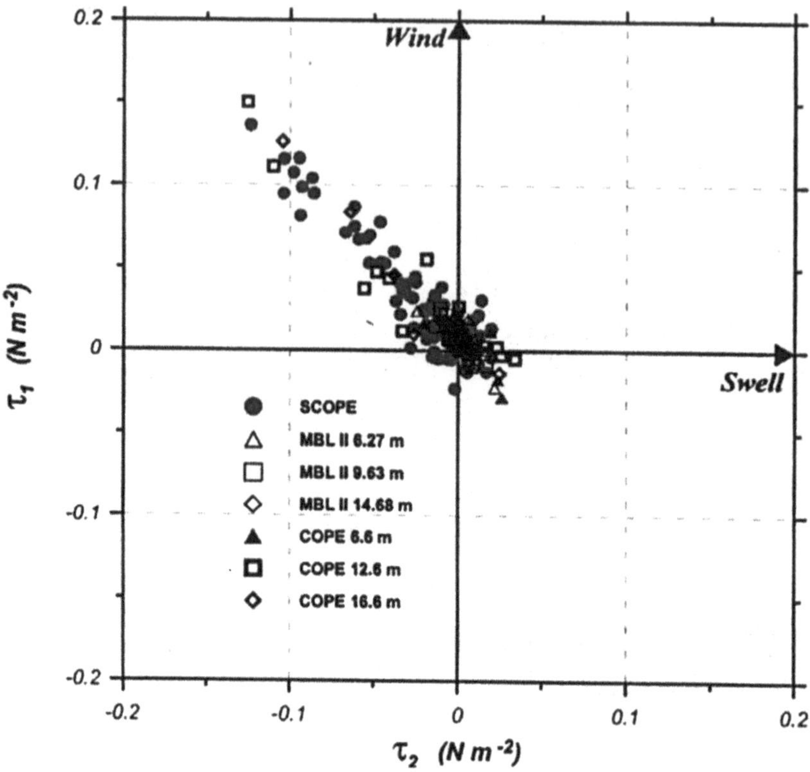

FIGURE 5. Same as Figure 3 for the case when swell propagates in the wind direction ($\cos\theta_2 > 0$) and light winds (2 ms^{-1} $< U <$ 6ms^{-1}).

$\tau_2 < 0$ and $\tau_1 > 0$ ($U \approx 9$ ms^{-1}). The counter swell regime for the SCOPE data was considered by Grachev & Fairall (2001).

4. Summary

This study focuses on the directional characteristics of the wind stress in different wind-wave regimes. The surface stress is a vector sum of the (i) pure shear stress (turbulent and viscous) aligned with the mean wind shear, (ii) wind wave-induced stress aligned with the direction of the pure wind sea waves, and (iii) swell-induced stress aligned with the swell direction. The direction of the wind wave-induced stress and the swell-induced stress components may coincide with, or may be opposite to, the direction of wave propagation (pure wind waves and swell, respectively). As a result, during different wind-wave situations the stress vector may lie in the different sectors created by the wind and swell directions. The determination of the wind stress vector is important in many applications, especially those related to radar remote sensing of the ocean surface. Airborne and spaceborne scatterometers may detect the surface stress direction rather than the mean wind direction (Colton *et al.* 1995, Rufenach *et al.* 1998, Cornillon & Park 2001).

REFERENCES

BELCHER, S.E., & HUNT, J.C.R. 1998 Turbulent flow over hills and waves. *Ann. Rev. Fluid Mech.* **30**, 507 - 538.

COLTON, M.C., PLANT, W.J., KELLER, W.C. & GEERNAERT, G.L. 1995 Tower-based measurements of normalized cross section from lake Ontario: Evidence of wind stress dependence. *J. Geophys. Res.* **100**(C5), 8791 - 8813.

CORNILLON, P., & PARK, K.-A. 2001 Warm core ring velocities inferred from NSCAT. *Geophys. Res. Let.* **28**(4), 575 - 578.

DRENNAN, W.M., KAHMA, K.K. & DONELAN, M.A. 1999 On momentum flux and velocity spectra over waves. *Boundary-Layer Meteorol.* **92**(3), 489 - 515.

EDSON, J.B., HINTON, A.A., PRADA, K.E., HARE, J.E. & FAIRALL, C.W. 1998 Direct covariance flux estimates from mobile platforms at sea. *J. Atmos. Oceanic Tech.* **15**, 547 - 562.

FAIRALL, C.W., WHITE, A.B., EDSON, J.B. & HARE, J.E. 1997 Integrated shipboard measurements of the marine boundary layer. *J. Atmos. Oceanic Tech.* **14**, 338 - 359.

GEERNAERT, G.L., HANSEN, F., COURTNEY, M. & HERBERS, T. 1993 Directional attributes of the ocean surface wind stress vector,*J. Geophys. Res.* **98**(C9), 16,571 - 16,582.

GRACHEV, A.A. & FAIRALL, C.W. 2001 Upward momentum transfer in the marine boundary layer. *J. Phys. Oceanogr.* **31**(7), 1698 - 1711.

HARE, J.E., HARA, T., EDSON, J.B. & WILCZAK, J.M. 1997 A similarity analysis of the structure of airflow over surface waves. *J. Phys. Oceanogr.* **27**, 1018-1037.

HWANG, P.A. & SHEMDIN, O.H. 1988 The dependence of sea surface slope on atmospheric stability and swell conditions. *J. Geophys. Res.* **93**(C11), 13903 - 13912.

MAKIN, V.K., & KUDRYAVTSEV, V.N. 1999 Coupled sea surface-atmosphere model. 1. Wind over waves coupling. *J. Geophys. Res.* **104**(C4), 7613 - 7623.

RIEDER, K., SMITH, J.A. & WELLER, R.A. 1994 Observed directional characteristics of the wind, wind stress, and surface waves on the open ocean. *J. Geophys. Res.* **99**(C11), 22,589 - 22,596.

RIEDER, K., SMITH, J.A. & WELLER, R.A. 1996 Some evidence of colinear wind stress and wave breaking. *J. Phys. Oceanogr.* **26**(9), 2519 - 2524.

RUFENACH, C.L., BATES, J.J. & TOSINI, S. 1998 ERS-1 Scatterometer measurements - Part 1: The relationship between radar cross section and buoy wind in two oceanic regions. *IEEE Trans. Geosci. Remote Sensing*, **36**(2), 603 - 622.

SMEDMAN, A.-S., TJERNSTRÖM, M. &. HÖGSTRÖM, U. 1994 Near-neutral marine atmospheric boundary layer with no surface shearing stress: A case study. *J. Atmos. Sci.* **51**, 3399-3411.

SULLIVAN, P.P., MCWILLIAMS, J.C. & MOENG, C.-H. 2000 Simulation of turbulent flow over idealized water waves. *J. Fluid Mech.* **404**, 47-85.

An Improved Parameterization for Energy Exchange from Wind to Stokes Waves

S.G. Sajjadi and M.T. Bettencourt
CHL, John C. Stennis Space Center, Building 1103, Suite 103, Mississippi 39529, USA.

Abstract

Previous analysis for the generation of non-linear surface waves by shear flow (Croft & Sajjadi 1993, Sajjadi *et al.* 1997 and Sajjadi 1998) is extended by: (i) presenting results on Stokes wave; (ii) imposing the boundary condition at the surface wave, rather than at the mean surface; and (iii) including the dominant viscous term in the complete Orr-Sommerfeld equation. The inclusion (i) yields an energy transfer that is larger than those predicted for monochromatic waves, while (ii) has no real effect and (iii) only a small effect for gravity waves. The present analysis is mainly confined to the second-order Stokes wave but its extension to higher-order Stokes waves is straightforward and suggested. Results are also generalized for fully non-linear Stokes waves (nth order) which confirms the findings of previous studies by Sajjadi and co-workers. The energy transfer parameters are obtained via numerical integration of the full Orr-Sommerfeld equation, and it is found the results agree well with that of Conte & Miles (1959) and confirms that the extra energy associated with the growth of non-linear waves is due to higher harmonics of Stokes wave. An expression is derived for parameterized form of this energy exchange from wind to Stokes waves which is could be used in spectral wave models. The present parameterization is compared with other models formulations and its is shown to agree well with the numerical simulation of full Reynolds stress equations, Sajjadi (2002b).

1. Introduction

The first theory for the generation and growth of non-linear surface water waves was offered by Croft & Sajjadi (1993) who demonstrated that application of Miles' (1957) critical layer mechanism to a sharp-crested Stokes wave leads to a growth rate that is two or three times greater than that previously reported.

Croft & Sajjadi's (1993) theory considered a uniform shear flow $U(y)$ that is perturbed by a Stokes wave

$$y = y_0(x,t) = \Re \sum_{n=1}^{\infty} a_n e^{ik_n(x-c_nt)} \qquad (1.1)$$

in a Cartesian coordinate system (x, y) that is moving in the positive x-direction. They expressed the Stokes wave profile at $t = 0$ by the representation based on that originally proposed by Longuet-Higgins (1981)

$$y = K^{-1} \ln|\sec K(x - x_0)| \qquad (1.2)$$

where $x_0 = \frac{1}{2}$, $|K(x - x_0)| \leqslant \pi/3$, $K = 2\pi/L$ is the wavenumber and L is the wavelength. They then represented (1.2) by Fourier series and evaluated the amplitude of each harmonics by numerical integration.

They referred the velocity to the surface current (which must exist in consequence of the mean shear stress acting on the water), so that $U(0) = 0$ and the mean velocity in the moving reference frame was $U(y) - c$. They formulated the boundary-value problem for the perturbed flow in $y > y_0$, and then calculated the resulting forces on $y = y_0$ to obtain the energy transfer to Stokes wave.

However, there remained a fundamental question with Croft & Sajjadi's (1993) theory which was frequently asked but remained unanswered and that is, what mechanism is responsible for such increased energy transfer rate? Croft & Sajjadi (1993) argued that this increase energy transfer is due to non-linear nature of surface Stokes wave. Indeed in the later communications, Sajjadi *et al.* (1997) and Sajjadi (1998) improved upon the earlier theory by introducing the effect of turbulence in the wind blowing over Stokes wave.

The boundary-value problem posed by Croft & Sajjadi (1993) was the solution of the Orr-Sommerfeld equation subject to the boundary conditions at infinity and at the mean (undisturbed) position of the surface. The aim of this paper is to: (i) present results on Stokes wave; (ii) impose the boundary condition at the surface wave, rather than at the mean surface; (iii) include the dominant viscous term in the complete Orr-Sommerfeld equation; and (iv) confirm the earlier conjecture by Croft & Sajjadi (1993) that this increased energy transfer is due to higher harmonics that are present in Stokes wave and show how much of this increased energy is transferred to higher harmonics. We shall find that: (i) yields an energy-transfer coefficient that is smaller than, but of the same order of magnitude as, that previously estimated by Croft & Sajjadi (1993) but larger than those calculated for monochromatic waves; (ii) has no real effect on the end-results; (iii) shows that the viscous effects in the air just above the surface Stokes wave are small compared with those in the water, being of relative order $\sigma R_a^{-1/2} R_w$ (σ is the ratio of air to water density), where

$$R = c/k\nu \tag{1.2a}$$

or for gravity wave,

$$R = c^3/g\nu \tag{1.2b}$$

denotes a Reynolds number based on wave-speed, wavenumber, and the viscosity of either fluid (suffices a and w denote air and water, respectively); and (iv) shows that the damping ratio, given by

$$\zeta = \frac{2\Im\{\sqrt{\mathcal{F}_1 + k^2a^2\mathcal{F}_2}\}}{\Re\{\sqrt{\mathcal{F}_1 + k^2a^2\mathcal{F}_2}\}}$$

where the expressions for $\mathcal{F}_i$, $i = 1, 2$ is given in §4 below, has extra components proportional to k^2a^2 for Stokes wave compared with the corresponding expression for the monochromatic wave. This indicates that the extra energy is transferred to Stokes wave via the presence of $\mathcal{F}_2$ term whose magnitude is less than $\mathcal{F}_1$ but its overall effect is to increase the magnitude of ζ and hence yield an additional energy transfer from wind to Stokes wave.†

The model to be developed in §2 resembles that developed by Croft & Sajjadi (1993) in that it neglects perturbation Reynolds stresses‡ (associated with the interaction between turbulent fluctuations in the original and perturbed flows); it differs in that it includes perturbation viscous stresses and is based on the intrinsic equations of motion (in which the streamlines appears as coordinate lines) as oppose the orthogonal curvilinear coordinates which was adopted by Croft & Sajjadi (1993). We shall include only a boundary-layer approximation to the perturbation viscous stresses, however, anticipating that (for large R) these stresses can be significant only in the small neighbourhoods of the surface Stokes wave (*outer viscous layer*) and at $U = c$ (*inner viscous layer*).

Although we recognize that the turbulent stresses will profoundly affect the wave growth (Miles 1967, Sajjadi 1998), however, in the present model, the role of the Reynolds stresses is confined to the determination of the unperturbed mean velocity profile. For air flowing concurrently with the waves, there is a critical height where the unperturbed wind speed equals the wave phase speed. The upward motion of the air flow over the wave induces a pressure variation which leads to a vortex sheet of periodically varying strength forming at the critical height. Then the 'vortex force' on the wave leads to a transfer of energy from wind to the waves. We will show that the growth rate, for each harmonic of a non-linear wave, consist of three components. The first is essentially a positive energy-transfer from the shear flow, being essentially the same as the inviscid mechanism of Miles (1957). The second and the third components represent, respectively, additional energy transfer from viscous dissipation in the air and the water. We will further demonstrate that simple asymptotic expressions, used in the most large-scale wave models, leads to growth rate that is a factor of $O(1/\varepsilon)$, too large, where

$$\varepsilon \sim \frac{1 + c/U_*}{(1/\kappa)\ln(\lambda/z_0)}$$

Here we will offer an alternative expression which agrees better with both experimental and numerical data.

It might be objected that the neglect of the perturbation Reynolds stresses relative to perturbation viscous stresses is far more questionable than their neglect in an inviscid model, but the purpose of including the viscous stresses here is to show that they are indeed negligible compared with the terms arising from the inviscid part. Note that the outer viscous layer, which is expected to be important with regard to the interaction of viscous stresses in the air and water, is confined within the viscous sublayer of the undisturbed flow; it has indeed been amply confirmed experimentally that the flow near the water is aerodynamically smooth for wind-speeds as high as 800 cm/s at a height of 400 cm above the surface.

† Note that for monochromatic waves ($\mathcal{F}_2 \equiv 0$) the expression for ζ reduces to that found by Miles (1959).

‡ The inclusion of perturbation Reynolds stresses will be reported in part 2 of this paper.

Thus, with regard to the remark just made, it is appropriate to introduce the dimensionless shear parameter

$$S_a = U'(0+)/kc = U_*^2/kc\nu_a = R_a(U_*/c)^2 \qquad (1.3a, b, c)$$

Note that (1.3b) follows from (1.3a) through the equality of the shearing stresses $\rho_a U_*^2$ (where U_* is Prandtl's friction velocity) and $\rho_a \nu_a U'(0+)$, while (1.3c) follows from (1.3b) through (1.2a). We find that the outer viscous layer will be confined within the viscous sublayer if approximately $S_a < 10$ and that this inequality will be satisfied for those combinations of parameters for which viscous dissipation in the air might have been expected to be significant. Note however that S_a may be rather large at higher wind-speeds, c.f. Croft & Sajjadi (1993). In the former, analysis has been given for arbitrary values of S_a on the assumption that $U(y)$ is exactly linear, while in the latter, a fixed wave ($C = 0$ or $S_a = \infty$) was treated on the same assumption. Both findings showed that the phase shifts associated with the viscous stresses can lead to a positive energy transfer to the disturbance [c.f. also Lin (1954)].

We shall develop the equations of motion for the water in §3 on the assumption $S_w \ll 1$, where [based on $\rho_a U_*^2 = \rho_w \nu_w U'(0-)$]

$$S_w = U'(0-)/kc = \sigma U_*^2/kc\nu_w \qquad (1.4a, b)$$

and

$$\sigma = \rho_a/\rho_w \qquad (1.5)$$

This permits the neglect of the shear flow in the water and the derivation of our results directly from Lamb (1932, §349).

Having developed the equations of motion in §§2 and 3, we shall impose the boundary conditions of continuity of velocity and stress in §4 to obtain an approximation to the complex wave-speed. We assume that the magnitude of the wave-speed is closely approximated by its unperturbed, inviscid value [$c^2 = gk^{-1}(1 + k^2a^2)$ for Stokes gravity waves] and that this value may be used in the determination of the perturbation flows. In §§5 and 6 we shall present numerical results based on revised and extended analysis originally offered by Conte & Miles (1959).

One of the major component of next-generation operational models includes coupling interaction between meteorology and wave models. However, software tools to facilitate coupling are currently unavailable, requiring major computer coding of both models to include interactions during time steps and also developing new computer architecture, such as code parallelization, to accelerate the solution schemes.

Most wave models to date suffer from non-physically based source terms, often empirically derived, and sometimes yielding inaccurate solutions. For example, underestimation of wave growth due to wind-wave interaction is a well-known problem in current wave models, and attempts to address this problem are often crude. In recent years, Sajjadi *et al.* (1997, 1999); Sajjadi (1998, 2002a,b) have identified that the underestimation in wave models is due to the neglect of turbulent interaction between the atmosphere and ocean, and due to a lack of consideration for different phase speeds and nonlinearity in surface wave profiles. Nearly all parameterizations follow the original critical-layer contribution made by Miles (1957) which only accounts for wave growth due to inviscid shear-flow instability, and assumes the wind speed is low compared to the wave speed. In §7 we will compare the result of the present parameterization with those of Miles (1957), Janssen (1991) and experimental data.

2. Airflow over the Stokes wave

We choose, as independent variables, the coordinates s and n measured along and normal to the streamlines (see Figure 1) in a frame of reference moving with the wave-speed c and, as dependent variables $q(s,n)$ and $\theta(s,n)$, the velocity along a streamline and the inclination of the streamline, respectively. (In accordance with the procedure outlined above, c may be approximated as real throughout the following analysis except in (4.5i,j) and (4.6).) We begin with the intrinsic equations of motion (Milne-Thomson 1968, §21.39)

$$q\frac{\partial q}{\partial s}+\frac{1}{\rho}\frac{\partial p}{\partial s}=\nu\frac{\partial^2 q}{\partial n^2} \tag{2.1a}$$

$$q^2\frac{\partial \theta}{\partial s}+\frac{1}{\rho}\frac{\partial p}{\partial n}=0 \tag{2.1b}$$

$$\frac{\partial q}{\partial s}+q\frac{\partial \theta}{\partial n}=0 \tag{2.1c}$$

where ρ denotes density, p hydrodynamic pressure, ν kinematic viscosity (all parameters in this section referring to the air), and the right-hand side of (2.1a) represents the dominant shear term in a boundary-layer-type approximation. We shall seek the perturbation flow coupled with the second order Stokes wave

$$\eta(s)=ae^{iks}+\tfrac{1}{2}a^2ke^{2iks}\equiv\eta_1+\eta_2 \tag{2.2}$$

on $n=0$ in a uniform, parallel shear flow $q^{(0)}=U(n)-c$. (Following the usual convention, the imaginary parts of complex quantities proportional to $\exp(iks)$ and $\exp(2iks)$ are to be discarded in the final interpretation.) The final motion will be unstable if $\Im\{c\}>0$, exhibiting the time-growth factor $\exp(k\Im\{c\}t)$.

We first note that the unperturbed solution to (2.1a,b,c) implied by our assumption of a strictly parallel shear flow is

$$q=q^{(0)}(n),\qquad p=p^{(0)}(s),\qquad \theta\equiv 0. \tag{2.3a,b,c}$$

In fact, we shall use (2.1a,b,c) to describe perturbations with respect to a turbulent flow for which $U(n)$ is the mean flow and in which the viscous stress $\rho\nu U'$ is actually balanced by a Reynolds stress; the model provided by (2.1a,b,c) then neglects perturbation Reynolds stresses.

We may linearize (2.1a,b,c) in the independent variable $\theta(s,n)$ by differentiating (2.1a) with respect to both s and n and (2.1b) twice with respect to s, taking the difference between the results to eliminate p, and eliminating q_s through (2.1c) to obtain

$$\frac{\partial}{\partial s}\left[\frac{\partial}{\partial n}\left(q^2\frac{\partial\theta}{\partial n}\right)+\frac{\partial}{\partial s}\left(q^2\frac{\partial\theta}{\partial s}\right)\right]=\nu\frac{\partial^3}{\partial n^3}\left(q\frac{\partial\theta}{\partial n}\right) \tag{2.4}$$

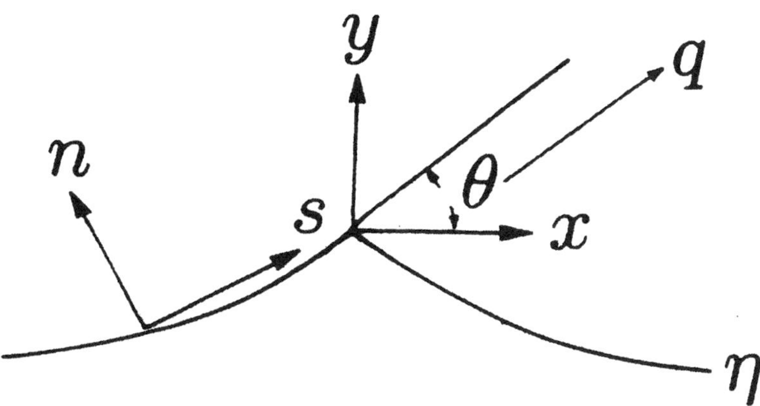

FIGURE 1. The coordinates for the intrinsic equations of motions (2.1a,b,c).

Now, to first order in θ, we may approximate q by its undisturbed value $U(n)-c$ and assume θ to exhibit the harmonic s-dependent of the form

$$\theta(s,n) = e^{iks}\Theta_1(n) + ake^{2iks}\Theta_2(n) \tag{2.5}$$

to obtain

$$\frac{\partial}{\partial n}\left(q^2\Theta_1'\right) - k^2\left(q^2\Theta_1\right) = \frac{\nu}{ik}\frac{\partial^3}{\partial n^3}\left(q\Theta_1'\right) \tag{2.6}$$

and

$$\frac{2}{k}\frac{\partial}{\partial n}\left(q^2\Theta_2'\right) - 8k\left(q^2\Theta_2\right) = \frac{\nu}{ik}\frac{\partial^3}{\partial n^3}\left(q\Theta_2'\right) \tag{2.7}$$

as the linearized equation of motion for $O(a)$ and $O(a^2)$, respectively. We remark that (2.6) differs from a boundary-layer approximation to the Orr-Sommerfeld equation (governing the perturbation stream function in Cartesian coordinates) in that it has a singularity at $U = c$; this implies that the linearized approximation to θ cannot be uniformly valid in the neighbourhood of $U = c$.† We shall find that this singularity introduces no essential difficulty (in so far as we require only the perturbation stresses at the interface $n = 0$), but it should be distinguished from the singularity that occurs at $U = c$ for the inviscid Orr-Sommerfeld equation of (2.11a) below; the latter singularity is a consequence of neglecting the viscous forces in a neighbourhood where the inertial forces tend to zero.

We shall present the full boundary condition in §4 below, but we note that

$$\theta(0,s) = \eta'(s) \quad \text{and} \quad \theta \to 0 \text{ as } n \to \infty \tag{2.8a, b}$$

Note that these boundary conditions are imposed at the displaced, rather than

† In our earlier calculations, Croft & Sajjadi (1993), formulation in orthogonal curvilinear coordinates avoided this difficulty. However, that formulation led to an inhomogeneous form of the Orr-Sommerfeld equation unless terms in U^{iv} and U''' were neglected.

the mean, position of the interface, thereby avoiding the assumption that the surface-wave displacement η must be small compared with a characteristic length (say c/U') for the shear profile; thus, we have only to assume $k|\eta| \ll 1$, rather than $U'(0+)|\eta|/c[= S_a k|\eta|] \ll 1$. It is for this reason that we choose a formulation in terms of $\theta(s,n)$, the streamline inclination in non-Cartesian coordinates, rather than the more conventional formulation in terms of a stream function in Cartesian coordinates.

We shall seek asymptotic solution to (2.6) and (2.7) in the limit as $R = c/k\nu \to \infty$. The formal procedure is essentially the same as that for the Orr-Sommerfeld equation (see Lin 1955, §§3.4 and 3.6) and yields two solutions for both equations (2.6) and (2.7) which satisfy the boundary condition (2.8b). The first of these, the *inviscid solution*, may be obtained by setting $\nu = 0$ in (2.6) and (2.7); the second or *viscous solution*, may be obtained by neglecting the second term on the left-hand side of (2.6) and (2.7) or equivalently, omitting the pressure gradient in (2.1a) and disregarding (2.1b). We find it convenient to solve for $(U - c)\theta$ and $(U - c)\theta_n$ (which are proportional to vertical velocity and perturbation shearing stress) in these two cases and to separate the s-dependence by introducing the factor $\eta'(s)$; defining the dimensionless variables

$$\xi = kn, \qquad f(\xi) = [U(n) - c]/c \tag{2.9a, b}$$

we then express $\Theta_i(n)$ as an inviscid plus viscous contributions in the form

$$\Theta_i(n) = \hat{\Theta}_i(n) + \tilde{\Theta}_i, \qquad i = 1, 2 \tag{2.10a}$$

where

$$f(\xi)\hat{\Theta}_i(n) = -ik\phi_i(\xi), \qquad f(\xi)\tilde{\Theta}_i'(n) = -ik^2\chi(R^{1/2}\xi) \tag{2.10b, c}$$

$$\Theta_i(n) = -ika\left[\frac{\phi_i(\xi)}{f(\xi)} + \int_\infty^\xi \frac{\chi_i(R^{1/2}\xi)\,d\xi}{f(\xi)}\right] \tag{2.10d}$$

and

$$f\phi_i'' - (f'' + f)\phi_i = 0 \tag{2.11a}$$

$$\frac{d^2\chi_i}{d\xi^2} - iRf\chi_i = 0 \quad \text{or} \quad \chi_i'' - if\chi_i = 0 \tag{2.11b, c}$$

as may be confirmed either by substituting (2.10d) in (2.6) and (2.7) and allowing R to tend to infinity or through the approximation described in the proceeding sentence.

The inviscid equation (2.11a) is identical with the inviscid Orr-Sommerfeld equation, which shows that our introduction of intrinsic coordinates and the imposition of the boundary condition (2.8a) at the displaced position of the interface have not altered the inviscid problem. The viscous equation (2.11b), on the other hand, differs from its counterpart in the asymptotic solution of the Orr-Sommerfeld equation in consequence of our choice of variables. However, the methods of asymptotic solution remain the same and, we may use the WKB approximation (Sajjadi 1988) to obtain

$$\chi_1 \sim f^{-1/4} \exp\left\{-\int_{\xi_c}^{\chi} \sqrt{iRf}\, d\xi\right\}\left[1 + O\left(R^{-1/2}\right)\right] \tag{2.12a}$$

$$\chi_2 \sim f^{-1/4} \exp\left\{-\int_{\xi_c}^{\chi} \sqrt{2iRf}\, d\xi\right\}\left[1 + O\left(R^{-1/2}\right)\right] \tag{2.12b}$$

where $f(\xi_c) = 0$ and the phase of the radical in (2.12a) or (2.12b) is $\pm\frac{1}{4}\pi$ as $\xi \gtrless \xi_c$, the path of integration being indented under the branch point at $\xi = \xi_c$ (Lin 1955, §3.4). Note also that the error factor in (2.12a) and (2.12b) is referred to the exact solution of (2.6) and (2.7), respectively, and not to (2.11a).

Furthermore, it should be noted that neither (2.12a) nor (2.12b) are uniformly valid near $\xi = \xi_c$, but they suffice for the present purpose in so far as $S \ll R^{1/2}$ (the condition that the inner and outer viscous layers be well separated), a condition that will be satisfied for those combinations of parameters for which viscous dissipation in the air is most significant (albeit still small).

It now remains to express the perturbation stresses on the interface in terms of ϕ_i and χ_i. Neglecting terms $O(R^{-1})$, in keeping with the present boundary-layer approximation, we may calculate the normal stress from (2.1a), (2.3a,b), (2.10) and (2.11b) according to

$$p^{(nn)} = -e^{iks}\left(P_1 - P_1^{(0)}\right) - ake^{2iks}\left(P_2 - P_2^{(0)}\right) \tag{2.13a}$$

$$\begin{aligned} &= -(ik)^{-1}\rho[\Theta_1' q^2 - (\nu/ik)(q\Theta_1')_{nn}]e^{iks} \\ &\quad -(2ik)^{-1}\rho[\Theta_2' q^2 - (\nu/2ik)(q\Theta_2')_{nn}]ake^{2iks} \end{aligned} \tag{2.13b}$$

$$= \rho c^2 k\{[f\phi_1'(0) - f'\phi_1(0)]\eta_1 + [f\phi_2'(0) - f'\phi_2(0)]\eta_2\} \tag{2.13c}$$

Evaluating (2.13c) at $n = 0$, where $f = -1$ and $f' = S$ [see (1.3)]. we may write

$$\left(p^{(nn)}\right)_{n=0+} = -\rho c^2 k[\varpi_1\eta_1\phi_1(0) + \varpi_2\eta_2\phi_2(0)] \tag{2.14}$$

where

$$\varpi_j = \frac{\phi_j'(0)}{\phi_j(0)} + S = (\alpha_j + i\beta_j)\left(\frac{U_1}{c}\right)^2, \quad j = 1, 2 \tag{2.15a, b}$$

the parameter α_j and β_j are related to the resulting pressure which acts on the boundary (where $\phi_j(0) = 1$)

$$p = \rho U_1^2 k \sum_{n=1}^{\infty}(\alpha_n + i\beta_n)\eta_n \tag{2.15c}$$

and the argument zero implies evaluation at $\xi = 0+$. Note that the result given (2.14) and (2.15a,b) for second-order Stokes wave can be generalized for higher-order Stokes wave in much the same way with difficulty. For example, for third-order Stokes wave

$$\eta = ae^{iks} + \tfrac{1}{2}a^2ke^{2iks} + \tfrac{3}{8}a^3k^2e^{3iks} \equiv \eta_1 + \eta_2 + \eta_3$$

(2.13) and (2.14) become

$$p^{(nn)} = -(ik)^{-1}\rho[\Theta_1' q^2 - (\nu/ik)(q\Theta_1')_{nn}]e^{iks}$$
$$-(2ik)^{-1}\rho[\Theta_2' q^2 - (\nu/2ik)(q\Theta_2')_{nn}]ake^{2iks}$$
$$-(3ik)^{-1}\rho[\Theta_3' q^2 - (\nu/3ik)(q\Theta_3')_{nn}]a^2k^2e^{3iks} \qquad (2.16a)$$
$$= \rho c^2 k\{[f\phi_1'(0) - f'\phi_1(0)]\eta_1 + [f\phi_2'(0) - f'\phi_2(0)]\eta_2 + [f\phi_3'(0) - f'\phi_3(0)]\eta_3\} \qquad (2.16b)$$

and where (2.16b) is evaluated at $n = 0$, we obtain

$$\left(p^{(nn)}\right)_{n=0+} = -\rho c^2 k[\varpi_1\eta_1\phi_1(0) + \varpi_2\eta_2\phi_2(0) + \varpi_3\eta_3\phi_3(0)] \qquad (2.17)$$

Thus for higher-order Stokes wave (2.14) and (2.15) generalize to

$$\left(p^{(nn)}\right)_{n=0+} = -\rho k^2c^2\sum_{j=1}^{\infty}\left(\frac{\phi_j'(0)}{\phi_j(0)} + S\right)\eta_j = -\rho kc^2\left(\frac{U_1}{c}\right)^2\sum_{j=1}^{\infty}(\alpha_j + i\beta_j)\eta_j \qquad (2.18a, b)$$

The expressions (2.18) are in full agreement with earlier works of Sajjadi and co-workers.

The tangential stress is given by (within the boundary-layer approximation)

$$p^{(sn)} = \rho\nu(q_n - U') \qquad (2.19a)$$
$$= (-ik)^{-1}\rho\nu[q(\Theta_1'\eta_1 + \Theta_2'\eta_2)]' \qquad (2.19b)$$
$$= \rho c^2 R^{-1/2}[\chi_1'(R^{1/2}\xi)k\eta_1 + \chi_2'(R^{1/2}\xi)k\eta_2] \qquad (2.19c)$$

Substituting χ_i from (2.12) and setting $\xi = 0+$, we obtain

$$\left(p^{(sn)}\right)_{n=0+} = -e^{\pi i/4}\rho c^2 R^{-1/2}[\chi_1(0)k\eta_1 + \sqrt{2}\chi_2(0)k\eta_2] \qquad (2.20)$$

Note that the viscous solution enters the calculation of the normal stress and the inviscid solution that of the tangential stress only through the boundary conditions, which relate $\phi_i(0)$ and $\chi_i(0)$.

3. Equations of motion for the Stokes wave

We shall now proceed on the assumption that the shear flow in the water (which is induced by the traction of the shear flow in the air) may be neglected. As was stated in §1, this will be a good approximation if $S_w \ll 1$, where S_w is defined by (1.4). We must stress at this point that this assumption does not preclude the existence of a surface current, since our velocities are defined relative to such a current; all that $S_w \ll 1$ implies is that the water moves approximately uniformly with the surface current to a depth of the order $1/k(= \lambda/2\pi)$. In addition to this, the small shear flow that is present must be oppositely directed to the airflow, for example $U_w < 0$ if $U_a > 0$; it follows that $U - c$ could not vanish in $n < 0$ for wave traveling downwind, whence the shear flow in the water could not transfer energy to the surface wave through the critical layer mechanism. Note that such an energy transfer would be predicted for a wave traveling upwind, but it generally would be much smaller than the energy absorbed by viscous dissipation, primarily because $U_w \ll 1$ at the depth where $U_w - c = 0$.

Assuming small perturbations with respect to a uniform flow $-c$ (relative to the moving frame of reference), we follow Lamb (1932, §349) and construct a solution for a surface wave of the form (2.2) moving over a viscous liquid. If we omit the hydrodynamic static pressure from $P^{(nn)}$ and note that $v = -c\theta$, we may pose the streamline inclination and perturbation stresses in the forms

$$\theta = \left(a_1 e^{\xi} + b_1 e^{\kappa\xi}\right) ik\eta_1 + \left(a_2 e^{\xi} + b_2 e^{\kappa\xi}\right) ik\eta_2 \tag{3.1}$$

$$\begin{aligned} p^{(nn)} &= -\rho c^2 \left[(1+2iR^{-1})a_1 e^{\xi} + 2i\kappa R^{-1} b_1 e^{\kappa\xi}\right] k\eta_1 \\ &\quad -\rho c^2 \left[(1+2iR^{-1})a_2 e^{\xi} + 2i\kappa R^{-1} b_2 e^{\kappa\xi}\right] k\eta_2 \end{aligned} \tag{3.2}$$

$$\begin{aligned} p^{(sn)} &= \rho c^2 \left[2R^{-1} a_1 e^{\xi} + (2R^{-1} - i) b_1 e^{\kappa\xi}\right] k\eta_1 \\ &\quad \rho c^2 \left[2R^{-1} a_2 e^{\xi} + (2R^{-1} - i) b_2 e^{\kappa\xi}\right] k\eta_2 \end{aligned} \tag{3.3}$$

$$\kappa = (1 - iR)^{1/2}, \qquad \Re\{\kappa\} > 0 \tag{3.4a, b}$$

where a_i and b_i are constants to be determined by the boundary conditions at the interface, and ρ, ν, and R are to be evaluated for the water. Note that the boundary-layer approximation is not applicable to the water (since the interface is approximately free for the heavy fluid below the interface) and that (3.1)–(3.4) are based on the full, linearized equations of viscous flow; subsequently, we shall assume $R_w \gg 1$, but this approximation has yet to be invoked.

4. Energy transfer to the Stokes wave

In order to evaluate the energy transfer from shear flow to Stokes wave we must first proceed with determination of wave-speed. Thus, we may infer the boundary conditions at the interface from the considerations that both θ_a and θ_w must be equal to the slope of the surface wave, that the velocity (or θ_n) be continuous, that the shear stress be continuous, and that the discontinuity in normal stress be prescribed. Accordingly

$$\theta_a = ik\eta, \qquad \theta_w = ik\eta, \qquad \Delta\theta_n = 0 \tag{4.1a, b, c}$$

$$\Delta p^{(sn)} = 0, \qquad \Delta p^{(nn)} = L\eta \tag{4.1d, e}$$

where

$$\Delta(\ ) = (\ )_{n=0+} - (\ )_{n=0-} \tag{4.2}$$

and $L\eta$ the (static) restoring viscous stress of the interface. We may relate the operator L to the inviscid wave-speed in the absence of the upper-fluid according to

$$L\eta = \rho_w c_w^2 k\eta \tag{4.3}$$

for second-order Stokes gravity waves, we have

$$c_w^2 = gk^{-1}(1 + k^2a^2) = c_{w0}^2 + c_{w1}^2 \tag{4.4}$$

Moreover, we need not pose the boundary conditions at infinity, assuming them to be satisfied implicitly.

Substituting (2.10d), (2.14), (2.20), (3.1)–(3.3) and (4.3) in (4.1a–e), setting $f(0) = -1$ and $f'(0) = S_a$, and cancelling common factors, we may place the results in the form

$$\phi_1(0) - e^{\pi/4}R_a^{-1/2}\chi_1(0) = 1 \tag{4.5a}$$

$$\phi_2(0) - e^{\pi/4}R_a^{-1/2}\chi_2(0) = \tfrac{1}{2} \tag{4.5b}$$

$$a_1 + b_1 = 1 \tag{4.5c}$$

$$a_2 + b_2 = 1 \tag{4.5d}$$

$$\phi_1'(0) + S_a\phi_1(0) + \chi_1(0) = a_1 + \kappa b_1 \tag{4.5e}$$

$$\phi_2'(0) + S_a\phi_2(0) + \chi_2(0) = \tfrac{1}{2}(a_2 + \kappa b_2) \tag{4.5f}$$

$$-e^{-\pi i/4}\sigma R_a^{-1/2}\chi_1(0) = 2R_w^{-1}(a_1 + b_1) - ib_1 \tag{4.5g}$$

$$-e^{-\pi i/4}\sigma R_a^{-1/2}\chi_2(0) = \tfrac{2}{\sqrt{2}}R_w^{-1}(a_2 + b_2) - ib_2 \tag{4.5h}$$

$$c_{w0}^2 = c_0^2[(1 + 2iR_w^{-1})a_1 + 2i\kappa R_w^{-1}b_1 - \sigma\varpi_1\phi_1(0)] \tag{4.5i}$$

$$c_{w1}^2 = c_1^2[(1 + 2iR_w^{-1})a_2 + 2i\kappa R_w^{-1}b_2 - \sigma\varpi_2\phi_2(0)] \tag{4.5j}$$

Solving (4.5a–j) for $\phi_i(0), \chi_i(0), a_i$ and $b_i, i = 1, 2$, and substituting in (4.5i) and (4.5j), we obtain

$$c^2 = c_0^2 + c_1^2 = c_{w0}^2[\mathcal{F}_1 + k^2a^2\mathcal{F}_2] + O[R_w^{-3/2}, \sigma R_a^{-1/2}R_w^{-1/2}, \sigma S_a R_a^{-1}, \sigma^2] \tag{4.6}$$

where

$$\mathcal{F}_1 = 1 - 4iR_w^{-1} + \sigma\varpi_1 - \sigma(1 - \varpi_1)^2 e^{\pi i/4}R_a^{-1/2}$$

and

$$\mathcal{F}_2 = 1 - (2 + \sqrt{2})iR_w^{-1} + \sigma\varpi_2 - \sigma(1 - \varpi_2)^2 e^{\pi i/4}R_a^{-1/2}$$

Substituting φ_i from (2.15b) and R_w and R_a from (1.2b) in (4.6) and neglecting higher-order terms, we obtain the damping ratio for second-order Stokes wave

$$\zeta = \frac{2\Im\{c\}}{\Re\{c\}} = \underbrace{\sigma\beta_1\left(\frac{U_1}{c}\right)^2}_{\text{(i)}} \underbrace{-\frac{4g\nu_w}{c^3}}_{\text{(ii)}}$$

$$\underbrace{-\sigma\left(\frac{g\nu_a}{2c^3}\right)^{1/2}\left[1 - 2(\alpha_1 + \beta_1)\left(\frac{U_1}{c}\right)^2 + (\alpha_1^2 + 2\alpha_1\beta_1 - \beta_1^2)\left(\frac{U_1}{c}\right)^4\right]}_{\text{(iii)}}$$

$$+\underbrace{\sigma\beta_2\left(\frac{U_1}{c}\right)^2}_{\text{(iv)}}\underbrace{-\frac{(2+\sqrt{2})g\nu_w}{c^3}}_{\text{(v)}}$$

$$\underbrace{-\sigma\left(\frac{g\nu_a}{2c^3}\right)^{1/2}\left[1-2(\alpha_2+\beta_2)\left(\frac{U_1}{c}\right)^2+(\alpha_2^2+2\alpha_2\beta_2-\beta_2^2)\left(\frac{U_1}{c}\right)^4\right]}_{\text{(vi)}} \quad (4.7)$$

where the terms (i)–(iii) on the right-hand side represent, respectively, the positive energy-transfer from the shear flow, the viscous dissipation in the water, and the viscous dissipation in the air for the fundamental harmonic (being essentially the same as that for a monochromatic wave), whereas terms (iv)–(vi) represent the same for the second harmonic.

5. Numerical solution

The energy transfer parameters α_j and β_j are evaluated by numerical integration of the Orr-Sommerfeld equation, derived in §2†

$$(f^2\Theta')' - f^2\Theta = (iR)^{-1}(f\Theta')''' \quad (5.1)$$

Following the notations in §2, letting $\Theta = -ik\phi f'$ and following Croft & Sajjadi (1993), we may express the solution of (5.1) as a combination of inviscid plus viscous solutions $\phi + h \equiv F$ where $\phi(\eta)$ satisfies the following boundary value problem

$$f\left(\frac{d^2\phi}{d\xi^2} - \phi\right) - \left(\frac{d^2 f}{d\xi^2}\right)\phi = 0, \qquad (0 < \xi_0 \leqslant \xi < \infty) \quad (5.2)$$

subject to the boundary conditions

$$\phi_0 = f_0, \qquad (\xi = \xi_0) \quad (5.2a)$$

$$\frac{d\phi}{d\xi} + \phi \to 0, \qquad (\xi \to \infty) \quad (5.2b)$$

As was shown by Croft & Sajjadi (1993), the asymptotic analysis for large R indicates that the viscous solution approximately obeys the equation

$$\frac{d^4 h}{d\xi^4} - i\gamma(\xi)\frac{d^2 h}{d\xi^2} = 0 \quad (5.3)$$

subject to the boundary conditions

$$h_0 = 0, \qquad h_0'' = 1 \quad (5.4)$$

$$h' = h''' = 0 \quad \text{as} \quad \xi \to \infty \quad (5.5)$$

where

$$f(\xi) = \log(\xi/\xi_c), \qquad (\xi_c > \xi_0) \quad (5.6)$$

† For clarity sake, we will describe the procedure for the fundamental harmonic $j = 1$.

and

$$\gamma(\xi) = Rf(\xi) \tag{5.7}$$

To integrate equation (5.3) we first reduce it to a pair of second order differential equations

$$\left.\begin{aligned} &h'' = \chi, && \text{subject to } h_0 = 0 \text{ and } h' \to 0 \text{ as } \xi \to \infty \\ &\chi'' - i\gamma\chi = 0, && \text{subject to } \chi_0 = 1 \text{ and } \chi' \to 0 \text{ as } \xi \to \infty \end{aligned}\right\} \tag{5.8}$$

The integration of equations (5.8) pose no real difficulty, however the solution of equation (5.2) pose a major difficulty due to the singular behaviour of the equation around $\xi = \xi_c$.

Note that ξ is confined to the real axis except in the neighbourhood of the regular singular point $\xi = \xi_c$, where the path of integration must be taken under the singularity. The components of this singularity are 0 and 1, and there exists only one analytic solution, $\phi_1 = O(f)$, in this neighbourhood. The second linearly independent solution has a logarithmic branch point there and may be posed as

$$\phi_2 = \phi_1 \log f + \phi_3 \tag{5.9}$$

where ϕ_3 is analytic and $O(1)$ in the neighbourhood of $f = 0$ and ϕ_1 is the regular solution of (5.2).

Substituting (5.9) into (5.2) we obtain the following inhomogeneous equation

$$\begin{aligned} &f\left(\frac{d^2\phi_3}{d\xi^2} - \phi_3\right) - \left(\frac{d^2 f}{d\xi^2}\right)\phi_3 = -f\log f\frac{d^2\phi_1}{d\xi^2} - 2\frac{df}{d\xi}\frac{d\phi_1}{d\xi} - \\ &\left[\frac{d^2 f}{d\xi^2} - \frac{1}{f}\left(\frac{df}{d\xi}\right)^2 - \left(f - \frac{d^2 f}{d\xi^2}\right)\log f\right]\phi_1 \end{aligned} \tag{5.10}$$

where ϕ_1 is the regular solution of (5.2), namely

$$f\left(\frac{d^2\phi_1}{d\xi^2} - \phi_1\right) - \left(\frac{d^2 f}{d\xi^2}\right)\phi_1 = 0 \tag{5.11}$$

The procedure adopted here is to determine ϕ_1 and ϕ_3 successively and then to combine ϕ_1 and ϕ_2 to satisfy (5.2a,b).

Substituting (5.9) into (5.2a,b), we obtain the following boundary conditions

$$\left.\begin{aligned} &\phi_1 = G && \text{and} \quad \phi_2 = G(1 - \log f) && \text{at} \quad \xi = \xi_0 \\ &\frac{d\phi_1}{d\xi} + \phi_1 = 0 && \text{and} \quad \frac{d\phi_2}{d\xi} + \phi_2 = -H && \text{as} \quad \xi \to \infty \end{aligned}\right\} \tag{5.12}$$

where

$$G = \phi_1 \log f + \phi_2 \qquad \text{evaluated at } \xi = \xi_0$$

$$H = \phi_1\left(\frac{1}{f}\frac{df}{d\xi} + \log f\right) + \left(\frac{d\phi_1}{d\xi}\log f\right) \qquad \text{as } \xi \to \infty$$

We discretise the above system of equation using central differences and seek their solutions over a Stokes wave defined by (4.2) and then calculate the resulting pressure acting on the surface, from (2.15c), where the energy transfer parameters α and β are calculated according to

$$\alpha + i\beta = f_0 \left(\frac{dF}{d\xi} - \frac{df}{d\xi} \right)_0 \tag{5.13}$$

Note that, the energy transfer to the Stokes wave at the surface is proportional to β, namely

$$\beta = -\pi |F|_c^2 (f_c''/f_c') \tag{5.14}$$

where the subscript c implies evaluation at the singular point $\xi = \xi_c$,

$$\xi_c = \Omega f_0^{-2} e^{-f_0}, \qquad \frac{c}{U_1} = -f_0 = \log(\xi_c/\xi_0), \qquad \Omega = g\xi_0/U_1^2 \tag{5.15}$$

and $\Omega = O(10^{-3}\text{–}10^{-2})$ is Charnock's constant.

6. Numerical solution about the singular point

The solution of the boundary value problem (5.2) posed in the previous section requires us to:

(a) solve (5.2) for the regular solution ϕ_1;
(b) substitute (5.9) in (5.2) and solve the resulting equation for ϕ_3;
(c) combine ϕ_1 and ϕ_2 to satisfy the boundary conditions (5.2a,b); and
(d) determine β through either (5.13) or (5.14).

In carrying out this program numerically for the solution of equation (5.2) the velocity f provides a more convenient scale than the independent variable ξ. Under the change of independent variable (5.6), the boundary value problem (5.2), (5.2a,b) becomes

$$L\phi = f \left(\frac{d^2\phi}{df^2} - \frac{d\phi}{df} \right) + (1 - \xi_c^2 f e^{2f}) = 0, \quad (-\infty < f_0 \leqslant f \leqslant \infty) \tag{6.1}$$

$$\phi_0 = f_0 \tag{6.1a}$$

$$\frac{d\phi}{df} + \xi_c e^f \phi \to 0, \qquad (f \to \infty) \tag{6.2b}$$

where (6.1) defines the operator L, and the singularity now appears at $f = 0$.

Applying the method of Frobenius we can now obtain series expansions valid about the singularity $f = 0$ for two linearly independent solutions. If we let

$$\phi = \sum_{k=1}^{\infty} a_k f^{k+c}, \qquad a_0 \neq 0$$

and expand the exponential in (6.1) in powers of f, we obtain

$$\sum_{k=1}^{\infty} a_k(k+c)(k+c-1)f^{k+c-1} - \sum_{k=1}^{\infty} a_k(k+c)f^{k+c} +$$
$$[1-\xi_c^2 f(1+2f+2f^2+...)]\sum_{k=1}^{\infty} a_k f^{k+c} = 0$$

Equating coefficients of f^{c-1} we find that $c = 0$ or $c = 1$. Now equating coefficients of $f^c, f^{c+1}, f^{c+2}, ...$ for $c = 1$ we obtain a_k's for $k \geqslant 1$ in terms of a_0. Without loss of generality we set $a_0 = 1$, thus the series for the regular solution ϕ_1 is

$$\phi_1 = f + \frac{1}{6}\xi_c^2 f^3 + \frac{7}{36}\xi_c^2 f^4 + \frac{\xi_c^2}{240}(31+2\xi_c^2)f^5 + \frac{\xi_c^2}{5400}(333+101\xi_c^2)f^6 + ... \tag{6.2}$$

To obtain the logarithmic solution we substitute (5.9) into (6.1) and, observing that $L\phi_1 = 0$, we obtain

$$L\phi_3 = \left(1+f^{-1}\right)\phi_1 - 2\frac{d\phi_1}{df} \tag{6.3}$$

as the inhomogeneous differential equation for ϕ_3. Substituting (6.2) into (6.3) and again expanding we obtain

$$\phi_3 = -1 + \tfrac{1}{2}(1-\xi_c^2)f^2 + \tfrac{5}{180}(3-20\xi_c^2)f^3 + \tfrac{1}{432}(6-137\xi_c^2-18\xi_c^4)f^4$$
$$+\tfrac{79}{900}\left(\tfrac{45}{1896} + \tfrac{179}{632}\xi_c^2 - \xi_c^4\right)f^5 + ... \tag{6.4}$$

The general solution of (6.1) is

$$\phi = A\phi_1 + B\phi_2 \tag{6.5}$$

where A and B are constants to be chosen so as to satisfy the boundary conditions (6.1a,b). The equations for determining A and B are:

$$\left.\begin{aligned} &A\phi_1(f_0) + B\phi_2(f_0) = f_0 \\ &A\left[\frac{d\phi_1}{df} + \xi_c e^f\phi_1\right]_{f_+} + B\left[\frac{d\phi_2}{df} + \xi_c e^f\phi_2\right]_{f_+} = 0 \end{aligned}\right\} \tag{6.6}$$

where f_+ is, for the purpose of numerical integration, infinite; the actual value of f_+ varies, depending upon the parameters ξ_c and Ω.

The numerical integration for boundary value problems, defined in §6, was performed using a five point central difference approximation. Writing any of these equations in the form

$$\Phi'' - u\Phi = w$$

where $\Phi \equiv \{\phi_1, \phi_3, h, \chi\}$, and discretizing them using following approximations

$$\Phi = \tfrac{1}{360}\left(\Phi_{i+2} + 56\Phi_{i+1} + 246\Phi_i + 56\Phi_{i-1} + \Phi_{i-2}\right)$$

$$\Phi'' = \ell^{-2}\left(\Delta_i^2 + \tfrac{1}{12}\Delta_i^4\right)\Phi$$

here Δ_i is the central difference operator and ℓ is the uniform step length, we obtain quindiagonal matrix equations

$$\mathcal{A}\boldsymbol{\Phi} = \boldsymbol{w}$$

These equations are then solved using the standard Gauss elimination technique.

The presence of the singularity at $f = 0$ necessitated the evaluation of the series (6.2) and (6.4) at some point f_1 sufficiently removed from the singularity before the numerical solution could be carried out. From these series we obtained ϕ_1 and ϕ_3 and their derivatives at f_1. The value of f_1 chosen varied from 10^{-4} to 10^{-7} depending upon the value of ξ_c. The lower limit of integration was fixed at f_0. The numerical integration for the viscous equation (5.3) was then performed using the central finite difference approximation of the coupled system (5.8) and the result was combined with the numerical solution of (5.2) to yield the solution to the Orr-Sommerfeld equation (5.1). Having obtained the numerical solution to (5.2), β was calculated using equation (5.14).

7. Application of the Theory to Spectral Wave Models

Spectral wave models, such as WAM, describe the evolution of a two-dimensional ocean wave variance spectrum through integration of the transport equation

$$\frac{D\mathsf{F}}{Dt} + \frac{\partial}{\partial\varphi}(\dot{\varphi}\mathsf{F}) + \frac{\partial}{\partial\mu}(\dot{\mu}\mathsf{F}) + \frac{\partial}{\partial\vartheta}\left(\dot{\vartheta}\mathsf{F}\right) = \mathsf{S}$$

where F represents the spectral density with respect to directions ϑ, latitudes φ and longitudes μ (the dot over symbols denotes the rate of change of the position and propagation of a wave packet).

The source function S is represented as a superposition of the wind input S_{in}, white capping dissipation S_{dis}, and non-linear transfer S_{nl}, i.e.

$$\mathsf{S} = \mathsf{S}_{\text{in}} + \mathsf{S}_{\text{dis}} + \mathsf{S}_{\text{nl}}$$

Here we shall only be concerned with S_{in} and offer an alternative formulation for it.

The wind input is usually given by

$$\mathsf{S}_{\text{in}} = \zeta\mathsf{F}$$

where ζ represents the growth rate of the waves. According to Miles (1957) mechanism, the growth rate, when normalized with frequency ω, is expressed as

$$\frac{\zeta}{\omega} = \beta\sigma\left(\frac{U_*}{c}\right)^2 \tag{7.1}$$

In the third generation model of WAM, the energy transfer parameter appearing in (7.1), taken from the asymptotic expression derived by Janssen (1989, 1991), is given by

$$\beta_{\text{asym}} = \frac{\beta_m}{\kappa^2} kz_c \ln^4(kz_c), \qquad kz_c < 1 \tag{7.2}$$

where κ is the von Karman's constant, $\beta_m = 1.2$ a constant and

$$kz_c = \min\left(1, kz_0 e^{\kappa/(U_*/c+0.011)}\right)$$

is the dimensionless critical height (k is the wave number and z_c the critical height at which $U = c$).

We shall now apply the foregoing theory to the growth of water waves by wind on the hypothesis that the air mean velocity in the turbulent boundary layer may be regarded as parallel shear flow. We shall assume that the mean velocity profile in the vicinity of the air-water interface is asymptotically logarithmic according to

$$U(n) = \frac{U_*}{\kappa}\log(n/n_0) \tag{7.3}$$

where U_* is the wind friction velocity and κ is Karman's constant. We perform calculations for waves with wavelength $\lambda = 0.64$ m, steepness $ak = 0.01$. We choose the value of 10^{-4} for the non-dimensional roughness length kn_0 and evaluate the energy transfer parameter over a range of U_*.

In Figure 2 we have plotted the expression for $\beta_{\rm asym}$, given by (7.2), as a function of c/U_* (the dashed line). As can be seen, the peak value of $\beta_{\rm asym}$ is approximately 34 which occurs at $c/U_* = 16$. Also in Figure 2 we have plotted β calculated with the present numerical integration of the Orr-Sommerfeld equation for the first 3 harmonics of Stokes wave, namely β_i, $i = 1, 3$ (the dotted lines). Note incidentally that β_1 corresponds to the monochromatic wave and the results obtained by Janssen (1989, 1991) (also for monochromatic waves) leads to growth rate that is a factor of

$$O\left(\frac{\kappa^{-1}\ln(\lambda n_0)}{1 + c/U_*}\right)$$

too large compared to the growth rate obtained from β_1. Note that, β_1 agrees very well with $\beta_{\rm num}$ (pluses), obtained from the numerical integration of full Reynolds stress transport equations over water waves using the two-component limit of turbulence (Sajjadi 2002b), over the range $7 \leqslant c/U_*/ \leqslant 18$. However, in the range $2 \leqslant c/U_* \leqslant 5$ both the present model and the asymptotic theory of Janseen (1989,1991) fails to capture the saturated trend of β. This is not surprising since neither of these models take into account the implicit nature of turbulence other than in prescribing the logarithmic mean wind velocity profile over the surface wave. It can be seen that β tends to an approximately constant value of 15 in the range $2 \leqslant c/U_* \leqslant 7$, being the range in which U_* is large (normally referred to as the slow moving wave regime).

Recently by invoking a closure model, based on Townsend (1976), Sajjadi (2002a) constructed a model for the specification of turbulent Reynolds stresses over water waves where the resulting equation together with its corresponding boundary conditions was solved numerically. The energy transfer parameter was then calculated from the derived expressions for the momentum flux for slow wind-wave regime ($c/U_* \leqslant 5$) which agreed well with the numerical integration of Sajjadi (2002b). The result of these calculations are also shown in Figure 2 (crosses) for comparison.

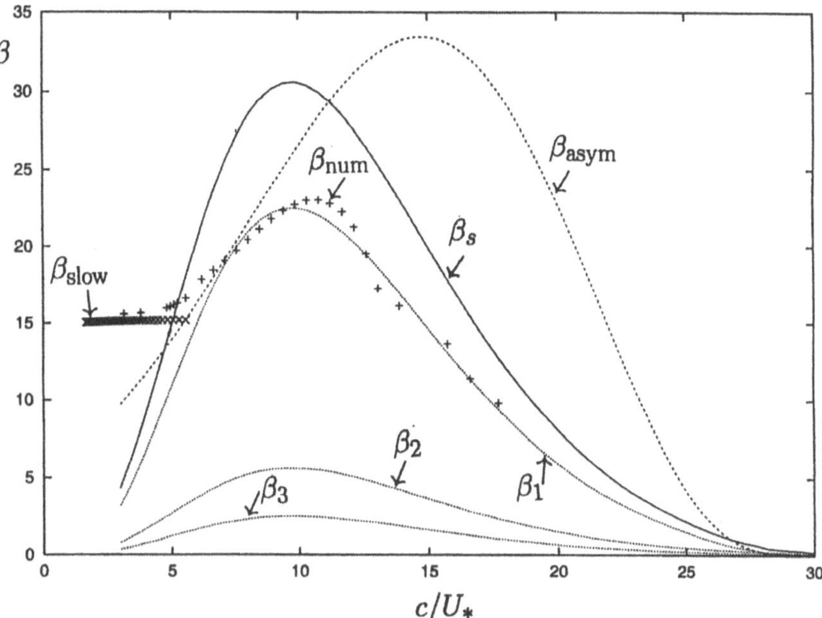

FIGURE 2. The plot of β against c/U_*. Dotted lines (...) are the result obtained by the present model for the first three harmonics of Stokes wave, pluses (+) are numerical results of Sajjadi (2002b), crosses (×) is the numerical results obtained by Sajjadi (2002a) for slow moving waves, solid line (—) is the result obtained by the present model for a third-order Stokes wave, and the dashed line (- - -) is Janssen's (1991) asymptotic result.

Figure 2 also shows the plot of β_s, given by

$$\beta_s = \sum_{n=1}^{N} (ak)^{n-1} \beta_n \tag{7.4}$$

for $N = 3$ which corresponds to energy transfer from wind to a third-order Stokes wave. From (7.4) it is evident that for $ak \ll 1$, $\beta_1 > \beta_2 > \beta_3$ and thus the increased energy transfer to non-linear waves is due to the presence of the additional terms, namely β_i, $i \geqslant 2$, compared to the monochromatic counterpart. Note that in the present calculations, the magnitude of $\beta_4 \ll 1$ and therefore its inclusion to series (7.4) will have an insignificant effect to the overall magnitude of β_s.

The result for energy transfer parameter, β, can be parameterized so that it can be incorporated in large scale wave models. The new parameterization, based on the present theory and that of Miles (1996), is given by (for full detail, see Sajjadi 2002c).

$$\beta = \beta_v + \beta_c, \qquad \beta_v = \frac{5\kappa^2}{\varepsilon}, \qquad \beta_c = \frac{5}{2}\pi \hat{W} \ell_0^4 \left[1 - \left(4 - \frac{\pi^2}{3}\right)\varepsilon^2\right]$$

$$\varepsilon = \ell_0^{-1}, \quad \ell_0 = -\gamma - \log \hat{W}, \quad \hat{W} = kz_0 e^{c/U_\lambda}(U_\lambda/c)^2, \quad U_\lambda = 2U_* \tag{7.5}$$

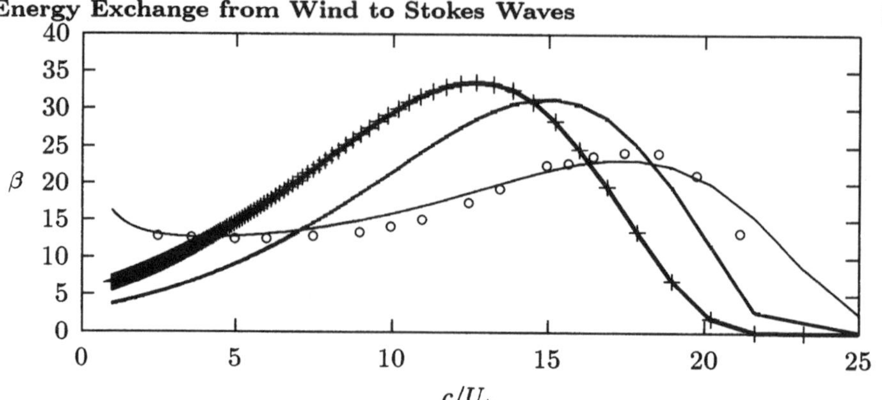

FIGURE 3. The plot of β against c/U_*. — present parameterization, eqn (7.5); — Miles (1957), eqn (7.6); +++ WAM model, eqn (7.2); o numerical simulation, Sajjadi (2002b).

Note in arriving at this parameterization the present theory is extended in order to take into account the implicit nature of turbulence over surface waves. The extra term β_v, appearing in (7.5), represents the contribution from viscoelastic effect of turbulence above the surface wave (see Sajjadi 2002c) and in its absence (7.5) reduce to the parameterization for β which would be valid only for the quasi-laminar model considered in this paper.

The result of the parameterization for β, given by (7.5), is shown in the Figure 3. The circles are the experimental data, obtained from the numerical integration of full Reynolds stress equations, which as can be seen agrees well with the parameterization. Like the data, obtained from the numerical integration, the peak in the energy transfer parameter β occurs at $c/U_* \approx 18$. Also in Figure 3 Miles'(1957) theory given by

$$\begin{aligned}\beta &= \pi\xi_c\left\{\frac{1}{6}\pi^2 + \log^2(\gamma\xi_c) + 2\sum_{n=1}^{\infty}\frac{(-1)^n\xi_c^n}{n!n^2}\right\}^2, \qquad \xi_c < 1 \qquad (7.6)\\ &= 4\pi\xi_c^{-3}e^{-2\xi_c}\left(1 - 6\xi_c^{-1} + 31\xi_c^{-2}\right), \qquad \xi_c \geqslant 1\end{aligned}$$

$$\xi_c = kz_0 e^{(\kappa/U_*/c)}$$

and Janssen's (1991) parameterization, given by (7.2), are compared with (7.5). Note that, Miles' theory predicts the peak value for β at $c/U_* = 15$ and Janssen's parameterization yields a peak value at $c/U_* \approx 12.5$. Note also that both Miles' (1957) theory and Janssen's (1991) parameterization predict a different variation of β over the entire range of c/U_* and clearly do not agree with the data obtained by numerical integration.

8. CONCLUSIONS

In this paper we have extended the previous analysis for the growth of non-linear surface waves by shear flow by: (i) presenting results on Stokes wave; (ii) imposing the boundary condition at the surface wave, rather than at the mean surface; and (iii) including the dominant viscous term in the complete Orr-Sommerfeld equation. The inclusion (i) yields an energy transfer that is larger than those predicted for monochromatic waves, while (ii) has no real effect and (iii) only a small effect for gravity waves. The present analysis is mainly confined to the second-order Stokes wave but its extension to higher-order Stokes waves is straightforward and suggested; though the result of a single calculation for third-order Stokes wave is included. The nondimensional energy transfer parameters are obtained via numerical integration of the full Orr-Sommerfeld equation, and it is found the results agree well with the recent numerical integration of full Reynolds stress transport equations over water waves using the two-component limit of turbulence performed by Sajjadi (2002b), over the range $7 \leqslant c/U_*/ \leqslant 18$.

In this paper we have shown that the growth of non-linear waves (comprising of several harmonics) consists of three components for each harmonic, namely; the positive energy-transfer from the shear flow (being essentially the classical critical layer mechanism), the viscous dissipation in the water, and the viscous dissipation in the air. Note that the asymptotic result obtained by Janseen (1989, 1991) leads to growth rate that is a factor of $O(1/\varepsilon)$ too large (see the second paragraph of page 3). We conclude that the increased energy transfer to non-linear waves obtained here is due to the presence of the additional components β_i, $i \geqslant 2$ which are not present for monochromatic waves. We further conclude that the disagreement between the present results and the result obtained by full numerical integration of the Reynolds stress transport equations is due to the exclusion of implicit effect of turbulence in slow moving wave regimes.

ACKNOWLEDGEMENT: This work is supported by DoD-HPCMO under the contract number N62306-99-D-B004.

REFERENCES

CONTE, S. & MILES, J.W. 1959 On the numerical integration of the Orr-Sommerfeld equation. *J. Soc. Indust. Appl. Math.*, **7**, 361.

CROFT, A.J. & SAJJADI, S.G. 1993 A mechanism for the transferring of energy from wind to Stokes wave.*Math. Engng. Ind.*, **4**, 23.

JANSSEN, P.A.E.M. 1989 Wave-induced stress and the drag of air flow over sea waves. *J. Phys. Oceanogr.*, **19**, 745.

—1991 Quasi-linear theory of wind wave generation applied to wave forecasting. *J. Phys. Oceanogr.*, **21**, 1631.

LAMB, H. 1932 *Hydrodynamics*. 6th edn. Cambridge University Press.

LIN, C.C. 1954 On the stability of two-dimensional parallel flows. Parts I, II and III. *Quart. Appl. Math.*, **3**, 117, 218, 277.

—1955 *The Theory of Hydrodynamic Stability*. Cambridge University Press.

LONGUET-HIGGINS, M.S. 1981 Oscillating flow over steep sand ripples. *J. Fluid Mech.*, **107**, 1.

MILES, J.W. 1957 On the generation of surface waves by shear flows. *J. Fluid Mech.*, **3**, 185.

—1959 On the generation of surface waves by shear flows. Part 2. *J. Fluid Mech.*, **6**, 568.

—1967 On the generation of surface waves by shear flows. Part 5. *J. Fluid Mech.*, **30**, 163.

—1996 Surface-wave generation: a visoelastic model. *J. Fluid Mech.*, **322**, 131.

MILNE-THOMSON, L.M. 1968 *Theoretical Hydrodynamics.* 5th edn. Macmillan.

SAJJADI, S.G. 1988 Shearing flow over Stokes waves. Technical Report, Coventry Polytechnic.

—1998 On the growth of a fully non-linear Stokes wave by turbulent shear flow. Part 2. Rapid distortion theory. *Math. Engng. Ind.*, **6**, 247.

—2002a Effect of turbulence due to tropical storm and hurricane on the growth of the Stokes wave. Submitted for publication to *Proc. Roy. Soc.* A.

—2002b Numerical simulation of turbulent flow over Stokes wave: Two-component limit of turbulence. Submitted for publication to *Int. J. Numer. Meth. Fl.*

—2002c A parameterization of Miles' viscoelastic theory of turbulence for energy exchange from wind to Stokes wave. Under preperation.

SAJJADI, S.G., THOMAS N.H. & HUNT, J.C.R. 1999 *Wind-Over-Wave Couplings: Perspectives and Prospects.* Oxford University Press.

SAJJADI, S.G., WAKEFIELD, J. & CROFT, A.J. 1997 On the growth of a fully non-linear Stokes wave by turbulent shear flow. Part 1. Eddy-viscosity model. *Math. Engng. Ind.*, **6**, 185.

TOWNSEND, A.A. 1976 *The Structure of Turbulent Shear Flow.* 2nd edn. Cambridge University Press.

Wind-Generated Water Waves: Two Overlooked Mechanisms?

M.E. McIntyre
Centre for Atmospheric Science† at the
Department of Applied Mathematics and Theoretical Physics,
Centre for Mathematical Sciences, Wilberforce Road, Cambridge CB3 0WA, UK
http://www.atm.damtp.cam.ac.uk/people/mem/

ABSTRACT

From research on stratospheric dynamics there has emerged what might prove to be some useful fresh ideas about the problem of water-wave generation by wind. As pointed out in an earlier contribution (1993), water waves can be systematically amplified by two irreversible, ratchet-like mechanisms that depend on spatio-temporal inhomogeneities, such as wind gustiness and wave groupiness. Neither mechanism is represented in the models used for wind-wave forecasting. The first mechanism is simply the drag from what might be called Rossby or vorticity lee waves in the airflow downstream of water-wave groups. The second is intermittent Rossby-wave breaking or vertical mixing of spanwise horizontal vorticity in the airflow, a highly nonlinear, non-Fourier-superposable mechanism. Both mechanisms can amplify non-breaking water waves, as well as contributing to the amplification of breaking water waves.

1. INTRODUCTION

I want to suggest that the mechanisms by which wind generates water waves can be illuminated by today's understanding of the large-scale fluid dynamics of the Earth's stratosphere. Persistent, irreversible momentum transport, involving a phase-coherent interaction between waves and turbulence, enters both the wind-wave problem and the stratospheric problem in an essential and fundamentally similar way. The ideas — which both subsume, and also go beyond, recent extensions of the Miles theory, including that of Belcher *et al.* (1999) presented in the predecessor to this wind-wave conference proceedings — can be traced back to G.I. Taylor's classic paper on eddy motion in the atmosphere (Taylor 1915) and its nonlinear extension by Killworth & McIntyre (1985), hereafter KM. The relevance of these ideas to the wind-wave problem were first pointed out in a Sectional Lecture (McIntyre 1993) to the Minisymposium *Sea Surface Mechanics and Air–Sea Interaction* at the 18th International Congress of Theoretical and Applied Mechanics.

In the case of the stratosphere we have a secure and well-developed understanding of the highly inhomogeneous 'wave–turbulence jigsaw puzzle' with

† The Centre for Atmospheric Science is a joint initiative of the Department of Chemistry and the Department of Applied Mathematics and Theoretical Physics (http://www.atm.damtp.cam.ac.uk/).

which the stratosphere confronts us when viewed on a global scale. For the history of ideas leading to that understanding the reader may consult, for instance, the review by Hoskins *et al.* (1985) and two recent reviews of mine (2000, 2002). The history takes on a new interest when compared to the history of ideas in wind-wave generation, especially if one takes as a conceptual starting point the linear-theoretic wave ideas embodied in the Miles theory — as further developed by, for instance, Lighthill (1962), Janssen (1982, 1991) and Belcher *et al.* (1999) — and then goes on to recognize various forms of strong nonlinearity that operate in reality (e.g. Banner 1990, McIntyre 1993 & refs., Belcher & Hunt 1998 & refs.), especially those nonlinearities associated with the critical layer or matched layer in the airflow above the water waves, with the implication that Fourier superposition must generally fail in descriptions of water-wave generation. Over the years, our understanding of stratospheric dynamics has undergone a closely analogous transition from linear to nonlinear thinking.

2. A glimpse of the real stratosphere

Concerns about stratospheric ozone have produced highly sophisticated observing systems that allow us to see the real stratosphere in considerable detail. Figure 1, of which an animated version can be seen on my website,† is reproduced here by courtesy of Dirk Offermann and Martin Riese of the CRISTA space-based remote-sensing project; see Riese *et al.* (2002). It shows the distributions at about 31 and 37 km altitude of a chemical tracer, nitrous oxide, which for present purposes is accurately an inert material tracer — the pictures are like laboratory dye pictures — because photochemical timescales for nitrous oxide are of the order of years, far longer than the relevant fluid-dynamical timescales. White regions are data gaps.

Aside from the overall pole-to-equator gradient, with high equatorial tracer values (red) and low polar values (purple), the pattern seen in both pictures is shaped by nearly-horizontal, layerwise-two-dimensional air motion on timescales of days to weeks. It reveals distinct airmasses such as the well-mixed region (blue) on the right, with almost perfectly uniform tracer values, sandwiched between relatively isolated polar and tropical airmasses with very different tracer values, and with steep gradients in transition zones between. It is clear, especially from the animated version of figure 1 and from many other data studies and from numerical model simulations of the air motion, all of which tend to produce generically similar tracer pictures (e.g. Norton 1994, Lahoz *et al.* 1996), that horizontal fluid motion is causing strong mixing in an extensive midlatitude region between the polar and tropical airmasses.

The early stages of a typical mixing event can be seen on the left. A long tongue of subtropical air (light orange–yellow) is being drawn around the edge of the vortex, crossing the tip of South America, and the animation clearly shows that air is circulating within the part of the midlatitude region overlying the Atlantic and South America. Air is also circulating and mixing over the

† at www.atm.damtp.cam.ac.uk/people/mem/papers/LIM/index.html#crista-movie

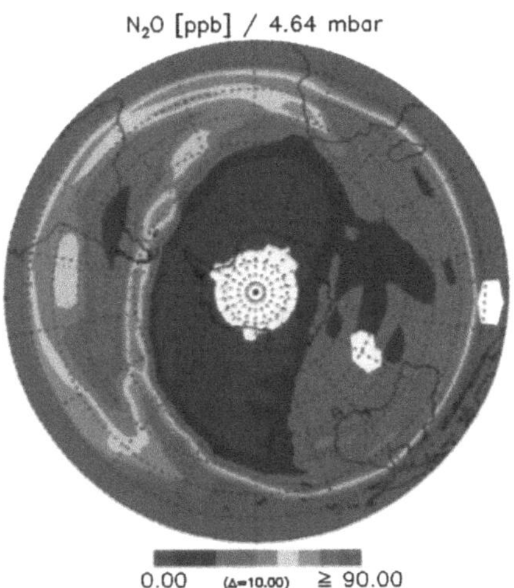

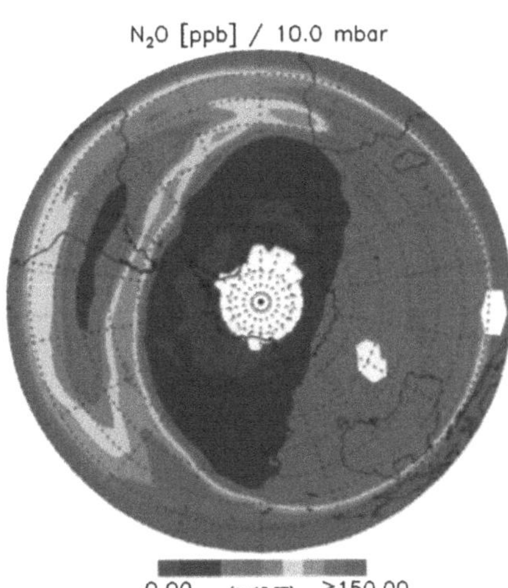

FIGURE 1. Nitrous oxide (N_2O) mixing ratios observed at two stratospheric altitudes on 11 August 1997 by the CRISTA instrument, from Riese *et al.* (2002). White regions are data gaps. On Rossby-wave timescales of days and weeks N_2O is an accurate passive tracer, though destroyed photochemically on Brewer–Dobson timescales of years (see end of §2 below). In the right half of each picture the N_2O mixing ratios increase equatorward nearly monotonically or stepwise monotonically (being nearly constant over the large blue regions on the right). The mixing ratios increase equatorward from polar-vortex values close to zero (purple) to large tropical values imported from the troposphere by the Brewer–Dobson upwelling (red). **At left and right respectively,** pressure-altitudes are 4.64 hPa and 10 hPa, roughly 37 km and 31 km; ranges of mixing ratios in parts per billion by volume are 0–90+ and 0–150+ with contour intervals 10 and 16.67, where '+' signifies that maximum values may slightly overshoot the plotted range. The light band at the subtropical edge of the surf zone (yellow) highlights the ranges 60–70 and 100–116.67 ppbv. CRISTA (Cryogenic Infrared Spectrometers and Telescopes for the Atmosphere) detects a number of chemical species through their infrared spectral signatures and is a large (1350 kg) helium-cooled instrument flown from the Space Shuttle.

right-hand part overlying the Indian and Southern Oceans and Australia but, because that part has already been fairly well mixed, the circulating motion there does not show up clearly in the animation.

We shall see that the midlatitude mixing region, which for reasons to be explained is called the *stratospheric surf zone*, is the stratospheric counterpart of the critical layer in the airflow above water waves, turned on its side. It is only the dimensionless parameters that matter and not, of course, the absolute spatial scale, nor the superficial geometrical differences. But one point to note is that the stratospheric surf zone is anything but narrow. Critical layers have often been thought of as narrow or thin, a mental picture that tended to feature prominently in the early history of thinking about stratospheric fluid dynamics (e.g. Dickinson 1969), just as has tended to be the case in other parts of the research literature on theoretical fluid dynamics.

In the real stratosphere the situation shown in figure 1, with its broad and extensive surf zone, is very typical. The whole picture is not only well simulated by numerical models, but is also well understood dynamically, using ideas like that of 'potential-vorticity inversion' (§4 below; also, e.g., Hoskins *et al.* 1985, Ford *et al.* 2000, 2002). There is an intimate interplay of wavelike and quasi-turbulent dynamics leading, as we shall see, to a persistent momentum transport, outside the scope of classical turbulence theory. It gives rise to a westward force on the stratosphere and an eastward reaction on the denser tropospheric air below. It is this momentum transport that is fundamentally similar to the momentum transport involved in the wind-wave problem.

3. Breaking Rossby waves

The animated version of figure 1 makes the wavelike aspect conspicuous. It shows the long axis of the central purple region rotating clockwise through an angle of about 70° longitude in 5 days, 10–15 August 1997. The purple region is the so-called 'polar vortex', more precisely, the core of the polar vortex, characterized by large negative values of the relevant measure of vorticity — in this case the so-called Rossby–Ertel 'potential vorticity' (e.g. Hoskins *et al.* 1985) — which behaves approximately like an advected tracer and has a distribution like that of the nitrous oxide, up to an additive constant. The rate at which the long axis of the purple region rotates is determined by a competition between the mean winds on the one hand — which blow clockwise, in high latitudes such as 60°S, at speeds of the order of $80\,\mathrm{m\,s^{-1}}$, about nine times faster than 70° longitude in 5 days — and a powerful wave propagation mechanism on the other, which rotates the long axis anticlockwise relative to the mean winds. Today this is usually called the 'Rossby-wave' mechanism, even though (by the usual etiquette honoured more in the breach than in the observance) it 'should', probably, be called the Kelvin–Kirchhoff–Rayleigh–Rossby mechanism.†

The wave mechanism operates, as is well known, whenever the relevant measure of vorticity, Q say, has a cross-stream mean gradient $\bar{Q}_y$, such as the atmosphere's strong pole-to-equator gradient caused by the Earth's rotation. The counterpart in the airflow above water waves is the vertical gradient of ordinary vorticity, associated with the curvature $\bar{u}_{yy}$ of the mean velocity profile. Material displacements across that gradient $\bar{Q}_y$, whatever its cause, give rise to a pattern of fluctuating Q anomalies, Q' say, that alternate in sign downstream. They do so every 90° of longitude in the case of figure 1. Inversion of those Q anomalies to obtain the fluctuating velocity field produces north–south velocities a quarter wavelength out of phase with the displacements, hence phase propagation. This is *one-way* phase propagation, and the signs make it anti-

† Of course we can't call these vorticity waves 'Kelvin waves', because that means something else in today's terminology, namely Coriolis-trapped gravity waves. 'Kirchhoff waves' would suggest a restriction to the wave-2, quasi-elliptical case. Carl-Gustaf Rossby was a great pioneer in atmospheric dynamics, discovering among other things one of its most crucial concepts, that of potential vorticity (Rossby 1936, 1940), which in stratified systems plays a role like that of ordinary vorticity in aerodynamics through the concept of 'potential-vorticity inversion'.

clockwise in figure 1, and upstream in the airflow above water waves. For recent work on the concept of potential-vorticity inversion and its ultimate limitations see Ford *et al.* (2000, 2002).

The resulting wavelike, quasi-elastic resilience of the edge of the vortex core shows up in figure 1 in another way, through the lack of mixing across the edge, as evidenced by the steep tracer gradients at the edge and the very different tracer values inside and outside the vortex, now routinely observed by stratospheric researchers and much studied because of the significance for ozone-layer chemistry. 'Shear sheltering' is also involved (Juckes & McIntyre 1987; Hunt & Durbin 1999). The vortex core is largely isolated chemically from its surroundings, as evidenced by its purple colour in figure 1 (as with the smoke in traditional smoke rings familiar to Kelvin, Kirchhoff and Rayleigh). Yet just next to the edge, in the midlatitude surf-zone region, horizontal mixing is strong, as already noted. It is strong for well-understood reasons associated with flow unsteadiness, hyperbolic points, and so on (e.g. Polvani & Plumb 1992). We may say that the surf zone is chaotically advective. It is often described as two-dimensionally or 'geostrophically' turbulent, more aptly 'layerwise-two-dimensionally turbulent', under the constraint of the strong stable stratification. The stratification is strong in the sense that its timescale, i.e. the timescale of internal gravity waves or buoyancy oscillations, minutes to hours, is far shorter than the days, weeks and years already mentioned.

The evidence for horizontal mixing shows that the Rossby waves must be considered to be *breaking*. The flow unsteadiness, hyperbolic points, etc., hence the turbulence itself, are due to the Rossby-wave motion, just as the turbulence in ocean-beach surf can be said to be due to the water-wave motion. That is the reason for the term 'stratospheric surf zone'. The Rossby-wave breaking is part of what makes the momentum transport persistent and irreversible, in the same sense as the momentum transport generating longshore currents on ocean beaches is rendered persistent and irreversible, by the breaking of incoming water waves. The next section will analyse the Rossby-wave case in more detail, in the simplest possible way, showing the essential role of nonlinearity.

As for the stratosphere's reponse to the momentum transport under discussion, that is a separate question, and irrelevant to the wind-wave problem; but the essence of what happens is noted briefly because it explains the pole-to-equator gradient that makes nitrous oxide such an effective material tracer in figure 1. The response depends on Coriolis forces, which are immensely strong on the timescales of interest; the reader interested in the mathematical details may consult the review by Holton *et al.* (1995). Because stratospheric air is persistently pushed westward, Coriolis forces persistently deflect it poleward, giving rise to a pumping action. This may be called 'gyroscopic pumping', the familiar 'Ekman pumping' being just the special case of it where the persistent force happens to be frictional rather than wave-induced.

The gyroscopic pumping in the stratosphere, due chiefly to momentum transport by breaking Rossby waves like those seen in figure 1, drives a slow but very persistent global-scale circulation, known as the Brewer–Dobson circulation, in which tropospheric air is drawn up into the tropical stratosphere and is then pumped poleward and eventually downward while undergoing photochemical transformation, on timescales of years. That is part of why northern pollution causes a southern ozone hole, and is also the reason for the strong pole-to-equator gradient of nitrous oxide seen in figure 1. Nitrous oxide is produced by biological processes and is well mixed throughout the troposphere. It is drawn up into the tropical stratosphere by the Brewer–Dobson circulation and then destroyed photochemically as it drifts poleward.

4. The dynamics of the wave–turbulence jigsaw

How does the wave–turbulence jigsaw suggested by figure 1 actually work to produce the persistent momentum transport? Part of the answer is that there are highly specific phase relations and phase shifts among the wavelike and quasi-turbulent elements, and that the spatial inhomogeneity conspicuous in figure 1, the juxtaposition of wavelike and turbulent regions, is an essential feature. The spatial inhomogeneity takes the problem well outside the scope of classical turbulence theories, as does the wave propagation itself. A thorough discussion of this last point, from a wider historical and theoretical-physics perspective, can be found in the first of my recent pair of reviews (2000). Here we sketch only what is most basic and essential for the Rossby-wave case and the wind-wave problem. The essentials include the advective nonlinearity.

All the essentials are contained in a simple and elegant model whose workings are fully understood, and which illuminates both the stratospheric problem and the wind-wave problem. This is the Rossby-wave critical layer problem first solved in the complementary papers of Stewartson (1978) and Warn & Warn (1978), hereafter 'SWW problem'. Fundamentally relevant, too, are the relations concerning momentum transport found by G. I. Taylor (1915), displayed in (4.6) below, and extended by KM, who also carefully reviewed and extended the SWW work. Our understanding of the SWW problem was brought to completion in the definitive work of Haynes (1989).

The starting point is the simplest relevant dynamical system, two-dimensional frictionless, incompressible motion. Here Q is the ordinary vorticity, so that inverting Q to obtain the velocity field demands nothing more than solving a Poisson equation of the form $\nabla^2\psi = Q$ under suitable boundary conditions, such as evanescence of $|\nabla\psi|$ at infinity. Coriolis effects are being ignored for the moment. Symbolically,

$$\psi = \nabla^{-2}Q\,, \tag{4.1}$$

where the stream function $\psi(x,y,t)$ is defined such that

$$\mathbf{u} = (u,v)\,, \qquad u = -\,\partial\psi/\partial y\,, \quad v = \partial\psi/\partial x\,, \tag{4.2}$$

(x,y) being Cartesian coordinates and (u,v) the corresponding components of the velocity $\mathbf{u}(x,y,t)$. The two-dimensional Laplacian $\nabla^2 = \partial^2/\partial x^2 + \partial^2/\partial y^2$, and the Green's function to invert it has a logarithmic kernel. In this dynamical system there is just one evolution equation,

$$\mathrm{D}Q/\mathrm{D}t = 0\,, \tag{4.3}$$

where $\mathrm{D}/\mathrm{D}t$ is the two-dimensional material derivative,

$$\mathrm{D}/\mathrm{D}t \;=\; \partial/\partial t + \mathbf{u}\cdot\nabla \;=\; \partial/\partial t + u\partial/\partial x + v\partial/\partial y \,. \qquad (4.4)$$

The single time derivative in (4.3) is the essential reason for the one-way character of Rossby phase propagation. Equation (4.3) with its single time derivative is common to this simple system and its more realistic generalizations, including realistic models of the stratosphere, with suitable redefinitions of Q and the associated inversion operator.

A step toward such generalizations is to introduce the Coriolis parameter $f(y)$. This is the vertical component of the Earth's absolute vorticity picked out by the atmosphere's stable stratification, and is a strong function of latitudinal distance y. Then (4.1) is replaced by

$$\psi \;=\; \nabla^{-2}(Q-f) \,. \qquad (4.5)$$

When f is made to depend linearly on y, with, say, $df/dy = \beta = \text{constant}$, (4.2)–(4.5) becomes the Rossby 'beta plane' or 'flat earth' model. With its constant background vorticity gradient $\bar{Q}_y = \beta$, this well known model provides the simplest textbook examples of Rossby-wave propagation, and in addition the framework for the SWW problem.

If we assume that averaging in the x-direction is well defined, with vanishing mean pressure gradient $\partial\bar{p}/\partial x$, then it is straightforward to derive from (4.2)–(4.5) and the associated x-momentum equation the following relations for the airflow:

$$\frac{\partial\bar{u}}{\partial t} \;=\; -\frac{\partial}{\partial y}\left(\overline{u'v'}\right) \;=\; \overline{v'Q'} \;=\; -\frac{\partial}{\partial t}\left(\tfrac{1}{2}\bar{Q}_y\overline{\eta'^2}\right) , \qquad (4.6)$$

where the overbars denote the (Eulerian) x-average and the primes fluctuations about it, and where $\eta'(x, y, t)$ is the fluctuating material displacement in the y-direction, satisfying $Q' = -\bar{Q}_y\eta'$. The middle and last equalities are Taylor's relations. The middle equality is an exact consequence of (4.2) and ∇^2(4.5) alone, and is often referred to as *Taylor's identity*.† The last equality holds in general for small disturbances only.

Taylor's expression $\frac{1}{2}\bar{Q}_y\overline{\eta'^2}$ has an exact, finite-amplitude counterpart, which KM discovered and exploited in their work on the SWW problem. Not surprisingly, the exact expression, omitted here for brevity, implies that the approximate expression $\frac{1}{2}\bar{Q}_y\overline{\eta'^2}$ remains qualitatively correct when a disturbance grows from small to finite amplitude, as long as the mean vorticity profile $\bar{Q}(y)$ is monotonic, i.e. $\bar{Q}_y(y)$ is one-signed, and is not too much changed from the profile in the initial state. The exact expression has been shown to be related to a Hamiltonian 'Casimir invariant' involved in the stability theorems of V. I. Arnol'd, and is often referred to today as the Rossby-wave 'activity' or minus the 'pseudomomentum' per unit mass (e.g. McIntyre 1981, Shepherd 1990).

† Taylor's identity generalizes to more realistic models of the stratosphere, with $-\partial\left(\overline{u'v'}\right)/\partial y$ replaced by the divergence of the so-called Eliassen–Palm flux (e.g. Andrews *et al.* 1987), which has an extra term representing vertical momentum transport.

Notice, now, the special case of a small disturbance growing exponentially, with all fluctuating quantities proportional to $\exp(\sigma t)$, say, with constant growth rate σ. This is precisely the case analysed in Belcher *et al.* (1999), in which the above equations represent two-dimensional frictionless airflow over two-dimensional frictionless water waves, with y vertical and with $\beta = 0$ and $\bar{Q}_y = -\bar{u}_{yy} > 0$. The last equality in (4.6) holds quantitatively, because of the small-disturbances assumption, and $\partial/\partial t$ is replaced simply by 2σ. We see immediately that if $\bar{Q}_y > 0$ then there is a momentum-transport divergence, $\partial\left(\overline{u'v'}\right)/\partial y > 0$, and, with the assumption of vanishing mean pressure gradient $\partial\bar{p}/\partial x$, an associated mean rate of change of momentum $\partial\bar{u}/\partial t < 0$ in the airflow. This momentum change has to be accompanied by an equal and opposite reaction on, and momentum change in, the water, manifesting itself (in this frictionless model) entirely as an amplification of the water waves driven by a nonvanishing correlation between surface-pressure fluctuations and surface displacement gradients $\partial\eta'/\partial x$.

As Belcher *et al.* point out, the original Miles theory is merely one case of this, in which σ is the growth rate of the Miles instability. The relations (4.6) generalize the result still further: it is now plain that the growth does not need to be exponential either. The result is far more robust than that. Any growth will do. As long as we stay in the realm of small-amplitude disturbances, the expression on the right of (4.6) tells us that the result is not at all dependent on the analytical devices associated with Cauchy's theorem and the calculus of residues, nor on any other analytical device applicable only to disturbances that grow precisely exponentially. The airflow suffers a momentum deficit, and the water waves are amplified — in this frictionless, two-dimensional model — whenever Taylor's wave-activity expression $\frac{1}{2}\bar{Q}_y\overline{\eta'^2}$ increases for any reason at all, such as the arrival of a wave group.

However, as figure 1 reminds us, it is essential to take into account the finite-amplitude effects of the advective nonlinearity if one wants to describe what happens beyond early growth. Then it is not enough to restrict attention to infinitesimal amplitude. This point is well illustrated by the SWW and related solutions, and by numerical simulations of situations like that of figure 1 (e.g. Haynes 1989, Norton 1994), in which the constant-Q contours in the critical-layer or surf-zone region wrap round in complicated ways, grossly violating the condition of small sideways slope necessary for the validity of small-disturbance theory. This again is Rossby-wave breaking. Taylor's wave-activity expression $\frac{1}{2}\bar{Q}_y\overline{\eta'^2}$ applies during the early stages of the process, when the sideways slopes of the constant-Q contours are still small. Then the expression becomes increasingly inaccurate as the slopes steepen and the contours begin to wrap round, leading to the irreversible mixing of the Q distribution. The Miles theory predicts its own breakdown in a fundamentally similar way.

The defining property of wave breaking — if one wants a concept general enough to apply to wave-induced momentum transport for a wide variety of transverse waves, including Rossby waves as well as gravity waves — is the rapid and irreversible deformation of those material contours that would be described by linear wave theory as sloping gently and undulating reversibly, if the linear theory is self-consistently applied. A careful justification of this definition was given in McIntyre and Palmer (1984, 1985). A central consideration is the role of Kelvin's circulation theorem in wave–mean interaction theory, as further discussed in the abovementioned review (2000). Examples of the relevant material contours include the constant-Q contours for Rossby waves, and, for water waves, material contours lying in the free surface.

A side benefit of the definition is that it avoids using "zoological" ideas of wave breaking, based on features of special cases such as whether or not there is air entrainment, whether or not instabilities are involved, and to what extent and in what sense the resulting flow should be considered turbulent.

Solutions to the SWW and related problems provide not only examples of Rossby-wave breaking as a phenomenon but also, in exquisite detail, descriptions of precisely how wave breaking mediates momentum transport. The SWW solution precisely represents, through the technique of matched asymptotic expansions, the interplay between a nonlinear critical layer or surf zone and its wavelike surroundings in a background shear flow $\bar{u}(y)$, including a mathematical description of all the phase shifts involved and their impact on quantities like $\overline{u'v'}$ and $\overline{v'Q'}$ in the first two, exact, equalities in (4.6). Part of the interplay involves the fact that the wavefield, through the PV inversion operator, can remotely sense the nonlinear rearrangement of the Q distribution within the surf zone.

For a first analysis of what happens, it is possible to leave the phase-shift details implicit if one is prepared to assume what is plain from cases like that of figure 1, and from the SWW analysis as well, namely that the result of the Q contours deforming irreversibly in some surf zone $-\frac{1}{2}b < y < \frac{1}{2}b$, say, is an irreversible rearrangement of the Q distribution that weakens the mean gradient $\bar{Q}_y$ in that zone, as suggested in figures 2(a) and 2(b).

Figure 2(a) depicts an idealized strongly-nonlinear scenario in which the surf zone is perfectly mixed; figure 2(b) shows an actual $\bar{Q}(y)$ profile from a complete, dynamically consistent solution due to Haynes (1989 and personal communication), within the SWW matched-asymptotics framework but for a case less simple than the original SWW solution, and in some ways more typical. The detailed flow is more chaotic, because of the onset of secondary instabilities. Yet figure 2(b) is qualitatively similar to figure 2(a), even though the effects of peripheral weak Rossby-wave breaking, in the form of Kelvin 'sheared disturbances', involved in shear sheltering, are also noticeable. To get the corresponding zonally averaged momentum changes, we need the inversion implied by (4.5). In the x-averaged view, this inversion is trivial: one merely has to integrate the change $\delta\bar{Q}$ in $\bar{Q}$ once with respect to y, since $\delta\bar{Q} = -\delta\bar{u}_y$:

$$\delta\bar{u}(y) = \int_y^\infty \delta\bar{Q}(\tilde{y})\,\mathrm{d}\tilde{y}\,, \tag{4.7}$$

which is well defined because $\int_{-\infty}^{\infty}\delta\bar{Q}(y)\mathrm{d}y = 0$.

The result is shown in the right-hand graph, figure 2(c). The momentum change corresponding to the more idealized, left-hand $Q(y)$ profile in figure 2(a) is a simple parabolic shape (not shown) qualitatively similar to the right-hand

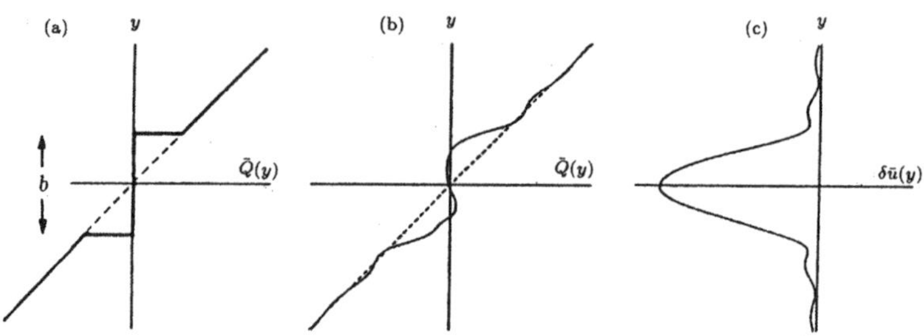

FIGURE 2. The relation between nonlinear Q-rearrangement and momentum transport in the simplest relevant model system, the dynamical system (4.2)–(4.5), when used to describe a breaking Rossby wave. Courtesy P. H. Haynes; for mathematical details see KM and Haynes (1989). Plot (a) shows idealized Q distributions before and after perfect mixing in some y-interval or latitude band $-\frac{1}{2}b < y < \frac{1}{2}b$. Plot ($b$) shows the x-averaged Q distribution, $\bar{Q}(y)$, in an actual model simulation using equations (4.2)–(4.5) within the SWW matched-asymptotics framework. Plot (c) shows the resulting mean momentum deficit, equation (4.7), whose profile would assume the simple parabolic shape (4.8) in the idealized case corresponding to (a).

graph, as is evident from a moment's consideration of (4.7):

$$\delta\bar{u}(y) \;=\; \begin{cases} \frac{1}{2}\bar{Q}_y(y^2 - \frac{1}{4}b^2) & (|y| < \frac{1}{2}b)\,, \\ 0 & (|y| > \frac{1}{2}b)\,\cdot \end{cases} \tag{4.8}$$

In this idealized case the total momentum change is precisely

$$\delta M \propto \int_{-\infty}^{\infty} \delta\bar{u}(y)\,\mathrm{d}y \;=\; -\tfrac{1}{12}\bar{Q}_y b^3\;. \tag{4.9}$$

It is plain from (4.7) that *any* Q rearrangement that creates a surf-zone-like feature with weakened $\bar{Q}_y$ will robustly, and irreversibly, give rise to a net momentum deficit. Integrating by parts, we have $\delta M = \int_{-\infty}^{\infty} y\delta\bar{Q}(y)\mathrm{d}y$; that is, δM is equal to the first moment of $\delta\bar{Q}(y)$. This is negative for any N-shaped $\delta\bar{Q}(y)$ profile representing a weakening of the gradient $\bar{Q}_y$ by mixing. Since a momentum deficit in the air corresponds to amplification of the water waves beneath, we may expect a very robust, ratchet-like, irreversible driving of the water waves whenever spanwise vorticity is mixed vertically in the airflow for whatever reason, perfectly or imperfectly. Such mixing is likely to result from, for instance, the arrival of a water-wave group, or the arrival of a wind gust over pre-existing water waves. It would be interesting to have more observational information about vorticity profiles in real gusts.

Notice, finally, how misleading it would be to suppose that small-amplitude theory holds good, with Taylor's wave-activity expression $\frac{1}{2}\bar{Q}_y\overline{\eta'^2}$ remaining valid on the right of (4.6), throughout the formation of a surf zone in the airflow caused by, for instance, the arrival and departure of a long water-wave group. Integrating (4.6) with respect to t, one would find zero net momentum change in the air and zero net driving of the water waves. The resolution of this paradox is, of course, the fact that the advective nonlinearity, responsible for producing the irreversible rearrangement of the Q distribution in the airflow, vitiates the small-amplitude expression $\frac{1}{2}\bar{Q}_y\overline{\eta'^2}$ as KM showed in detail.

5. Concluding remarks

Of course the real airflow over real water waves is three-dimensional, and three-dimensionally turbulent. KM §7 discusses some of what is involved in adapting the above theory, and argues that a spanwise-averaged two-dimensional representation will still capture the essence of what happens, which is intrinsically robust.

One of the peculiar features of the wind–wave problem is that the quasi-linear theory presently used in wave forecasting models (Janssen 1982, 1991) gives, in a sense, the right answer for the wrong reasons. The quasi-linear theory assumes that real critical layers are infinitesimally thin and behave as in the Miles theory, and that an ensemble of them is distributed over a finite depth because there is a broad spectrum of water-wave phase speeds. The analysis summarized in figure 2, which applies equally well to a narrow, monochromatic spectrum — to say nothing of the stratospheric example with its single, monochromatic Rossby wave having exactly 2 full wavelengths around a latitude circle, and a nearly constant phase speed — underlines the point that real critical layers or surf zones may well be thick, even in a monochromatic case.

The right answer is obtained by the quasi-linear theory because it *diffuses* vorticity in the y-direction through a truly Fickian or Brownian process, in small steps via the ensemble of infinitesimally thin, linearized critical layers. The real process of mixing, in more or less one large step rather than in many small ones, is far more powerful, as Figure 1 reminds us, unless of course the diffusion coefficient D in the diffusion model is 'tuned' to be artificially big.

It is, of course, vorticity that is diffused or mixed, not mean velocity or momentum. An equation like $\bar{u}_t = D(y)\bar{u}_{yy}$ is not a diffusion equation; rather, it is the integral with respect to y of a diffusion equation. The difference is crucial for any case of nonuniform diffusion or mixing, and indeed crucial for the entire scenario summarized by equations (4.7)–(4.9) and figure 2 with its implication that net momentum is exported from the air to the water. Readers may recall the old argument, overlooking the role of wave propagation mechanisms (e.g. Stewart & Thomson 1977), that scenarios like that in figure 2 are 'impossible because they would violate momentum conservation'. For another telling example, see the footnote on page 482 of KM §7.

The wave breaking featuring in the above discussion is that of the Rossby or vorticity waves in the airflow only. If the water waves are also breaking, then flow separation will deepen the layer over which vorticity is mixed in the airflow, further enhancing the irreversible momentum transport and consistent with the discussion of Banner (1990).

As pointed out in my original (1993) discussion, the Rossby-wave-breaking mechanism is only one of *two* mechanisms suggested by the analogy with the global-scale stratosphere. Space precludes more than a brief mention of the second mechanism here. The ratcheting-up of the water waves by irreversible momentum transport can take place not only through local changes in airflow Rossby-wave activity — the increase in Taylor's wave-activity expression $\frac{1}{2}\bar{Q}_y\overline{\eta'^2}$ in the early stages, made irreversible by Rossby-wave breaking — but also by any other process that causes Rossby-wave activity in the airflow to increase irreversibly. Another such process is the formation of Rossby lee waves in the airflow *downstream* of a water-wave group. Then the volume integral of $\frac{1}{2}\bar{Q}_y\overline{\eta'^2}$ increases simply through the rate of increase of the length of the lee-wave train. This may be relevant to explaining the 'energy front' of Chu *et al.* (1992). Further progress, which Stephen Belcher and I hope to pursue, will depend on the analysis of unpublished data for the actual $\bar{u}$ and $\bar{u}_{yy}$ profiles measured in the Chu *et al.* experiments.

ACKNOWLEDGEMENTS: I thank participants in the Newton Institute Programme and in the subsequent IMA Wind-over-Waves conference for many stimulating discussions and for comments on this work, including Mike Banner, Tetsu Hara, Julian Hunt, Ken Melville, and most especially Stephen Belcher in collaboration with whom I am pursuing, in particular, the lee-wave and energy-front problems. Thanks also to Dirk Offermann and Martin Riese of the CRISTA team for allowing me to quote figure 1 in advance of publication, to the UK Natural Environment Research Council for atmospheric modelling support in past years, and to the UK Engineering and Physical Sciences Research Council for support in the form of a Senior Research Fellowship.

REFERENCES

ANDREWS, D. G. *et al.* 1987 *Middle Atmosphere Dynamics.* Academic Press, 489 pp.

BANNER, M. L. 1990 The influence of wave breaking on the surface pressure distribution in wind–wave interactions. *J. Fluid Mech.*, **211**, 463–495.

BELCHER, S. E. & HUNT, J. C. R. 1998 Turbulent flow over hills and waves. *Ann. Rev. Fluid Mech.*, **30**, 507–538.

BELCHER S. E. *et al.* 1999 Turbulent flow over growing waves. In: Wind-over-Wave Couplings: Perspectives and Prospects, ed. S. G. Sajjadi, N. H. Thomas, J. C. R. Hunt; Oxford, Clarendon Press and IMA, 19–29.

DICKINSON, R. E. 1969 Theory of planetary wave zonal flow interaction. *J. Atmos. Sci.*, **26**, 73–81.

CHU, J. S. *et al.* 1992 Measurements of the interaction of wave groups with shorter wind-generated waves. *J. Fluid Mech.*, **245**, 191–210.

FORD, R. *et al.* 2000 Balance and the slow quasimanifold: some explicit results. *J. Atmos. Sci.*, **57**, 1236–1254.

FORD, R. *et al.* 2002 Reply to Comments on "Balance and the Slow Quasimanifold: Some Explicit Results". *J. Atmos. Sci.*, **59**, 2878–2882.

HAYNES, P. H. 1989 The effect of barotropic instability on the nonlinear evolution of a Rossby-wave critical layer. *J. Fluid Mech.*, **207**, 231–266.

HOLTON, J. R., *et al.* 1995 Stratosphere–troposphere exchange. *Revs. Geophys.*, **33**, 403–439.

HOSKINS, B. J. *et al.* 1985 On the use and significance of isentropic potential-vorticity maps. *Q. J. Roy. Meteorol. Soc.*, **111**, 877–946. *Corrigendum and further remarks*, **113**, 402–404.

HUNT, J. C. R. & DURBIN, P. A. 1999 Perturbed vortical layers and shear sheltering. *Fluid Dyn. Res.*, **24**, 375–404.

JANSSEN, P. A. E. M. 1982 Quasilinear approximation for the spectrum of wind-generated water waves. *J. Fluid Mech.*, **117**, 493–506.

JANSSEN, P. A. E. M. 1991 Quasi-linear theory of wind-wave generation applied to wave forecasting. *J. Phys. Oceanog.*, **21**, 1631–1642.

JUCKES, M. N. & MCINTYRE, M. E. 1987 A high resolution, one-layer model of breaking planetary waves in the stratosphere. *Nature*, **328**, 590–596.

KILLWORTH, P. D. & MCINTYRE, M. E. 1985 Do Rossby-wave critical layers absorb, reflect or over-reflect? *J. Fluid Mech.*, **161**, 449–492. [KM]

LAHOZ, W. A. *et al.* 1996 Vortex dynamics and the evolution of water vapour in the stratosphere of the southern hemisphere. *Q. J. Roy. Meteorol. Soc.*, **122**, 423–450.

LIGHTHILL, M. J. 1962 Physical interpretation of the mathematical theory of wave generation by wind. *J. Fluid Mech.*, **14**, 385–398.

MCINTYRE, M. E. 1981 On the "wave momentum" myth. *J. Fluid Mech.*, **106**, 331–347.

MCINTYRE, M. E. 1993 On the role of wave propagation and wave breaking in atmosphere–ocean dynamics (Sectional Lecture). In: *Theoretical and Applied Mechanics 1992*, Proc. XVIII Int. Congr. Theor. Appl. Mech., Haifa, ed. S. R. Bodner, J. Singer, A. Solan, Z. Hashin, 459 pp; Amsterdam, New York, Elsevier, 281–304.

MCINTYRE, M. E. 2000 On global-scale atmospheric circulations. In: *Perspectives in Fluid Dynamics: A Collective Introduction to Current Research*, ed. G. K. Batchelor, H. K. Moffatt, M. G. Worster; Cambridge, University Press, 631 pp., 557–624. Paperback edition with corrections to appear 2003.

MCINTYRE, M. E. 2002 Some fundamental aspects of atmospheric dynamics, with a solar spinoff. In: *Meteorology at the Millennium*, ed. R. P. Pearce; London, Academic Press and Royal Meteorol. Soc., 330 pp., 283–305. Academic Press International Geophysics Series, vol. **83**, series editors Renata Dmowska, J. R. Holton, H. T. Rossby.

MCINTYRE, M. E. & PALMER, T. N. 1984 The "surf zone" in the stratosphere. *J. Atm. Terr. Phys.*, **46**, 825–849.

MCINTYRE, M. E. & PALMER, T. N. 1985 A note on the general concept of wave breaking for Rossby and gravity waves. *Pure Appl. Geophys.*, **123**, 964–975.

NORTON, W. A. 1994 Breaking Rossby waves in a model stratosphere diagnosed by a vortex-following coordinate system and a technique for advecting material contours. *J. Atmos. Sci.*, **51**, 654–673.

POLVANI, L. M. & PLUMB, R. A. 1992 Rossby wave breaking, microbreaking, filamentation and secondary vortex formation: the dynamics of a perturbed vortex. *J. Atmos. Sci.*, **49**, 462–476.

RIESE, M. *et al.* 2002 Stratospheric transport by planetary wave mixing as observed during CRISTA-2. *J. Geophys. Res.*, **107** (D23) special CRISTA issue, paper no. 8179, doi: 10.1029/2001 JD000629.

ROSSBY, C.-G. 1936 Dynamics of steady ocean currents in the light of experimental fluid mechanics. *Mass. Inst. of Technology and Woods Hole Oc. Instn. Papers in Physical Oceanography and Meteorology*, **5(1)**, 1–43. See Eq. (75).

ROSSBY, C.-G. 1940 Planetary flow patterns in the atmosphere. *Q. J. Roy. Meteorol. Soc.*, **66** (Suppl.), 68–97.

SHEPHERD, T. G. 1990 Symmetries, conservation laws, and Hamiltonian structure in geophysical fluid dynamics. *Adv. Geophys.*, **32**, 287–338.

STEWART, R. W. & THOMSON, R. E. 1977 Re-examination of vorticity transfer theory. *Proc. Roy. Soc. Lond.*, **A 354**, 1–8.

STEWARTSON, K. 1978 The evolution of the critical layer of a Rossby wave. *Geophys. Astrophys. Fluid Dyn.*, **9**, 185–200.

TAYLOR, G. I. 1915 Eddy motion in the atmosphere. *Phil. Trans. Soc. Lond.*, **A215**, 1–23.

WARN, T. & WARN, H. 1978 The evolution of a nonlinear critical level. *Studies in Applied Math.*, **59**, 37–71.

On the Relative Importance of Wind Forcing and Nonlinear Interactions in the Downshift of a Gravity Wave Wavenumber Spectrum

J.F. Willemsen

Rosenstiel School of Marine and Atmospheric Science, University of Miami, Miami, Florida, USA.

Abstract

Under wind forcing, the deep water gravity wave amplitude spectrum experiences a shift of the spectral peak and its neighboring wave numbers toward lower wave numbers (downshift). This downshifting as the waves respond to a wind field is widely believed to be caused by the nonlinear interactions that govern the intrinsic wave dynamics. We investigate this process by integrating the Krasitskii equations in the presence of driving and dissipation. These equations make no *a priori* assumptions regarding the wave amplitude probability distribution function. Numerical results indicate that while nonlinear interactions play a role, the dominant cause of the downshift is the functional dependencies on wave number which are at the very least required for such a wind forcing term to have the correct dimensions.

1. Introduction

It has been understood for some time now that the deepwater gravity wave nonlinear interactions (NLI) play an important role in determining the shape of the wave spectrum. In particular Hara & Mei (1991) and Komen *et al.* (1994), e.g., have drawn attention to the role of NLI in shifting the spectral peak to lower frequencies in response to wind driving. However, characterizations of the wind driving source term which go beyond a u_*^2 parametrization (Plant 1982) invariably lead to $(U/c_k - 1)$ structures in the laboratory frame. Examples have been provided by, e.g., Al-Zanaidi & Hui (1984); Donelan & Pierson (1987) and Donelan (1999). Theoretical underpinnings may be found in recent work (e.g., Cohen & Belcher 1999). Any such structure in the growth term implies that different wave number components of the spectrum grow at different rates. Furthermore, even a u_*^2 parameterization requires k-dependent terms in order for the growth term to be dimensionally correct. The quantitative importance of such k-dependencies in the growth term *vis-a-vis* the nonlinear interactions is assessed below, and found to be entirely non-negligible. It is, in fact, the dominant effect causing downshifting.

2. Deterministic modelling of wave dynamics

2.1. *The basic nonlinear model*

The dynamics of gravity waves in deep water is governed by the Krasitskii (1994) equations to an excellent degree of approximation (because wave slopes are limited due to breaking). These are derived from a Hamiltonian and differ from the Zakharov (1966, 1968) equation by a canonical transformation and in the degree of symmetrization. For our purposes, these equations may be abbreviated as

$$\frac{\partial \zeta_k}{\partial t} = |k|\Psi_k + NL_\zeta(k) \tag{2.1}$$

$$\frac{\partial \Psi_k}{\partial t} = -g\zeta_k + NL_\Psi(k) \tag{2.2}$$

in which ζ represents the Fourier amplitude of the wave amplitude; Ψ represents the Fourier amplitude of the velocity potential; g represents gravitational acceleration; and k is a two (horizontal) component wavenumber. The precise forms of the nonlinear kernels denoted NL are readily obtained elsewhere (Krasitskii, 1994) and while of course they are used in the numerical work, their forms are not germane to this discussion. Their role is to redistribute energy across the wave spectrum. In this paper, wave statistics and the so-called "transport" formalism which focuses on the time evolution of the autocorrelation function are not discussed.

2.2. *A Model for the Driving*

The key elements of a plausible model for driving are that: it should be a reasonable representation of results from first principles studies, albeit these are most often done for a single wavenumber in the absence of others; it should explicitly state its assumptions regarding the parameterization of the windspeed; it should parameterize the directional properties of the wind; and it should damp rather than grow waves when $U/c_k < 1$.

Based on these considerations we will use the Donelan-Pierson model,

$$\frac{\partial/\zeta_k}{\partial t} \supset (U\cos\phi/c_k - 1)|U\cos\phi/c_k - 1|\omega_k\zeta_k \tag{2.3}$$

in which U is a characteristic wind speed such as U_{10}, while the angular frequency factor ω renders the expression dimensionally correct. The dimensionless parameter α is adjustable. It will be discussed further below. Quantitatively, the term on the right hand side (RHS) behaves very much like the Cohen-Belcher model, but it is computationally more efficient. Qualitatively, note that it has the desired features listed earlier. Without doing any computations, it is clear that the e-folding growth time is non-negligibly dependent on k. For example in the limit $k \gg 1$, the coefficient of ζ on the RHS scales as $k^{3/2}$ so the time constant decreases as $k^{-3/2}$. Conversely, low wave numbers grow continuously more slowly as a function of k until $U = c_k$.

2.3. *A Model for the Dissipation*

There is no doubt within the modeling community that much remains to be learned regarding modeling of the dissipation term. Here we follow the pragmatic approach espoused in e.g., Komen *et al.* (1994) and Tolman & Chalikov (1996). We construct a nonlinear broadband dissipation term that is tuned to produce a pre-selected asymptotic power law tail. For the numerical work described here, the following functional form has been selected:

$$\frac{\partial \Psi_k}{\partial t} \supset -\beta\omega_k (k^7|\Psi|^2/g)^\chi \tag{2.4}$$

Here β is an adjustable strength parameter, and the exponent $\chi = 2/3$ has been chosen to produce k^{-4} asymptotics in the downwind direction as suggested by Banner (1990) and others. The rationale is basically that one should try a dimensionless quantity raised to a power, as in Phillips (1985), although his specific reasoning does not apply to the Krasitskii equations. Physically, a term such as this one is meant to model energy loss to the wavefield via wave breaking. (Tolman & Chalikov introduce two separate dissipation terms, one in the region of the spectral peak, and another which applies to the asymptotic regime. The rationale for this is mentioned below.)

2.4. *Results from the Complete Model*

The basic equations (2.1) and (2.2) have been supplemented with the driving term (2.3) and the dissipation term (2.4), respectively. Ordinary viscous dissipation has been included as well, although it is a very small correction. The ensuing pair of integro-differential equations has been integrated using Fast Fourier techniques documented in Willemsen (1998, 2001a). Inclusion of driving and dissipation was considered in Willemsen (2001b). In the present work the domain is in 2 horizontal dimensions with a k-space grid consisting of 64×64 equally spaced values. The parameters have been assigned values $\alpha = 0.025$, while $\beta = 7.8 \times 10^{-3}$, for these runs, but the qualitative results are not sensitive to these choices. The same is true for wind speed dependencies: in this work a nominal $U = 10$ m/s has been applied, but qualitative results are entirely insensitive to this choice within reasonable limits.

The initial condition for the work reported here is a cosine function modulated by a Gaussian envelope with different degrees of fall-off in the x and y directions,

$$\cos(\kappa x)\cos(\kappa y)\exp(-\gamma_1 x^2)\exp(-\gamma_2 y^2),$$

with numerical parameter values $\kappa = \pi/20$, $\gamma_1 = 0.0015$, $\gamma_2 = 0.006$. The main motivation for this choice is that initial conditions such as these led Banner and Tian (1998) to observe strong sensitivity to initial slope as a parameter governing likelihood of wave breaking. Other cases have been investigated but the full results will not be discussed here. It has been found that past very early times there is little sensitivity to the precise parameterization of such functions.

Figure I contains the results relevant to this Letter. The upper panel was computed using driving and dissipation but with the nonlinear interactions

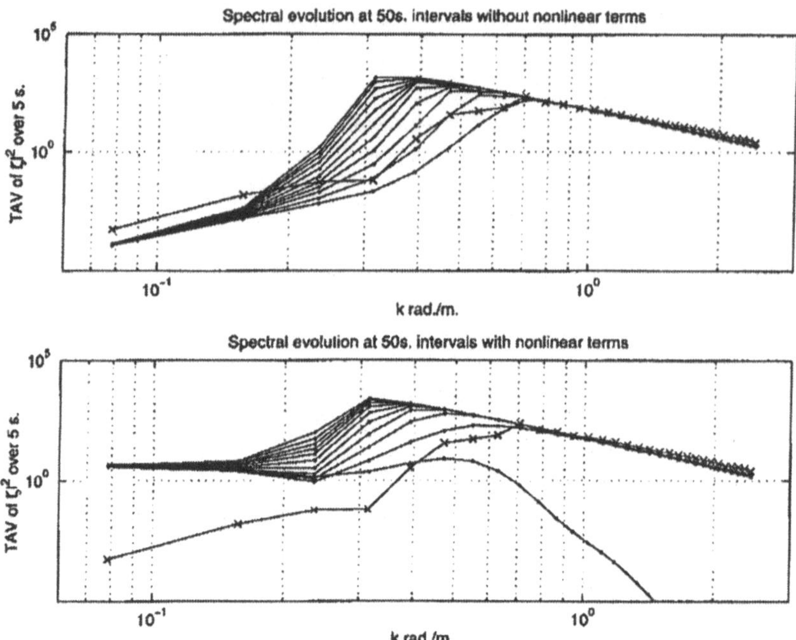

FIGURE 1. Both panels denote the temporal evolution of the time averaged (TAV) wave amplitude spectrum $|\zeta|^2$ 50 s intervals. For example, the lowest curves containing (.) symbols refer to the time average of the spectrum from 0 to 5 s. The next curve is the average of the spectrum from 50 to 55 s, the next 100 to 105 s, etc. The curves denoted by symbols ($\times$) represent the initial condition. The upper panel depicts the evolution when driving and dissipation are present but the nonlinear interactions are "turned off". The lower panel includes the complete model, with driving, dissipation, and nonlinear interactions.

turned off (NL = O in (2.1) and (2.2)). The second panel displays the results of the full model. In both panels the initial condition is displayed as a line with symbols "$\times$".

It is useful to study the panels in 3 distinct k-regimes. First, examine the lowest pair of k-space bins. Notice that in the linear (L) version the initial condition becomes suppressed. This is due to the "damping" nature of the driving term when $U/c_k < 1$. In the NL version, the initial condition becomes enhanced by several orders of magnitude. This is caused by NL cascading into this regime. The cascade is seen to be strong, defeating the aforementioned damping; and it is seen to be fast, operating to enhance the initial condition within the first 5 s of the run.

Passing now to the central regime surrounding the spectral peaks, it is evident that the peak migrates toward lower wave numbers in both the L and NL cases. The migration in the L case can only be due to the driving term. The migration ends in both cases due to a lack of resolution in the present model. Eventually the peak will move even further to the left. It ends sooner at the 4 left-hand

k rad/m	τ–calc	τ–obs L	τ–obs NL
0.16	455.9	455.6	436.5
0.24	86.9	87.2	90.9
0.31	36.5	37.1	39.7
0.39	20.3	20.9	24.5
0.47	13.1	13.8	15.9

TABLE 1. Calculated and observed growth constants in seconds

point for the NL case than the L case because the NL interactions are cascading in this region as well, yet not so much as to further shift the peak.

Finally, the high wavenumber regime is also markedly different in the L and NL cases. While both cases settle into an asymptotic tail, the L case achieves this by gradual dissipation. The NL interactions, on the other hand, initially very quickly and strongly "rob" the high wavenumber end of energy, supplying it to the lower wavenumbers. Yet after only 5 s more, the energy is recovered due to wind driving, which is balanced by dissipation to produce a stable tail. Note that the growth time constant for $k \approx 0.5$ rad/m is only 13 s, and for larger k values it is even smaller.

The table below records the calculated (calc) vs. observed (obs) growth time constants (τ) of $|\zeta(k,t)|^2$ for a range of k-values.

The observed L time constants are in excellent agreement with the calculated values, given uncertainties (due to selecting the precise range to fit) in the fit values. It is interesting that the NL time constants are significantly larger than the L values. Since dissipation is present in both models, the only logical conclusion is that the NL time constants are larger than the L values because nonlinear interactions are cascading energy into yet lower wave numbers. Thus the NL interactions are not so much responsible for causing a downshifting of the spectral peak as they are for *delaying* this process, which is dominated by the growth term.

A word of caution is in order. Parameter α determines the calculated growth rates, and the values cited above are rather fast for real ocean conditions under windspeeds of 10 m/s. However, it is easy to show that the time rate of change of the spectral density scales $S(k) = |\zeta(k)|^2$ with this coefficient in the following manner:

Based on the equation of motion satisfied by ζ it follows immediately that (at each k, and abbreviating the full functional dependence of the driving term)

$$\frac{dS}{dt} = 2\left\{\alpha\Delta(k,U)\omega_k S + \mathrm{Re}\left[\zeta\frac{\delta K}{\delta\psi}\right]\right\} \tag{2.5}$$

This expression reflects the fact that the potential energy is independent of Ψ. (It also reflects the fact that within the undriven linear theory, $\alpha = 0$, $S(k,t)$ oscillates with frequency $2\omega_k$ for a general solution of the form $\zeta = \zeta_+ e^{i\omega t} + \zeta_- e^{-i\omega t}$). The nonlinear interactions are contained in the second term on the RHS.

Let the numerical coefficient α refer to that used in the numerical integration of the equations of motion, which give rise to $\zeta(k,t)$ and $\Psi(k,t)$ dependencies. Introduce a more realistic coefficient which is much smaller than the one used for reasons explained earlier, say $\alpha = \lambda\alpha^0$ with $\lambda \gg 1$. Now, since the kinetic energy functional is homogeneous of degree 2 to all orders of perturbation theory, it follows that if $\Psi = \lambda\Psi^0$, while ζ remains unscaled, $\zeta = \zeta^0$, we obtain

$$\lambda^{-1}\frac{dS^0}{dt} = 2\left\{\alpha^0\Delta(k,U)\omega_k S^0 + \mathrm{Re}\left[\zeta^0\frac{\delta K^0}{\delta\psi^0}\right]\right\} \tag{2.6}$$

Physically, the meaning of the above equation is clear. The realistic growth coefficient leads to temporal evolution of the spectral density at a stretched time $\tau = \lambda t$, provided the Hamiltonian terms in $d\zeta/dt$ are scaled by λ. Thus our choice of values, driven by computational limitations that preclude extremely long runs, does not in any way artificially diminish the importance of the nonlinear interactions. Similar arguments may be invoked for the time evolution of the energy density functional. They similarly require a scaling of the parameter β.

Finally, notice that while the spectra exhibit downshifting, they do not show signs of "overshoot", which is present both in WAM models and in data (see, e.g., Komen *et al.* 1994). It may be necessary to correct for this by introducing a second dissipation term, which acts in the vicinity of the spectral peak, as in Tolman & Chalikov (1996). However, such a modification would have only a mild impact on the present qualitative results.

3. CONCLUSIONS

The above observations clearly demonstrate the important role played by the driving term in causing spectral downshift during wind-driven evolution of the sea surface. The principal cause of downshifting is the driving function. Its effects are influenced by the nonlinear interactions, but these are not the dominant causes.

These model results can be improved by introducing growth terms such as those of Cohen & Belcher (1999) upon properly shifting to the laboratory frame of reference. However, the question of wave damping by the wind in unfavorable conditions ($U\cos\phi/c_k < 1$) probably will require more careful attention at the theoretical level. Finally, it should be noted that if the Donelan-Pierson model may be justified according to the drag law on a solid surface, $\mathbf{F} \propto \mathbf{U}|U|$, then the angular dependence after a shift to the laboratory frame should really be $(U\ \cos\phi/c_k - 1)\sqrt{1 + (U/c_k)^2 - 2U\cos\phi/c_k}$ rather than the Donelan-Pierson form displayed in Eq.(2.3).

ACKNOWLEDGMENTS: I thank Mike Brown for stressing the relevance of these results.

REFERENCES

AL-ZANAIDI, M.A. & HUI, W.H. 1984 Turbulent air flow water waves, *J. Fluid Mech.* **148**, 225.

BANNER, M.L. 1990 Equilibrium spectra of wind wave. *J. Phys. Oceanogr.* **20**, 966.

BANNER, M.L. & TIAN, X. 1998 On the determination of the onset of breaking .for modulating surface gravity water waves, *J. Fluid Mech.* **367**, 107.

COHEN, J.E. & BELCHER, S.E. 1999 Turbulent shear flow over fastt-moving waves, *J. Fluid Mech.* **386**, 345.

DONELAN, M.A. & PIERSON, W.J., JR. 1987 Radar scattering and equilibrium ranges in wind-generated waves with application to scatterometry, *J. Geophys. Res.*, **92**, C5, 4971.

DONELAN, M.A. 1991 Wind-induced growth and attenuation of laboratory waves, in *Wind-Over-Wave Couplings*, ed. Sajjadi, S.G. *et al.* Oxford University Press.

HARA, T. & MEI, C.C. 1991 Frequency downshift in narrowbanded surface-waves under the influence of wind, *J. Fluid Mech.* **230**, 429.

KOMEN, G.J. *et al.* 1994 *Dynamics and Modeling of Ocean Waves.* Cambridge University Press.

PHILLIPS, O.M. 1985 Spectral and statistical properties of the equilibrium range in wind-generated gravity waves, *J. Fluid Mech.* **156**, 505.

PLANT, W.J. 1982 A relation between wind stress and wave slope, *J. Geophy. Re.* C87, 1961.

TOLMAN, H.L & CHALIKOV, D. 1996 Source Terms in a Third-generation wind wave model, *J. Phys. Oc.* **26**, 2497.

WILLEMSEN, J.F. 1998 Enhanced computational methods for nonlinear hamiltonian wave dynamics, *Jour. of Oc. and Atm. Tech.* **15**, 1517.

WILLEMSEN, J.F. 2001a Enhanced computational methods for nonlinear hamiltonian wave dynamics. Part II: New results, *Jour. of Oc. and Atm. Tech.* **18**, 775.

WILLEMSEN, J.F . 2001b Deterministic modeling of driving and dissipation for ocean surface gravity waves, *Jour. of Geo. Res. Oc.* **106**, 27187.

ZAKHAROV, V.E. 1966 On instability of waves in nonlinear dispersive media, *Sov. Phys. JETP.* **51**, 1107.

ZAKHAROV, V.E. 1968 Stability of periodic waves of finite amplitude on the surface of a deep fluid, *J. Appl.Mech. Tech.* **2**, 190.

Search for Characteristics of Deterministic Dynamics in Wind Wave Data

S.I. Badulin[1], V.I. Shrira[2], A.G. Voronovich[3], N.G. Kozhelupova[1] and Y.S. Yurezanskaya[4]

[1] P.P.Shirshov Institute of Oceanology 36 Nakhimovsky pr., Moscow 117851, Russia
[2] Department of Mathematics, Keele University, Keele ST5 5BG, UK,
[3] COETL NOAA, 325 Broadway, Boulder CO, USA
[4] Moscow Institute of Physics and Technology, 141700, Dolgoprudny, Moscow region, Russia

Abstract

The problem of identifying and quantifying deterministic features of random wind wave field in experimental data is addressed. On the basis of weakly non-linear consideration of wave dynamics it is shown that the so called "slave" or "forced", or "bound" harmonics play key role in the higher-order wave statistics. We suggest a new approach to find the higher order correlators of wave field from spatio-temporal wave data. The correlators are used as indicators quantifying the presence of quasi-regular wave patterns. An analysis of laboratory data obtained by G.Caulliez and F.Collard (1999) showed that the patterns manifest themselves in pronounced non-gaussianity strongly localized in the Fourier space.

1. Introduction

The answer to the fundamental question to what degree and in what sense a wind wave field can be viewed as deterministic depends on the criterion dictated by a particular physical problem. The first motivation to address this problem is prompted by the patterns of wind waves resembling cells or horse-shoes often observed in laboratory (Su 1982; Caulliez & Collard 1999; Collard & Caulliez 1999) and in open water (Kinsman 1965). On the other hand, there is a compelling necessity to address this question, since such visual evidence is supported by analysis of electromagnetic (EM) scattering by sea surface. There is a class of situations where significant deviations from predictions based on the Random Phase Approximation (RPA) occur (Chen et al. 1993; Gilman 1997). Within the framework of RPA water wave field is considered as an ensemble of statistically independent Fourier components, that is, dynamical links of the components due to their intrinsic non-linearity are completely ignored.

In the approximation of small water wave slopes both EM scattering and dynamical features of a random wave field can be expressed in a universal form in terms of certain high-order correlators of spatial Fourier amplitudes and, thus, EM scattering can be linked to the well-known concept of resonant interactions of weakly nonlinear water waves (Badulin, Shrira & Voronovich

2002). Recall that the above mentioned high-order correlators are associated with non-gaussianity of the wave field. Relation of higher-order statistics to water wave non-linearity was the central theme of a number of papers (e.g. Guza & Elgar 1985; Leykin 1995; Joelson et al. 2000). However, the use of commonly available temporal data series taken in one or, at best, few points for the analysis of subtle non-linear features of wave field is somewhat questionable because of unavoidable directional averaging. Obviously, spatial or better spatio-temporal data on water waves would be more adequate for the problem under consideration, but the task of collecting, processing and interpretation of such data is incomparably more difficult and the appropriate methods of analysis have not been properly developed. In the present paper we develop a new approach to analysis of wind wave field deterministic features on the basis of spatial and spatio-temporal data. Specifically, we are investigating the presence of dynamically linked components in the unique wind wave spatio-temporal data set obtained in the large IRPHE-Luminy wind wave facility by G.Caulliez and F.Collard (1999) (further CC99) and kindly provided for our study.

Usually the relation between wave resonances and higher-order wave statistics is considered ignoring some important features. In fact the effect of weak non-linearity (here, for simplicity, we discuss weak non-linearity only) is two-fold. First, it slightly affects master harmonics adequately described within the framework of linear approach. Second, and the most important in the present context, is the effect of emergence of new harmonics, the so-called *forced* or *slave* ones which are *strongly linked to the master wave components.* Thus, statistically, the effect of weak non-linearity is indeed weak for weak coupling of the master harmonics but *it is often strong for the slave components!* The latter is the main idea of our approach to the wave data analysis.

We are looking for the presence of dynamically linked components in a subset of the G.Caulliez and F.Collard (1999) wave data set (CC99). The naked eye observations clearly show the presence of regular three-dimensional water wave patterns embedded into a chaotic wave background. These patterns have different characteristic scales and geometry depending on wind wave regime (fetch and wind speed). However, the traditional spectral analysis, unsurprisingly, proves to be a totally inadequate tool to detect the presence or absence of the patterns and to reveal the underlying dynamical processes. Indeed, the spectra merely represent energy scale distribution, which is essentially a *linear* characteristics of the wave field, while to catch manifestations of *nonlinear features* we should resort to higher order statistics. A constructive way of how this can be achieved is presented below.

In § 2 we discuss the theoretical background and use a simple model to elucidate the role of "master" and "slave" harmonics in water wave statistics. Apart from the well known fundamental distinction between the "master" and "slave" modes, the point we would like to emphasize is that interaction between master modes occurs at slow nonlinear scale $O(\text{steepness}^{-2})$ which usually, and our data are not an exception, is smaller or comparable to the "averaging time". Thus, on the time averaging time scale the master modes are practically independent. Due to the weakness of interaction depending of the averaging time we arrive at either the situation where the manifestations of interaction are too "smeared" to be noticeable in case of large time of averaging, or for small times, we pick the manifestations of the interaction randomly, at different phases of the evolution, which does not lead to any coherent picture. The slave modes, on the contrary, can be strongly linked to master modes at much shorter, down to wave period, scale. Some slave modes are "rigidly" linked to the master ones. † This makes these links potentially observable in a random field even at small observation times. The analysis proper of the experimental data on spatio-temporal water wave dynamics is carried out in § 3. Different approaches including conventional statistical methods and heuristic ideas for data processing were aimed at identifying the phase links in water wave field in a relatively wide range of scales. Due to somewhat restrictive quality of even the best experimental data, the choice of data processing method and the corresponding parameters appears to be of principal importance. We analyze results in terms of harmonic resonances of master and slave harmonics of water waves. The found dynamically linked harmonics can be grouped into triplets and quadruplets satisfying three- and four-wave resonant conditions. The indications that these harmonic combinations are related to dynamical features of weakly nonlinear wind wave field are presented. Discussion in §4 summarizes our findings.

2. Phase links in weakly nonlinear wave field

A weakly nonlinear description of water waves begins with the well-known linear harmonic solutions. The higher-order terms of the corresponding asymptotic series are usually found by satisfying some proper orthogonality conditions ensuring unique non-singular solutions for the perturbation problem. The procedure implies that while coupling of the leading order terms is weak, in full agreement with the weak non-linearity hypothesis, the links of the higher-order harmonics, often referred to as "forced","slave" or "bound" harmonics, with the lower-order ones, often, are not weak at all. As we already mentioned, part of the slave harmonics are rigidly linked to the master ones and this fact makes them observable despite their often smaller amplitude. This key point of our study is illustrated by examples of *exact solutions* of weakly nonlinear equations for water waves. We employ the Hamiltonian formulation of water wave dynamics (Zakharov 1968; Krasitskii 1994) and the technique proposed in (Shrira, Badulin & Kharif 1996; Badulin *et al* 1995).

2.1. *Quasi-linear solutions for weakly nonlinear equations*

We employ the fact that water wave equations can be cast in Hamiltonian formulation (Zakharov 1968);

$$i\frac{\partial b(\mathbf{k})}{\partial t} = \frac{\delta H}{\delta b^{*}(\mathbf{k})} \tag{2.1}$$

† The absolute "rigidity" of such links is the mathematical consequence of orthogonally of the higher order approximations to the lower order ones

where the Hamiltonian H is expressed in terms of integral power series in complex amplitudes $b(\mathbf{k})$ and their complex conjugates b^*. We confine our attention to weakly nonlinear waves which justifies truncation of the Hamiltonian. Following (Krasitskii 1994) we retain terms up to $O(b^5)$, i.e. quartet and quintet interactions, which leads to the so called Zakharov five-wave reduced equation

$$\begin{aligned} H = \int \omega_0 b_0 b_0^* \quad &+\frac{1}{2}\int V_{0123} b_0^* b_1^* b_2 b_3 \delta_{0+1-2-3} d\mathbf{k}_{0123} \\ &+\frac{1}{2}\int W_{01234} b_0^* b_1^* b_2 b_3 b_4 \delta_{0+1-2-3-4} d\mathbf{k}_{01234} \end{aligned} \tag{2.2}$$

We use the notation where the arguments of functions are replaced by corresponding indices of arguments. As it was shown (Badulin *et al* 1995; Shrira, Badulin & Kharif 1996) *exact solutions* of the system (2.1,2.2) can be found in the form

$$b(\mathbf{k}, t) = \sum_{i=1}^{N} B_i(t)\delta(\mathbf{k} - \mathbf{k}_i) \exp[\mathrm{i}\Psi_i(t)] \tag{2.3}$$

by proper choice of canonical transformation. These solutions can be referred to as *quasi-linear* (Zakharov 1968).

The key element the paper is based upon is: *normal variables* $b(\mathbf{k})$ are not linear in physical canonical variables which are surface wave elevation η and velocity potential at the surface ψ. They are related to these physical variables through special canonical transformation that annihilates non-resonant terms in the Hamiltonian. In terms of the special normal variables $b(\mathbf{k})$ these solutions look like a linear superposition of harmonics, while all non-linear features are described by the special canonical transformation to physical variables. General form of the transformation is given by an integro-power series

$$\begin{aligned} a_0 \quad = \quad & b_0 + \int A^{(1)}_{012} b_1 b_2 \delta_{0-1-2} d\mathbf{k}_{12} + \int A^{(2)}_{012} b_1^* b_2 \delta_{0+1-2} d\mathbf{k}_{12} \\ + \quad & \int A^{(3)}_{012} b_1^* b_2^* \delta_{0+1+2} d\mathbf{k}_{12} + \int B^{(1)}_{0123} b_1 b_2 b_3 \delta_{0-1-2-3} d\mathbf{k}_{123} \\ + \quad & \int B^{(2)}_{0123} b_1^* b_2 b_3 \delta_{0+1-2-3} d\mathbf{k}_{123} + \int B^{(3)}_{0123} b_1^* b_2^* b_3 \delta_{0+1+2-3} d\mathbf{k}_{123} \\ + \quad & \int B^{(4)}_{0123} b_1^* b_2^* b_3^* \delta_{0+1+2+3} d\mathbf{k}_{123} \\ + \quad & \int C^{(1)}_{01234} b_1 b_2 b_3 b_4 \delta_{0-1-2-3-4} d\mathbf{k}_{1234} \\ + \quad & \int C^{(2)}_{01234} b_1^* b_2 b_3 b_4 \delta_{0+1-2-3-4} d\mathbf{k}_{1234} \\ + \quad & \int C^{(3)}_{01234} b_1^* b_2^* b_3 b_4 \delta_{0+1+2-3-4} d\mathbf{k}_{1234} \\ + \quad & \int C^{(4)}_{01234} b_1^* b_2^* b_3^* b_4 \delta_{0+1+2+3-4} d\mathbf{k}_{1234} \\ + \quad & \int C^{(5)}_{01234} b_1^* b_2^* b_3^* b_4^* \delta_{0+1+2+3+4} d\mathbf{k}_{1234} + \ldots \end{aligned} \tag{2.4}$$

where Fourier components of surface elevation and velocity potential at the surface *are linear* in terms of normal variables $a(\mathbf{k})$ (see Krasitskii 1994 for the coefficients and kernels)

$$\eta(\mathbf{k}) = M(\mathbf{k})[a(\mathbf{k}) + a^*(-\mathbf{k})]; \quad \psi(\mathbf{k}) = -\mathrm{i}N(\mathbf{k})[a(\mathbf{k}) - a^*(-\mathbf{k})] \qquad (2.5)$$

and

$$M(\mathbf{k}) = \sqrt{\frac{|\mathbf{k}|}{2\omega(\mathbf{k})}}; \qquad N(\mathbf{k}) = \sqrt{\frac{\omega(\mathbf{k})}{2|\mathbf{k}|}}; \qquad \omega(\mathbf{k}) = \sqrt{g|\mathbf{k}|} \qquad (2.6)$$

The form of the solution (2.4) illustrates our idea of higher-order harmonics coupling in the most transparent way. Consider solutions (2.3,2.4) for random initial conditions, i.e. assume B_i and Ψ_i to be random in (2.3). The form (2.4) allows us to estimate statistical properties of waves of such an ensemble by direct calculation, if statistical distribution of the initial characteristics is known. While the "master" modes in (2.3) are, at a certain timescale, independent, *certain components of non-linear superposition (2.4) are strongly linked.* From the experimental viewpoint it is of prime importance to discriminate different statistical behaviour which unfolds at different timescales. This point can be demonstrated easily by the following example.

2.2. *Phase coupling of master and slave harmonics for the random Stokes wave*

Consider the simplest quasi-linear solution which is the plane Stokes wave. In terms of normal variables $b(\mathbf{k})$ it corresponds to a single δ−pulse of constant amplitude B_0

$$b_{Stokes} = B_0\delta(\mathbf{k}_0 - \mathbf{k})\exp(\mathrm{i}\Psi)$$

Consider an arbitrary initial phase Ψ_0, which is justified by invariance with respect to initial coordinate. After gathering all combinations of the single "master" harmonic in canonical transformation (2.5) we get

$$\eta(\mathbf{x}) = -B_0\cos(\Psi) + \frac{1}{2}|\mathbf{k}_0|B_0^2\cos(2\Psi) - \frac{3}{8}|\mathbf{k}_0|^2B_0^3\cos(3\Psi) + O(B_0^4) \qquad (2.7)$$

where $\Psi = \mathbf{k}_0\mathbf{x} - \Omega_0 t + \Psi_0$. Stress that the phases of all harmonics are fixed relatively to the phase of the master harmonic. Hence, the phase of the master harmonic can be used as a reference for calculations of relative phases of higher-order slave harmonics. For example, if we consider an ensemble of random Stokes waves and compute the bi-coherence of, say, two times the first ("master") and one second ("slave") harmonics in (2.7), i.e.

$$B = \frac{|\langle\eta_1\eta_1\eta_2^*\rangle|}{\langle|\eta_1||\eta_1||\eta_2^*|\rangle} \qquad (2.8)$$

it proves to be unity. It *does not depend on distribution of initial phase* Ψ_0 (angle brackets in the above formulae mean averaging in Ψ_0) and *does not depend on amplitude of the master component* of the solution (2.7)

$$B_{112} = \frac{\langle\eta_{\mathbf{k}_0}\eta_{\mathbf{k}_0}\eta_{2\mathbf{k}_0}^*\rangle}{\langle|\eta_{\mathbf{k}_0}||\eta_{\mathbf{k}_0}||\eta_{2\mathbf{k}_0}^*|\rangle} \equiv 1$$

Similarly, bi-phase

$$\beta_{112} = \mathrm{Im}(\langle \eta_{\mathbf{k}_0} \eta_{\mathbf{k}_0} \eta^*_{2\mathbf{k}_0} \rangle) \equiv 0 \qquad (2.9)$$

does not depend on Ψ_0. This reflects the fact that the Stokes wave solution preserves permanent form: the phase of the slave harmonic of double frequency is fixed with respect to the master harmonic $\mathbf{k}_0$. A similar conclusion can be made for general solutions (2.3): *there are combinations of master and slave harmonics in the series (2.4) that do not depend on phases of master harmonics in (2.3).* This result is instrumental in our treatment of experimental data. The presence of non-vanishing fast phase independent high order correlators is manifestation of nonlinearity of the field and the presence of particular nonlinear patterns.

The existence of strongly correlated triads evidently contradicts the hypothesis of gaussianity of wind wave field. The hypothesis postulates vanishing of all triple correlators of spatial Fourier harmonics. We see that at least for specific combinations of wave components tri-correlators can be essentially non-zero. However, such a non-gaussianity is strongly localized in wavevector space, which makes it a challenging task to detect it in experimental data. The difficulties are quite obvious. We have to discriminate between "master" and "slave" spatial harmonics which have *a priori* rather different scales and amplitudes. The wide range of scales where we apply our processing methods inevitably leads to low confidence limits of resulting estimates for high order statistics.

3. Search for phase coupling in spatial experimental data

To find the correlated harmonics in wind wave field we used a subset of the experimental data obtained in Luminy wind-wave tank (CC99). The data represent a number of series of snapshots of wave surfaces calculated from wave slopes in two orthogonal directions measured by a specially designed optical system (CC99) for different wind-wave regimes. Five series of 100 snapshots each were analyzed. Dimension of each snapshot was about 80 × 80cm and spatial resolution exceeded 1cm. (For detail see (CC99).)

The photo in fig.1*a* demonstrates the presence of regular wave structures for wind speed 3 m/sec and fetch 9 m/sec. An example of reconstruction of wave profile is presented for wind speed 5 m/sec and fetch 6 metres in fig.1*b*. Both snapshots suggest that a relatively low number of spatial harmonics dominate the wave field. The dominating scales and angles of wave propagation can be identified from these figures "by naked eye". At the same time, the serious difficulties in quantifying these properties in terms of statistical characteristics are also apparent. The dominant wavelength in figs.1*b* is about 8 cm, which corresponds approximately to 12 experimental points. The resolution of slave harmonics which are combinations of dominant wave scales is likely to be inferior to the resolution of the dominant waves. Thus, the data are somewhat restrictive for the problem under consideration. Both the low-frequency noise due to the finite size of the observation domain and the high-frequencies poor spatial resolution can affect our results. To be on the safe side, verification of

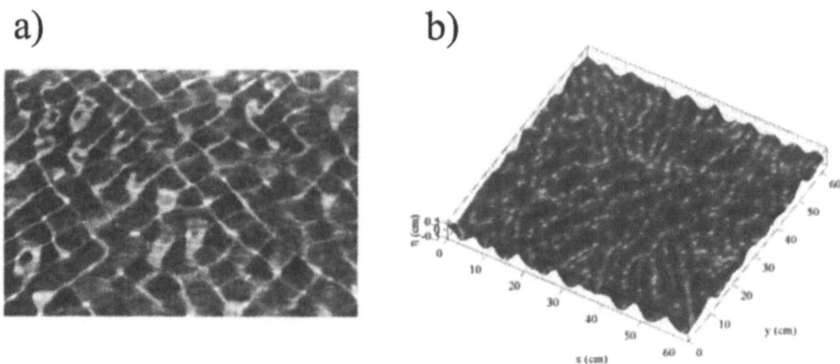

FIGURE 1. *a* — Photo of wave surface at wind speed 3 m/sec and fetch 9 m (see CC99). Rhomboid structures are seen quite well; *b* — An example of reconstruction of wave surface. Wind speed — 5 m/sec, fetch — 6 m

our ideas and results, checks for their consistency are carried out at each step of the data analysis.

3.1. *Spatial spectra and spatial harmonics resonances*

The first step of the analysis, calculations of the energy spectra, is a standard way for quantifying the dominating scales. Such calculations have been performed with great care in (CC99) to relate the features of the observed patterns and prevailing scales and angles of wave propagation.

We analyzed the spatial spectra of available reconstructions of water wave surfaces in order to specify the likely candidates for water wave master and slave harmonics. The compromise is to be reached between required high resolution of spectral components and relatively wide range of scales under consideration. We used both spatial and temporal (in series of snapshots) averaging in order to have representative spectral estimates. The well-known Welsh method (Marple-jr. 1985) for spectral estimates was used for calculation of spatial spectra for each snapshot of the series. The averaging of spectrum over snapshots' series was used for further analysis. Different processing parameters were specified for each series for better resolution of spectral peaks. The calculated spectra are presented in fig.3 in the Descartes and polar coordinates (see for refs. in CC99). As it was noted (CC99) the averaged (in series of snapshots) spatial spectra do not reflect the existence of the observed quasi-regular forms of wave surfaces. In two cases only (stages III and IV in terms of CC99) these oblique components can be identified as isolated spectral peaks in the range of dominating wave scales.

In fact, the classical spectral analysis is doomed to provide very poor information on the patterns. While the main peaks of average spectra can be related to the master modes of sample solutions (2.3), their slave counterparts are very likely to be masked by the wave background.

At the same time, spatial spectra of individual snapshots have a number of

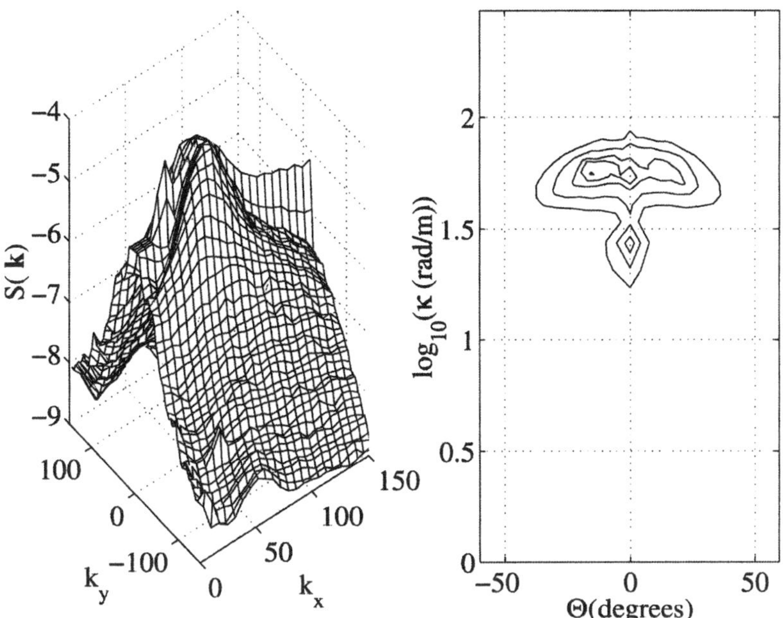

FIGURE 2. Mean spatial spectra of water wave elevation for the series of 100 snapshots (wind speed 5 m/sec, fetch 6 m). Peaks for the along-wind and oblique components can be identified. These peaks are shifted to larger scales as compared to spectral estimates for wave slope data (see fig.2 case 3 in CC99)

peaks that appear repeatedly at specific scales and, thus, are likely to be more relevant to the observed quasi-regular wave patterns. Regardless the rather low confidence limits for the individual spectral estimates we used parameters of these peaks to find possible links between different spatial components of the wave field.

3.2. *Resonance conditions for spatial harmonics*

We consider three- and four-wave resonant conditions for spatial harmonics of the type

$$\mathbf{k}_0 = \mathbf{k}_1 + \mathbf{k}_2 \tag{3.1}$$

$$\mathbf{k}_0 = \mathbf{k}_1 + \mathbf{k}_2 + \mathbf{k}_3; \quad \mathbf{k}_0 + \mathbf{k}_1 = \mathbf{k}_2 + \mathbf{k}_3; \quad \mathbf{k}_0 + \mathbf{k}_1 + \mathbf{k}_2 + \mathbf{k}_3 = 0 \tag{3.2}$$

and verify these conditions for coordinates of the peaks in the individual, i.e. for each snapshot, spectral estimates.

We define spectral peaks as local maxima of individual spectra that appear at the fixed wavevector in more then 50% of snapshot spectra of a series. The

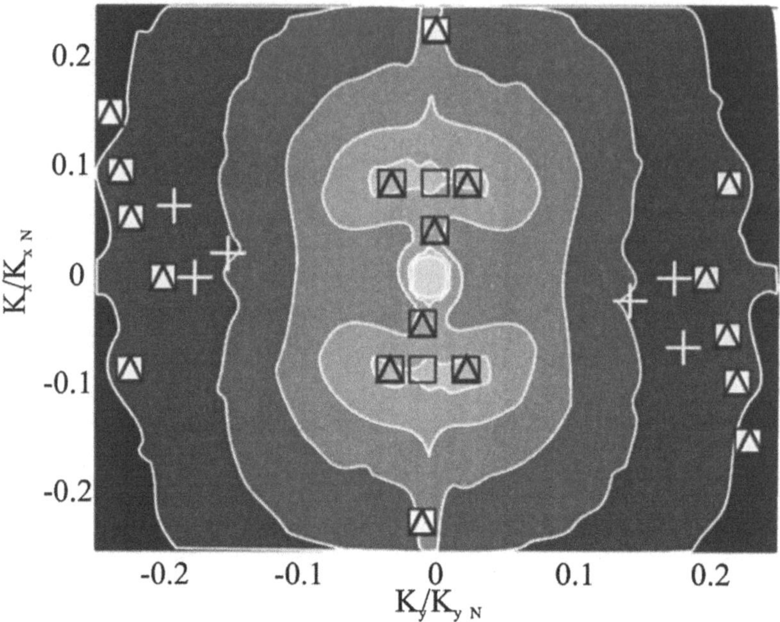

FIGURE 3. Spectral density distribution for wind speed 5m/s, fetch 6m/s. Symbols show local coordinates of instantaneous (based on the individual snapshots) spectral peaks. △ — wavevectors satisfying three-wave resonant conditions (3.1); □ — four-wave conditions (3.2); + — peaks non-attributed to resonances (3.1,3.2). The figure axes are normalized by the spatial Nyquist frequency $2/L$, where L is the snapshot scale)

wavevector of these peaks were considered as resonant when conditions (3.1) or (3.2) were satisfied with a mismatch of one point of Fourier transform †

An example of the distribution of spectral peaks is shown in spectra contours in Fig.3 for the experimental series at wind speed 5m/sec and fetch 6m. Vertical axis corresponds to along-wind direction ($x-$coordinate of wavevector). The relation of the peaks to four-, three- or both resonances (3.1 and 3.2 correspondingly) is shown in different symbols. The following features common for all processed series are worth noting here:

- Almost all of selected spectral peaks are involved into three-wave or four-wave resonances of the type (3.1) and (3.2);
- All resonant combinations contain at least one component of the main spectral peak;
- Resonant quartets always include components of resonant triplets.

† We take 5 points in Fourier domain arranged as a cross, and say that there is a peak for the snapshot if the amplitude in the maximum is two times larger then the mean value defined by averaging over the cross. If such a peak is found in at least 50% of snapshots we claim it to be statistically significant.

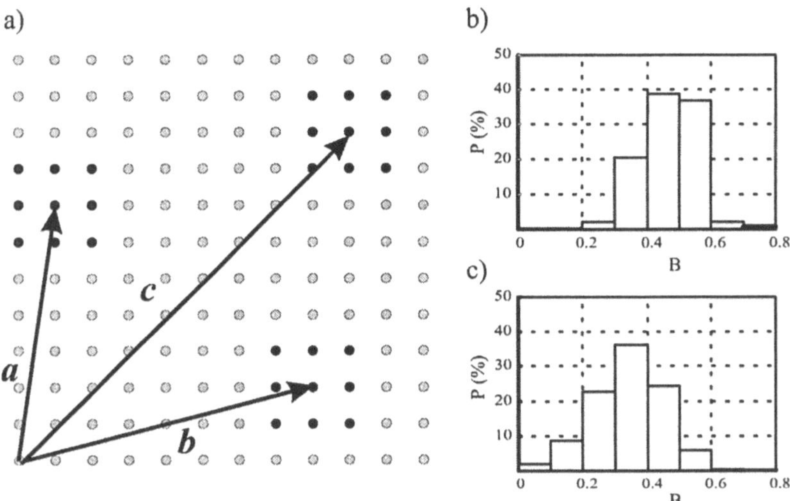

FIGURE 4. Calculations of bi-phase. *a* — Bi-coherence is calculated for combinations of different wavevectors near the found spectral peaks (shown in black); *b* — Probability density of bi-coherence for resonant triads $\mathbf{a}+\mathbf{b}=\mathbf{c}$ ($9^2=81$ combinations) as it is shown in sketch *a*; *c* — Probability density of bi-coherence for near-resonant triads $\mathbf{a}+\mathbf{b}\approx\mathbf{c}$ corresponding to mismatch in one point of the Fourier transform nested in black points of sketch *a)*.

Strictly speaking the relation of the found spectral peaks to dynamical features of the wave field is not proved due to the low statistical confidence and has to be verified in an alternative way. At the same time we can consider the above results as mere hints to possible links of the observed features and our theoretical anzats in the form of quasi-linear solutions (2.3). Moreover, the list of harmonics we selected by the above heuristic method has to be extended. In fact, a majority of the selected three- and four-wave combinations contains high-frequency components. Observability of these harmonics is enhanced by the low level spectral background in contrast to the harmonics of moderate scales which are contaminated by the high level background in the vicinity of main spectral peaks. Thus, in our further search of strongly correlated wave combinations we verify both the components of the found triads and quartets, as well as other triads containing some of the triads (quartets) components. For example, all cases of trivial resonances of the type $2\mathbf{a}=\mathbf{b}$, $3\mathbf{a}=\mathbf{b}$ *etc.* have been carefully examined. As a result a number of perspective candidates for master and slave components of the wave field was essentially extended. We stress that dynamically important modes do not necessarily manifest themselves as peaks.

3.3. *Phase coupling of resonant harmonics. Localized non-gaussianity*

The next step in our analysis is an attempt to answer the question: *Are the components of the found resonant combinations (3.1,3.2) dynamically linked?* The formal answer to this question could be formulated in terms of triple-

and quadruple-correlators corresponding to resonant conditions (3.1,3.2). High magnitudes of bi-coherence B (2.8) (or tri-coherence for four-wave resonances) would mean strong linkage of evolution of the involved resonant components. The relative phases of these components or bi-phase β (2.9) depend weakly on time (or, in terms of the available data, on the snapshot number). These relative phases (bi- and tri-phases) can be in principle easily calculated for all possible wavevector combinations, but their high dimension (four scalar arguments for bi-phase and six — for tri-phase) is an essential impediment for such an analysis.

An idea how to facilitate this analysis is demonstrated by fig. 4*a*. Bi-phases are calculated for a number of near-resonant combinations of wavevectors nested near the resonant ones. Bi-coherence B was calculated for all possible combinations of wavevectors $\mathbf{a}, \mathbf{b}, \mathbf{c}$ in squares 3×3 that satisfy resonant condition (3.1) approximately. Fig. 4*c* shows distribution of B for the resonant condition detuning by 1 point of Fourier transform. Fig. 4*b* is for the exact resonant condition (3.1) (closed triad of wavevectors $\mathbf{a}, \mathbf{b}, \mathbf{c}$). Two important conclusions can be acquired from these histograms. First, bi-coherence can reach rather high values lying far beyond the statistically predicted error limits. Secondly, resonant triads (fig. 4*b*) are definitely correlated, the small values of bi-coherence (up to the value 0.25) have zero probability in terms of the presented histogram. In fact, we have also demonstrated that closed combinations of wavevectors (exact resonance in terms of eqs. 3.1,3.2) are better correlated than the detuned ones.

Our next step is a logical continuation of the previous attempts to specify strongly correlated wave combinations. We show that resonant combinations of peak components selected in the previous section are better correlated than arbitrary triads (quartets) of wave components. Let us fix wavevectors $\mathbf{a}$, $\mathbf{b}$ of the spectral peaks found above and the corresponding resonant component $\mathbf{c}$. Calculate bi-coherence B as a function of detuning wavevector $\mathbf{d}$ for close wavevector combinations $\mathbf{a}$, $(\mathbf{b} + \mathbf{d})$ and $(\mathbf{c} + \mathbf{d})$. As fig.5 shows, the maximum of B is localized at $\mathbf{d} = 0$ that is, exactly for the wave components specified in §3.2 by means of spectral analysiss. Note, that the domain of high values of B is strongly localized in wavevector space. Thus, the *non-gaussianity* of water wave field we detected in our study is *strongly localized* in wavevector space.

3.4. *Dynamical links between wave components*

The final step of our analysis is aimed at demonstrating the underlying dynamical cause of the found strong correlations of specific wave components, that is, we have to show that the established features of correlations of wave amplitudes and phases are consistent with the concept of weakly nonlinear dynamics.

Consider a triad of wave harmonics with wavevectors $\mathbf{a}, \mathbf{b}, \mathbf{c}$ for a particular snapshot. The product of complex amplitudes A_i, B_i, C_i of these harmonics which contributes to bi-coherence B is characterized by instantaneous phase angle ϕ_i. Statistical independence of A_i, B_i, C_i implies, evidently, their independence on the relative phase ϕ_i and homogeneous distribution of ϕ_i itself. Fig. 5*b* shows quite clearly the existence of preferences in values of the relative phase angle ϕ in terms of its probability density (upper) and similar distribution of instantaneous amplitudes of wave components as functions of ϕ (middle

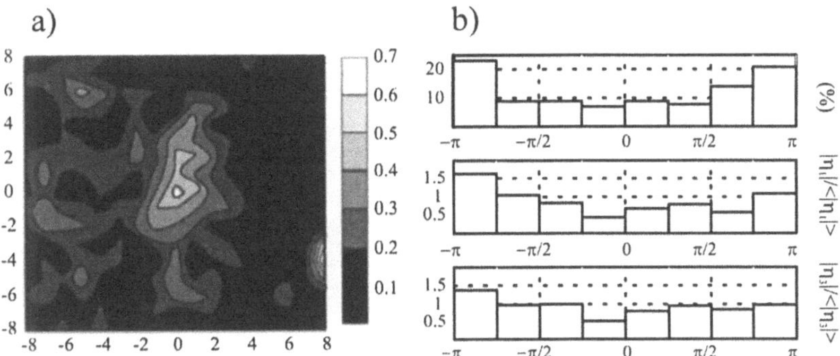

FIGURE 5. Wind speed 5 m/sec, fetch 6 m. *a)* Bi-coherence as a function of detuning wavevector $\mathbf{d}$ (shown in points of Fourier transform) for a particular exact three-wave resonance $\mathbf{a} + (\mathbf{b} + \mathbf{d}) = \mathbf{c} + \mathbf{d}$, $\mathbf{a} = \mathbf{b} = (-8, 4)$, $\mathbf{c} = (-16, 8)$. Maximum of B is reached for $\mathbf{d} = (0, 0)$; *b)* Partial phases and amplitudes contributing to the calculated bi-coherence and bi-phase for the resonant triad $\mathbf{a} = \mathbf{b} = (-8, 4)$, $\mathbf{c} = (-16, 8)$: *Upper histogram* — probability of phase for triad $\phi_{123} = \arg(\eta_1\eta_2\eta_3^*)$; *middle* — normalized amplitude of wave component η_1 as a function of ϕ_{123}; *bottom* — normalized amplitude of wave component η_3 as a function of ϕ_{123}

and bottom). Note, that the figure corresponds to the simplest case of nonlinear wave dynamics, that of appearance of double slave harmonics ($\mathbf{c} = 2\mathbf{a}$) similar to the Stokes solution (2.7). In full accordance with the form of this solution we have correct values of the most probable phase ϕ_{123}. Quite similarly, the amplitudes of master (η_1 — middle) and slave double (η_3 — bottom) harmonics evolve coherently. Small deviation of the slave component relatively to the master one is consistent with the Stokes solution as well (the slave harmonic is quadratic in the master harmonic amplitude).

4. DISCUSSION

The present study offers a complementary view of experimental data and findings by G.Caulliez & F.Collard (1999). Attempts to quantify visible quasi-regular patterns in water wave field in terms of traditional spectral analysis are doomed, since this "regularity" implies an essential role of intrinsic nonlinear water wave dynamics and, thus, strong statistical dependence of certain water wave components. In fact, the fundamental question we are trying to address, at least partially, is: "To what extent the hypothesis of Fourier components statistical independence is adequate to water wave description?" Our brief review of the theoretical background in §2 provided helpful clues for any attempt to extract dynamical features from experimental data. We have not presented a systematic analysis of all the available experimental data and results of their processing. The main aim of the paper was rather formulation and demonstration of possible ways to address it. The main result is a successful detection of strongly correlated water wave components by direct calculation of triple - and quadruple-correlators of wind wave Fourier harmonics. Thus, at least locally in Fourier space, the wind dynamics is essentially non-gaussian, although the overall field dynamics might be quasi-gaussian.

The main difficulty in the chosen approach is in the high dimension of the parameter space we have to analyze. There are two ways to overcome it: first, by brute computer force, which is certainly doable now; second, the one we adopted, is to use heuristic methods to simplify the task by localizing *a priori* the parameter sub-spaces of interest. Note, that the detection of strong correlations we demonstrated cannot be considered as statistically confident. At the same time, being consistent with other results, first of all, with the predictions of non-linear wave theory, this result can be considered as a trustworthy one. The suggested approach being applied to more extensive data sets with better spatial resolution will certainly produce quantitatively reliable results. The results are expected not only to clarify the fundamental questions regarding the non-gaussianity of the wind wave field but to have tangible implications for electro-magnetic scattering by sea surface. Thus, now one of the main challenges is in obtaining new experimental data on wind wave dynamics which would satisfy the necessary criteria and allow us to carry out the suggested analysis with the desired statistical confidence.

ACKNOWLEDGEMENT: The work was supported by INTAS grants 97-575, 2001-234 and Russian Foundation for Basic Research N01-05-64603, 01-05-64464.

REFERENCES

BADULIN, S.I., V.I. SHRIRA & A.G. VORONOVICH 2002 EM scattering from sea surface in the presence of wind wave patterns. Part I. Scattering by coherent wave patterns. *Internat.Journ.Remote Sensing* (submitted).

CAULLIEZ, G. & COLLARD, F. 1999 Three-dimensional evolution of wind waves from gravity capillary to short gravity range. *European Journal of Mechanics B/Fluids*, **18**, 389–402.

Collard F. & Caulliez G. 1999 Oscillating crescent-shaped water wave patterns, *Physics of fluids* **11**, 3195–3197.

BADULIN, S.I., SHRIRA, V.I., KHARIF, C. & IOUALALEN, M 1995 On two approaches to the problem of instability of short-crested water waves *J. Fluid Mech.* **303**, 297–325.

CHEN K.S., FUNG A.K., FAOUZI A., 1993 An empirical bispectrum model for sea surface scattering, *IEEE Transactions on geoscience and remote sensing*, **31**, N4.

GILMAN M.A. 1997 Bispectrum and analysis of the statistics of electromagnetic waves backscattered by sea surface, *Journ. of comp. and syst.sc.intern.* **36**, 972–980.

ELGAR S. & R.T.GUZA 1985 Observations of bispectra of shoaling surface gravity waves *J. Fluid Mech.* **161**, 425–448.

JOELSON M., TH.DUDOK DE WIT, PH.DUSSOLLIEZ & A.RAMAMONJIARISOA 2000 Searching for chaotic deterministic features in laboratory water surface waves *Nonlinear Processes in geophysics* **7**, 37–48.

KRASITSKII V.P. 1994 On reduced Hamiltonian equations in the nonlinear theory of water surface waves *J. Fluid Mech.* **272**, 1–20.

KINSMAN B. 1965 Wind Waves; their Generation and propagation on the Ocean Surface Englewood Cliffs, N.J.: Prentice-Hall Inc.

LEYKIN I.A. 1995 Asymmetry of wind waves studied in a laboratory tank *Nonlinear processes in geophysics* 280–289.

SHRIRA, V. I., BADULIN, S.I. & KHARIF C. 1996 A model of water wave 'horse-shoe' patterns, *J. Fluid Mech.* **318**, 375–404.

SU M.-Y., BERGIN M., MARLER P. & MYRICK R. 1982 Experiments on non-linear instabilities and evolution of steep gravity-wave trains *J. Fluid Mech.* **124**, 45–72

MARPLE JR., S.L. 1985, Digital spectral analysis with applications. Prentice-Hall,Inc., Englewood Cliffs,New Jersey

ZAKHAROV, V. E. 1968 Stability of periodic waves of finite amplitude on the surface of a deep fluid *J. Appl. Mech. Tech. Phys. (U.S.S.R.)* **9**, 86–94.

Vorticity Dynamics in the Water Below Steep and Breaking Surface Waves

J.C.R. Hunt[1,2], I. Eames[3] and S.E. Belcher[4]

[1]Departments of Space and Climate Physics and Geological Sciences, University College London, Gower Street, London WC1E 6BT, UK.
[2]J.M. Burgers Centre, Delft University of Technology, Delft, Netherlands.
[3]Department of Mechanical Engineering, University College London, Torrington Place, London WC1E 7JE, UK.
[4]Department of Meteorology, University of Reading, Earley Gate, Reading RG6 6BB, UK.

Abstract

This paper provides a new analysis of some aspects of vortical dynamics below surface water waves which are driven either by distant forcing or locally by a sheared turbulent wind. In the first situation, vorticity produced by the shear free condition applied at the surface, remains confined to surface layers as a result of vorticity cancellation in straining regions, even for finite amplitude water waves with significant movement of fluid to and from the surface. This justifies the irrotational flow assumption for finite amplitude water waves.

When the waves are generated locally by a wind shear, the mean and fluctuating vorticity generated in the water by the air flow over flat and undulating surfaces is analysed to provide a possible explanation of anomolous turbulent dissipation profiles below waves. We then suggest a new theoretical model for the mean water, air and two-phase velocity fields in quasi-steady breaking waves, building on the work of Banner & Melville (1976) and Longuet-Higgins & Turner (1974).

Interactions between finite amplitude three-dimensional waves and mean and fluctuating vortical fields in the water are analysed to differentiate between the mechanisms leading to Langmuir circulation, turbulence amplification by waves and downwards eddy bursting below waves observed by Komori *et al.* (1993). A previously overlooked mechanism for generation of vorticity in the water flow is identified, that corresponds to the 'fast' generation of vorticity by the drift (in contrast to the well-known slow generation by the Stokes drift). This mechanism may explain the observations of rapidly developing streamwise vortices by Veron & Melville (2001).

1. Introduction

The geophysical and practical reasons for studying ocean waves and the difficult fluid mechanical problems that still need to be resolved were reviewed by Melville (1996). When waves on a free surface are generated non-locally by forcing, or locally by wind stresses, it is generally assumed that the wave induced velocity field in the liquid is to first approximation irrotational. This is the basis of the powerful computational 'numerical wave tanks' now used for predicting waves in engineering and environmental flows (*e.g.* Grilli, Guyenne & Dias 2001). However, as is well-known, this assumption is not strictly valid everywhere in the flow because vorticity is generated at a free surface (Phillips

1977) and because the wind stresses produce a vorticity profile which in the mean is a maximum at the free surface and also fluctuates considerably as the eddies interact with the wave induced velocity field (Belcher *et al.* 1992) and the free surface.

We first consider, in §2, the vorticity dynamics of small amplitude and then steep non-breaking waves, without any wind stresses. This requires considering unsteady vortical flows near stagnation points, where vorticity generated at the surface can penetrate into the interior of the flow. Applying an analysis of vorticity cancellation in straining flows (in §2.2), recently developed for calculating wakes of obstacles in complex flows (Hunt & Eames 2002), shows quantitatively that vorticity generated near the surface does not penetrate far into the interior of the flow.

However, if the waves break, the surface topology, kinematics and vortical dynamics change completely, leading to vorticity generated by the surface layers being transported into the interior (Melville 1996). When a turbulent wind blows over a water surface, it drives a turbulent shear flow in the water with mean velocity $U_W(z)$. Following numerical simulations of turbulent air flow over a water surface by Lombardi, de Angelis & Bannerjee (1996) and more recent studies of turbulent boundary layers by Hunt & Morrison (2000), it has become clearer that the mechanisms which determine the turbulent structure in the surface layers of the water are similar to those in a turbulent boundary layer over a rigid surface, even though the surface velocity fluctuations are non-zero. These results lead to an estimate (in §3.1) of the vorticity in this 'wind-drift' layer $\mathrm{d}U_W/\mathrm{d}z$ in relation to the straining motion induced in the water by wave motion. The non-linear model of Harris, Belcher & Street (1996) shows that most of the flow below the waves is irrotational as assumed in most models for wind generation of waves (*e.g.* Makin & Kudryavstev 1999). However close to the water surface, in the 'inner layer', there is a significant change in the mean turbulent dissipation $\langle\epsilon(z)\rangle$ profile, which can be estimated theoretically and compared to experiments reviewed by Melville (1996). Hitherto these anomalies in $\langle\epsilon(z)\rangle$ have been attributed to breaking waves and Langmuir circulations.

The analyses of the turbulent air and water flow over and below non-breaking waves needs to be extended to calculate the flows in breaking waves where the main new element is the air-liquid two-phase fluid or foam layer with density ρ_{AW} that is created between the gas and liquid flows; a concept first discussed in detail by Longuet-Higgins & Turner (1974). The experiments and concepts reviewed by Banner & Peregrine (1993) and Melville (1996) show how the separated air flow, mediated by the two-phase flow layer, is related to the vorticity and turbulence in the water. Their work indicated the need to develop an approximate theory and order of magnitude model. This is what we attempt in §3.2 including a re-examination of the entrainment and detrainment gravity current model of Longuet-Higgins & Turner (1974) for the 'foam' layer.

These studies of the vorticity and energy dissipation rate in the water flow are a central element in modelling the dynamics of wind driven waves and thence the mean momentum transport from the air flow to the water. However, surface wave processes are also critical for mass and heat transport between the atmosphere and ocean. Most work has focussed on the layers close to the surface where waves generate intense small scale turbulence and gas bubbles are the main agents for scalar transfer (*e.g.* Melville 1996, Thorpe 1984). However, Komori, Nagaosa & Murkami (1993) showed experimentally that waves produce downward bursting eddy motion which greatly amplifies the transport of matter and heat away from the surface layers down into the water column. The theoretical study of these strong dynamical events (which can be seen in laboratory wind-wave channels) requires analysing how the mean and fluctuating motions under the waves interact with the mean and fluctating vortical motions of the wind driven turbulence flow. We review, in §4, the time averaged interactions over many wave periods that leads to Langmuir cell generation and turbulence amplification as analysed, respectively by Craik & Leibovich (1976) and Teixeira & Belcher (2002). Here a new order of magnitude analysis is developed for the interaction over short and long periods between the mean wind-drift vorticity and the non-linear straining by three-dimensional waves which generates a similar flow to that of Hawthorne & Martin (1955) for a shear flow–bluff body interaction. This requires analysing the mean Lagrangian displacement of fluid particles in waves using the recent results of Eames & McIntyre (1999) because this determines the vorticity from the distortion of fluid elements through Cauchy's formula (Batchelor 1967, p.276). Over long periods the displacements are equivalent to the Stokes drift.

2. Vorticity and free surface boundary layers

2.1. *Linear waves*

When the wave amplitude is much smaller than the wavelength $\lambda = 2\pi/k$ (where k is the wave number) *i.e.* $ak \ll 1$, the waves are said to be linear. It is important to briefly review the generation of the boundary layer flow beneath a linear wave because it is the key mechanism of vorticity generation in non-breaking waves.

The flow beneath a free surface may be decomposed as

$$\mathbf{u} = \nabla\phi + \mathbf{u}_r + \nabla\phi_i, \tag{2.1}$$

where $\nabla\phi$ is the leading order irrotational flow, $\mathbf{u}_r$ is the rotational component to the flow and $\nabla\phi_i$ is the image irrotational flow required to satisfy the kinematic boundary conditions on the free surface. The tangential shear stress exerted at the free surface is zero. Physically, a zero tangential shear stress at a surface translates to marked dye lines, which are initially perpendicular to the free surface, remaining perpendicular. An irrotational flow tends to rotate the dye lines at a deformed surface, and vorticity must be present to counteract this rotation. The strength of the vorticity $\omega(= \hat{\mathbf{y}} \cdot \nabla \times \mathbf{u}_r)$ at the surface is related to κ, the local curvature of the free surface, and q the local flow speed, through $\omega = 2\kappa q$. The sign of the vorticity at the free surface is related to the local curvature of the free-surface; surfaces with positive/negative curvature generate positive/negative vorticity (see figure 1(a)).

For a linear water wave, the vorticity at the free surface is

$$\omega = -\frac{\partial^2 \psi}{\partial x^2}. \tag{2.2}$$

Vorticity generated on the surface diffuses into the bulk of the fluid by the action of viscosity. The transport of vorticity ω is described by the advection-diffusion equation

$$\frac{\partial \omega}{\partial t} + \mathbf{u} \cdot \nabla \omega = \nu \left(\frac{\partial^2 \omega}{\partial x^2} + \frac{\partial^2 \omega}{\partial z^2} \right). \tag{2.3}$$

To leading order, vorticity is transported by the irrotational flow ($\nabla \phi$) so that (2.3) reduces, after applying the Boussinesq transformation, to

$$\frac{1}{J}\frac{\partial \omega}{\partial t} + \frac{\partial \omega}{\partial \phi} = \nu \left(\frac{\partial^2 \omega}{\partial \phi^2} + \frac{\partial^2 \omega}{\partial \psi^2} \right), \tag{2.4}$$

where $J = \partial(\phi, \psi)/\partial(x, z)$ is the Jacobian. The local balance reduces in the steady frame moving with the wave, to a balance between advection by tangential flow and diffusion perpendicular to wave surface,

$$\frac{\partial \omega}{\partial \phi} \approx \nu \frac{\partial^2 \omega}{\partial \psi^2} \tag{2.5}$$

Vorticity transport may be thus viewed as one-dimensional diffusion with ϕ/c^2 being interpreted as time. In the most general case, the stream function is described by

$$\psi(x, z) = \int_0^\infty a_n e^{ik_n x} e^{k_n z} \mathrm{d}k_n. \tag{2.6}$$

The vorticity at the free surface (from (2.2)) is $\omega = \int_0^\infty a_n k_n^2 e^{ik_n x} \mathrm{d}k_n$, so that fluid parcels at the free surface see on average a mean value of vorticity

$$\overline{\omega} = \frac{1}{\lambda} \int_{-\lambda/2}^{\lambda/2} \omega \mathrm{d}x = -\frac{1}{\lambda} \int_{-\lambda/2}^{\lambda/2} \frac{\partial^2 \psi}{\partial x^2} \mathrm{d}x = \int_0^\infty a_n k_n \sin(k_n \lambda/2) \mathrm{d}k_n. \tag{2.7}$$

For the case of a symmetric linear wave (symmetric about $x = 0$ and $z = 0$), *e.g.* $\psi = a \sin kx$, the average vorticity at the surface is zero. Thus vorticity cancellation generates a localised flow which does not grow with time,

$$\omega(x, z) = -2ak\sigma e^{z/l_v} \cos(kx - z/l_v), \tag{2.8}$$

where $l_v = (c/2\nu)^{-\frac{1}{2}}$ is the boundary layer thickness which tends to zero as the diffusivity $\lambda c/\nu \to \infty$. But, asymmetric waves (even in the linear case) generate a growing boundary layer beneath a wave because the mean vorticity at the surface is non-zero. Longuet-Higgins (1992) showed that the mean surface vorticity of capillary waves (which are non-symmetric about the average water level) is non-zero, and that far beneath the wave surface, the vorticity tends to twice the mean surface value. Sajjadi (2002) has elaborated more recently on this process.

2.2. *Non-linear non-breaking waves*

As the wave steepness increases, stagnation points appear on the surface of the wave in the unsteady frame moving with the wave (see figure 1(b)). From §2.1, the local value of vorticity is related to the curvature of the free surface, and that for non-linear (non-breaking) waves, the sign of the vorticity is predominately positive near the crest. The stagnation point on the wave surface occurs where the surface curvature changes sign, leading to opposite-signed vorticity being advected into the bulk of the fluid. The lengthscale of the irrotational straining flow at the stagnation region is $O(\lambda)$ and is much larger than the boundary layer thickness l_v. It is relevant to examine the transport of this vorticity advected into the bulk of the fluid, because as demonstrated by Hunt & Eames (2002), opposite-signed vorticity is rapidly cancelled out when strained. Here we elaborate on this important mechanism illustrating this effect with new calculations of the effect of a steady irrotational planar straining flow on boundary layer flows. The correction due to unsteadiness is also considered.

First consider the steady problem where the straining rate, $\alpha \sim c/\lambda$, is constant. A model of the flow near a stagnation point is described by the velocity potential ϕ' and streamfunction ψ',

$$\phi' \approx \frac{1}{2}\alpha(z'^2 - x'^2), \qquad \psi' \approx \alpha x' z', \tag{2.9}$$

where z' and x' are the local coordinates which are respectively perpendicular and parallel to the free surface. Since the boundary layer thickness l_v is small compared to the straining lengthscale, the transport of the vorticity (of strength $\pm\omega_W$) from the boundary layer into the interior is dominated by advection by the flow (2.9) and cross-stream diffusion. Thus the vorticity equation is locally described by

$$\alpha z' \frac{\partial \omega}{\partial z'} - \alpha x' \frac{\partial \omega}{\partial x'} = \nu \frac{\partial^2 \omega}{\partial x'^2}. \tag{2.10}$$

where the straining rate is $\alpha(t) \approx c/\lambda$. The vorticity is $\omega = \omega_W \mathrm{sgn}(x')$ at $x' = 0$. Solving (2.10), either by a similarity method or using the Boussinesq transformation (see Hunt & Eames 2002), shows that the vorticity magnitude Ω and wake width l varying according to

$$\frac{\Omega}{\omega_W} = \frac{l_v^2}{l^2}\left(\frac{z_0'}{z'}\right)^2, \qquad l^2 = \frac{\nu}{\alpha} + \left(l_v^2 - \frac{\nu}{\alpha}\right)\left(\frac{z_0'}{z'}\right)^2, \tag{2.11}$$

where z_0' is an initial distance $O(l_v)$ from the free surface. The effect of the straining field is to confine opposite-signed vorticity, which then rapidly diffuses into one another and is annihilated. When $z'/l_v \gg 1$, the effect of the strain on the wake is significant, with the width of the 'wake' tending rapidly to $(\nu/\alpha)^{\frac{1}{2}}$ and the mean vorticity decreasing as $\omega_W(l_v^2\alpha/\nu)(z_0'/z')^2$. A measure of the effect of wake flow on the ambient fluid is $Q(z')$ the volume flux it generates; for a straining flow, where $z'/l_v \gg 1$, the volume flux decreases as $Q(z')/Q(z_0) = (z_0'/z')^2 \sim (l_v/z')^2$, indicating that the flow beneath the wave is irrotational because the volume flux tends to zero. In the absence of the straining field, the vorticity magnitude would decrease rather more slowly as

$\sim 1/z'^{\frac{3}{2}}$ (for planar wakes). However, when the boundary layer thickness is not small compared to the straining lengthscale, $z'/l_v \sim 1$, the effect of straining field on vorticity annihilation is diminished, with boundary layer vorticity being advected into the main body of the fluid. The analysis has dealt with the case of a steady straining field. When the strain rate increases/decreases with time ($\dot{\alpha} > 0, < 0$), the above model will tend to under-estimate/over-estimate the effect of the straining field, which then depends on the integrated history of the strain rate.

The immediate consequence of this analysis is that while an unsteady stagnation flow transports vorticity into the ambient fluid, vorticity annihilation leads to such a large reduction in the magnitude of vorticity, that it can be neglected. This explains, to a large extent, the success of irrotational flow models in describing non-linear (non-breaking waves). However, when the boundary layer thickness (l_v) is comparable to, or larger, than the straining lengthscale ($\sim O(\lambda)$), vorticity is not cancelled out.

2.3. *Non-linear unsteady wave breaking*

Forced waves produced by a laboratory wave maker or ocean waves by a change in water depth, deform to such an extent that the tip of the wave crest plunges downwards into the wave surface and re-enters the water (*e.g.* Peregrine 1994). The topology of the water surface is changed significantly so as to become doubly connected (see figure 1(c,i)). The tongue of fluid (from the breaking wave) entering the ambient fluid generates vorticity by an inviscid mechanism described by Lamb (1932); the topological change means that the integral of velocity around a closed loop $\mathcal{C}$, $\int_{\mathcal{C}} \nabla\phi \cdot \mathrm{ds}$, is negative and negative vorticity is created at the point of intersection between the wave surface and crest tip. The fluid element at the other side leads to positive vorticity.

The high subsequent turbulent mixing generated by the plunging wave crest leads to vorticity in the surface layer of the water. A plunging jet impinging onto a water surface can behave in two different ways - either the velocity decreases and the jet spreads out (as if impinging onto an almost rigid surface) or it penetrates into the water surface and creates a submerged jet (just like a stream down a steep slope entering a reservoir). In the latter case vortex sheets are generated between the jet and the surrounding water, and air bubbles are entrained by an instability mechanism just where the jet enters (Thomas *et al.* 1983; Zhu, Oguz & Prosperetti 2000). This lack of uniqueness is analogous to that in inviscid flow around bluff bodies where the formation of free streamline vortex sheets is one possible solution (Batchelor 1965, p.494). The generation of vorticity at the edge of a plunging jet, which is subsequently distributed through the water is comparable to the vorticity generated by the collapsing wave cavity (Peregrine 1994).

The horizontal impulse associated with an inviscid (unsteady) wave, $I_x = \int_S \phi n_x \mathrm{d}S$, is conserved in time, before it breaks. The vortex sheet generated by the jet entering the liquid at an angle creates both a vertical and horizontal component of impulse in the interior of the liquid. The vorticity created during the plunging motion ultimately wraps up (in a perfect fluid) to create a vortex with a forward impulse Γd. The plunging jet provides a means of communicating

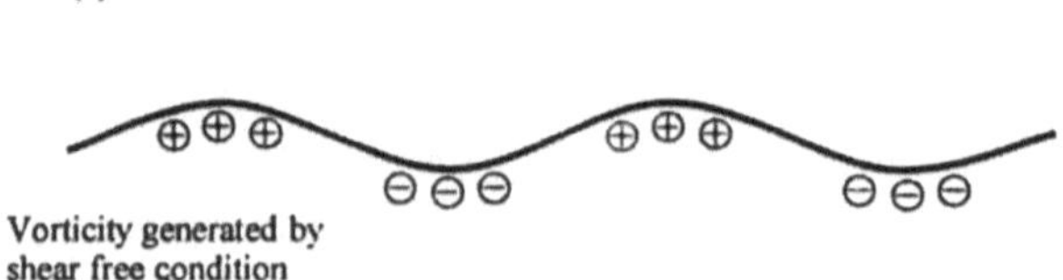

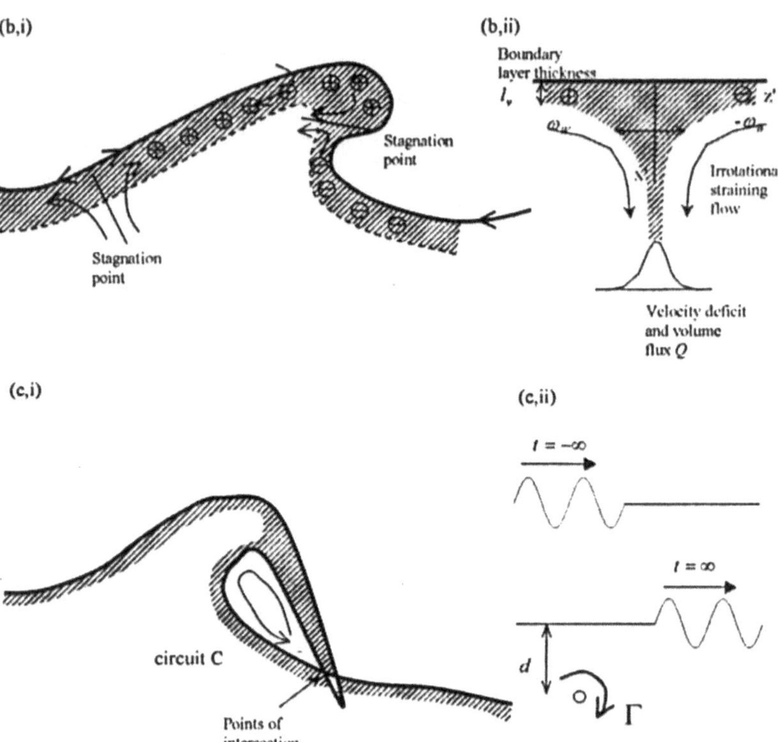

FIGURE 1. Schematic showing (a) linear waves, (b)(i) the stagnation points on the surface of a non-linear non-breaking waves, (ii) transport of boundary layer vorticity near by a stagnation flow, and (c) (i) wave crest re-entering the flow for an unsteady breaking wave and (ii) the generation of a vortex beneath the surface.

momentum from the wave to the ambient fluid, and leads to a net reduction of the horizontal impulse of the wave equal to

$$I_x(\infty) - I_x(-\infty) = \Gamma d, \tag{2.12}$$

where Γ is the circulation of the vortex created after the breaking process has finished and d is the separation between the free surface and the center of the vortex (see figure 1(c(ii))).

3. Wind driven flows and steady breaking waves

3.1. *Wind driven water flow under low amplitude waves*

When there is a turbulent wind over a nearly flat water surface, the mean velocity profile in the air has the approximate form

$$U_A(z) = U_{AW} - \frac{u_{*A}}{\kappa} \log\left(\frac{z}{z_{0A}}\right), \tag{3.1}$$

where U_{AW} is the mean velocity of the water at the air-water interface, u_{*A} is the friction velocity which is of the order of the r.m.s. velocity fluctuations in the turbulence, and z_{0A} is the roughness length of the air-sea interface presented to the air. It is assumed that the Reynolds number (based on the depth of the boundary layer) is very large so that there is a well-developed logarithmic layer, *i.e.* $U_{AW} l_v/\nu \gg 10^4$. Over a vertical scale of the order of the roughness length z_{0A}, the mean and fluctuating velocities and shear stresses at the surface in the air flow (U_A, u'_A and $\tau_A (= \rho_A u_{*A}^2), \tau'_A$) respectively match those in the water (U_W, u'_W and τ_W, τ'_W) which drives a turbulent shear flow in the water with a similar logarithmic profile

$$U_{AW} - U_W(|z|) = \frac{u_{*W}}{\kappa} \log\left(\frac{|z|}{z_{0W}}\right), \tag{3.2}$$

where the friction velocity in the water flow, u_{*W}, is related to the friction velocity in the airflow by the condition that the stress is continuous *viz.* , $\tau_A = \tau_W$ or $\rho_A u_{*A}^2 = \rho_W u_{*W}^2$, so that (Lombardi *et al.* 1996)

$$u_{*W} \approx \frac{u_{*A}}{\sqrt{\rho_W/\rho_A}} \approx \frac{u_{*A}}{30}. \tag{3.3}$$

Therefore the magnitude of the mean velocity in the water driven by a mean gradient dU_W/dz is only about 1/30 of that in the air flow. But, since the fluctuating velocities and stresses match, it follows that fluctuations in the shear stress driven by turbulence on the air side, τ'_A, which are of the order of τ_A, produce fluctuations in the water that are of the same order as those produced by the mean shear in the water. The shear-sheltering mechanism analysis of Hunt & Morrison (2000) implies that at very high Reynolds number the non-uniform shear $U_W(z)$ in the water below a flat surface acts to generate turbulence in approximately local equilibrium at each level so that it is independent of the 'slip' boundary condition at $z = 0$. Direct numerical simulations, even at quite low $Re \sim 10^4$ (Lombardi *et al.* 1996), show how the eddy structure of the water turbulence is broadly similar to that over a rigid surface, although the level of the surface velocity fluctuations is higher. The velocity fluctuations are higher beneath a surface wave are higher than beneath a rigid wall because the horizontal velocity fluctuations u'_W induced by the mean shear are not zero at $z = 0$.

What happens to the mean and turbulent flow and the energy dissipation in the water when non-breaking waves propagate along the surface? Since Harris *et al.* (1996) showed how a non-linear numerical model agreed well with linear theory, we can use linear theory to help resolve this question. In the water flow, within an inner layer of thickness $l_W \sim \lambda/z_{0W}$, the turbulence adjusts locally to the acceleration and deceleration associated with the wave. Within this layer, Belcher *et al.* (1996) showed that the additional changes to the peak shear stress induced by the wave varies in proportion to the wave slope,

$$\frac{\Delta\tau_W}{\langle\tau_W\rangle} \approx 10\frac{a}{\lambda}, \tag{3.4}$$

where $\langle\rangle$ denotes averaging over one wavelength. Below this inner layer the mean flow is determined by the wave field (since $(a/\lambda)c > \mathrm{d}U_W/\mathrm{d}z$): it is found that the turbulence adjusts 'rapidly' and the perturbation shear is much weaker.

The mean dissipation rate $\langle\epsilon\rangle$ near a flat surface decreases rapidly with distance as

$$\langle\epsilon\rangle \approx \frac{u_{*W}^3}{k|z|}, \tag{3.5}$$

for $|z| > z_{0W}$ (or $|z| > 10\nu/u_{*W}$ for a smooth surface). Since the dissipation rate, ϵ, varies with $(\tau_W + \Delta\tau_W)^{\frac{3}{2}}$, it follows that, within the inner region the mean dissipation $\langle\epsilon\rangle$ is significantly increased even when a/λ is only about 1/10:

$$\langle\epsilon\rangle \sim \frac{u_{*W}^3}{k|z|}\left(1 + 15\left(\frac{a}{\lambda}\right)^2 \exp(-|z|/l_w)\right) \tag{3.6}$$

for $|z| < l_w$. It is interesting to compare this calculation with measurements of Terray *et al.* (1995) for the dissipation rate in the upper layers of the ocean which yield

$$\langle\epsilon\rangle \sim \frac{u_*^3}{k|z|}(c|z|^{-p}), \tag{3.7}$$

where p varies over a considerable range. The conventional explanation is that the enhanced dissipation is due to wave breaking: here we have shown that even non-breaking waves significantly enhance the dissipation rate in the upper layers of the water flow.

3.2. *Non-linear steady wave breaking*

As the steepness increases further, the waves break, and vorticity and turbulence are generated through two different mechanisms. Firstly, as the tongue of a plunging breaker enters the water air is entrained and a vortical flow is formed (figure 1(c)), (Thomas *et al.* 1983). Secondly, the foam layer that forms in the breaking region significantly increases the free surface, which generates and diffuses vorticity in the water. Interstitial fluid in the foam layer, though small, is transported vertically downwards by drainage. With the use of high definition PIV, Melville (2001) has been able to examine the formation and dynamical influence of the vortical flow in the water generated by breaking.

Banner & Melville (1976) argue that in an equilibrium or quasi-steady breaking wave the position of airflow separation coincides with the position of flow reversal in the water (S_2). This implies that S_1 is also a stagnation point. (The shape of the surface is uncertain near S_2). Smith (1973) and Parkinson & Jandali (1970) have developed models of flow separation from streamline and

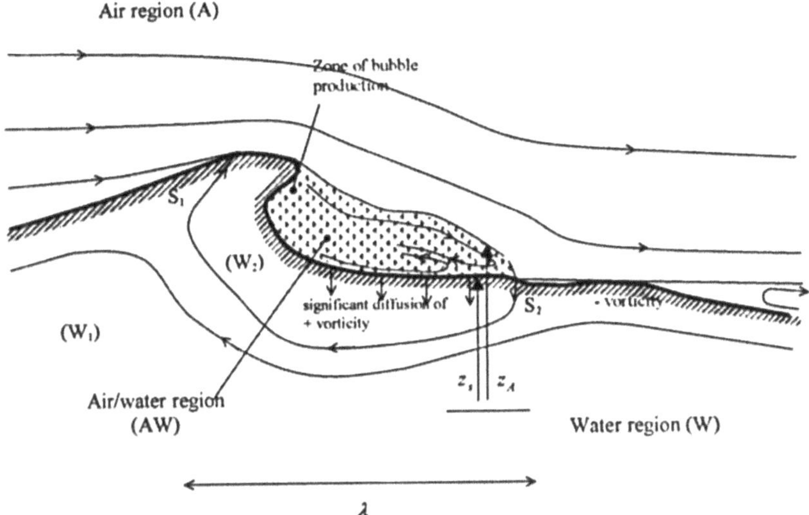

FIGURE 2. Schematic showing the flow in the frame moving with a breaking wave (following Banner & Melville 1976).

bluff bodies respectively, and this work may be relevant to understanding the processes near the separation streamline in a breaking wave, where the separation streamline is tangential to the surface. The other key aspect of the flow is the foam region or breaker which rides on the wave which Longuet-Higgins & Turner (1975) modelled as a gravity current flowing down an inclined wall. This produces a highly erratic interface between (AW) and (W_2), and induces a large diffusion across (W_2). Drainage of fluid from the foam region, which was neglected by Longuet-Higgins & Turner (1974) generates a Reynolds stress at the (AW) boundary, in a similar manner to that described by Maxworthy (1999).

In an inviscid flow without wind such a breaking event is inherently unsteady. But with a wind blowing, after overturning, a quasisteady breaking wave can persist. Figure 2 shows a schematic of the flow in the frame moving with the wave, where we distinguish between the flow regions: (W) water, (AW) air/water and (A) air and where the densities are respectively $\rho_W \geq \rho_{AW} \gg \rho_A$.

Foam region (AW)

Since there is a stagnation point at S_2, the dynamic pressure in the (A) air and (AW) air/water regions are comparable, so that $\rho_{AW} gh \sim \rho_A U_A^2(h)$, where $U_A(h)$ is the mean air speed at height h over the water surface. Waves break when the orbital speed at the crest is larger than the phase speed of the wave, so that U_w, the speed of the fluid (in the wave frame) is small and $\sqrt{gh} \approx c$. Since a gravity current in the presence of an external ambient flow generates a weak internal flow (Simpson & Britter 1979), the air-water zone behaves as a 'kind' of gravity current for which Longuet-Higgins & Turner (1977) developed a model. Here we elaborate on the particular form of the gravity current.

This also causes high dissipation in the water flow so that an overturning wave is in a steady state. The downwards vertical motion on (A)-(AW) interface drags air flow down to S_2, where the local positive pressure, as in high Re turbulent flows is associated with airflow separation downwind of S_2 (Hunt & Richards 1985).

Water region (W)

In the (W) region, we have, to a first approximation, inviscid stagnation points at S_1 and S_2. We need to calculate the velocity along the surface, u_s. Applying Bernoulli on the interface, $z = z_s$, where the surface pressure is constant $p_s = 0$, yields

$$-gz_s + \frac{1}{2}u_s^2 = 0 \tag{3.8}$$

For the case of irrotational flow near a stagnation region, we apply the free surface analysis of a stagnation point on a Taylor bubble (Batchelor 1965, p.475) shows that the local streamfunction is $\psi = Ar^2 \sin 2\theta$ and the velocity potential is $\phi = Ar^2 \cos 2\theta$. The difference between this analysis and the analysis described in §2.1, is that here the straining lengthscale λ is comparable to the thickness l_v of the vortical layers (see figure 2). Vorticity is advected into the ambient fluid without 'disappearing' as in the thin layer ($l_v \ll \lambda$) situation of §2.2. The vorticity field (ω_W) affects the flow near the stagnation points as the following analysis demonstrates. Expressing the vorticity field in terms of a streamfunction ψ, gives $\nabla^2\psi = -\omega$ or

$$\frac{\partial^2\psi}{\partial r^2} + \frac{1}{r}\frac{\partial\psi}{\partial r} + \frac{1}{r^2}\frac{\partial^2\psi}{\partial\theta^2} = -\sum_n a_n \sin 2n\theta, \tag{3.9}$$

where $\psi = 0$ on $\theta = 0, \pi$. The Fourier coefficients a_n are determined from the inlet condition $\omega = \omega_W \mathrm{sgn}(x)$,

$$a_n = \frac{4}{\pi}\int_0^{\pi/2} \sin 2n\theta\omega d\theta, \qquad \text{so that} \qquad a_{2k+1} = \frac{4\omega_W}{\pi^2(2k+1)}. \tag{3.10}$$

Expanding the streamfunction as a series in $\sin 2n\theta$ and determining the particular solution to the above equation, we obtain, as $r \to 0$, the leading order solution

$$\psi \approx \frac{a_1}{4} r^2 (\log r) \sin^2\theta = \frac{\omega_W}{\pi^2} r^2 \log r \sin 2\theta, \tag{3.11}$$

so that the tangential velocity on the surface of the wave near the stagnation region is

$$u_s = \pm\frac{\omega_W}{\pi^2} x \log|x|. \tag{3.12}$$

Thence, from (2.17), the local displacement of the wave surface is

$$z_s = -\frac{\omega_W^2 x^2 (\log|x|)^2}{2g\pi^4}. \tag{3.13}$$

The consequence of this local analysis is that a free surface above a stagnation point has a local minima either side of $x = 0$. The analysis breaks down far from $x = 0$ because the free-surface deformation (according to (3.25)) ultimately increases with distance because the boundary layer flow increases.

The broad pattern of the flow and surface deformation caused by the vorticity distribution in the wave can be understood by considering the ideal problem of a line vortex Γ a distance d below a free surface. When $\Gamma/\sqrt{gd^3} \sim 1$, the vortex moves at the same speed as the surface wave it induces. However, in a real wave problem, there is local enhanced dissipation, and a foam region with finite density so on the (W)-(AW) interface,

$$p_s - gz_s + \frac{1}{2}u_s^2 = -\epsilon(x), \tag{3.14}$$

where $p_s \approx (z_A - z_s)g\rho_{AW}/\rho_W$ (near S_2, p_s is determined by stagnation flow in the (AW) and (A)). Here $\epsilon(x)$ is the loss of pressure caused by dissipation under zone (AW). In the vortical zone of the water (W_2), there is rapid diffusion of vorticity because of turbulence, bubbles etc. But in (W_1), vorticity is confined to a narrow region close to the separation streamline. By understanding the diffusion of vorticity, we should be able to estimate diffusion of heat, mass etc.

4. Distortion of vorticity beneath waves

We now consider the interaction between the mean and fluctuating turbulent vorticity $\boldsymbol{\Omega}''(z)$, $\boldsymbol{\omega}''(\mathbf{x},t)$ (respectively) of wind-drift flows in the water (see §3), and the mean drift and fluctuating wave velocity field $\tilde{\mathbf{U}}, \tilde{\mathbf{u}}$ (respectively) produced by the waves. Note that $\tilde{\mathbf{U}} \sim (ak)^2 c$ is the Stokes drift and is proportional to the square of the wave slope. These interactions give rise to secondary mean and fluctuating vorticity fields ($\boldsymbol{\Omega}^{(S)}, \boldsymbol{\omega}^{(S)}$). These can best be understand by considering the equation for fluctuating vorticity in which the leading terms (which are labelled for future discussion) are

$$\frac{\mathrm{D}\boldsymbol{\omega}}{\mathrm{D}t} = \underset{(\Omega D)}{(\boldsymbol{\Omega}''\cdot\nabla)\tilde{\mathbf{U}}} + \underset{(\omega D)}{(\boldsymbol{\omega}''\cdot\nabla)\tilde{\mathbf{U}}} + \underset{(\Omega W)}{(\boldsymbol{\Omega}''\cdot\nabla)\tilde{\mathbf{u}}} + \underset{(SW)}{(\boldsymbol{\Omega}^{(S)}\cdot\nabla)\tilde{\mathbf{U}}} + \text{smaller terms.} \tag{4.1}$$

(The smaller terms correspond to vorticity diffusion and weaker non-linear interactions). The first and third terms describe the waves straining the mean vorticity of the wind drift and are zero for two-dimensional waves normal to the wind. But the third term (ΩW) leads to a significant mean effect when the waves are at angle to the wind giving rise to one form of Langmuir circulation (Craik & Leibovich 1976). The fourth (SW) term describes the secondary mean (vortical) field $\boldsymbol{\Omega}^{(S)}$ being strained by the gradients of the mean waves field ($\partial\tilde{U}_x/\partial z$) which induces an axial secondary vorticity $\boldsymbol{\Omega}^{(S)}$ which builds up the whole secondary field. This is now a well established mechanism in shear flows over wavy flows (both moving and fixed) (*e.g.* Craik & Lebovich 1976; Phillips 1998; Belcher & Hunt 1998). In the second term (ωD), vorticity of the wind drift turbulence ($\boldsymbol{\omega}''$) is strained by the mean wave motion ($\tilde{U}$) as the waves move (relatively rapidly) over the turbulent eddies. This mechanism was first proposed by Phillips (1957), recently analysed by Teixeira & Belcher (2002), and leads to energy passing slowly from the waves into the water-side turbulence where it is dissipated.

We now consider shorter timescale interactions of the order of a single wave period between the wind drift mean field and the wave for two- and three-dimensional waves. Laboratory experiments and numerical simulations of Komori *et al.* (1993) showed for flow of a deep turbulent boundary layer over two-dimensional waves, low speed fluid between the waves on the air side regularly bursts upwards into the higher speed air streams between the waves (probably helped by the Langmuir mechanism of the SW term). This process occurs less strongly in the water with weak downward bursts below two-dimensional waves. This is probably because unlike the air flow, there is no deep layer of large scale turbulent eddies in the water flow below the waves. However, it is observed that these bursts are greatly amplified (as seen from the movement of dyed surface layers) as soon as the waves become three-dimensional. It appears as if the mean drift vorticity $\boldsymbol{\Omega}''$ is distorted by the three-dimensional finite-amplitude wave motion ($\Omega''_y \partial \tilde{U}_x/\partial y$) so as to produce a trailing vorticity ω_x on the side of the wave and thence a downward velocity u_z. The mechanism is similar to the mechanism operating in a shear flow over an obstacle, such as a hemisphere on a surface, which induces a trailing vortex (horse-shoe vortex) and a vertical motion (Hawthorne & Martin 1955).

Quantitative analysis of the vorticity distortion requires use of Cauchy's relation between vorticity of the fluid element located at $\mathbf{X}$ at time t, $\boldsymbol{\omega}(\mathbf{X}, t)$, and its value at an earlier time t_0, $\boldsymbol{\omega}(\mathbf{a}, t_0)$, when the fluid particle was located at $\mathbf{a}(t_0)$, *viz.*

$$\omega_i(\mathbf{X}, t) = \gamma_{ij}\omega_j(\mathbf{a}, t_0), \tag{4.2}$$

where $\gamma_{ij} = \partial X_i/\partial a_j$ (Batchelor 1967, p.276) and the suffices 1,2 and 3 correspond to the x, y and z axis. Thus the framework for calculating vortex/turbulence interaction with water waves is based on a Lagrangian formulation where the motion of fluid particles are studied. Most research has previously exploited a linearised calculation of the Lagrangian drift velocity, namely

$$\frac{\mathrm{d}\mathbf{X}}{\mathrm{d}t} = \mathbf{u}(\mathbf{a}, t) + \left(\int_0^t \mathbf{u}(\mathbf{a}, t)\mathrm{d}t'\right) \cdot \nabla_a \mathbf{u}(\mathbf{a}, t), \tag{4.3}$$

(Batchelor 1967, p. 361). Although the above approximation to the drift velocity is valid for oscillatory (and rotational) flows, it does not capture the salient features of drift when the wave is non-linear or isolated. To calculate $\mathbf{X}$ over a shorter timescale around three-dimensional waves it is necessary to consider explicitly the local flow using the method originally developed for bluff bodies by Darwin (1953), extended by Eames *et al.* (1994) and applied to waves by Eames & McIntyre (1999). The latter paper showed that for potential flows, which is, (as we showed in §2 and 3), a good approximation for non-breaking waves (a majority according to Banner & Peregrine 1993), there is an explicit expression for X_1 in terms of the velocity potential ϕ and the total velocity $|\tilde{\mathbf{u}}|$ *viz.* for water side where the undisturbed velocity is $-c$;

$$X_1 = -\left[\frac{\phi}{c}\right]_0^t + \int_0^t \frac{|\tilde{\mathbf{u}}|^2}{c}\mathrm{d}t. \tag{4.4}$$

Eames & McIntyre (1999) showed formally that taking its mean value, $\langle \mathrm{d}X_1/\mathrm{d}t\rangle$ over many waves is equivalent to $\tilde{U}_1$ the streamwise Stokes drift velocity. The first term in (4.30) is the drift term and corresponds to a positive permanent displacement of fluid elements, while the second term is reflux and is generally positive for progressive waves.

To see the implications of these results to water waves, we proceed to calculate how shear generated by a wind stress is distorted by a three-dimensional wave field. For simplicity, we consider a three-dimensional wave field described by

$$\phi = ac\cos(k_x x)\cos(k_y y)e^{\sqrt{k_x^2+k_y^2}z} \tag{4.5}$$

moving with speed c, and shown schematically in figure 3(a). The wind stress creates a shear layer of thickness l_v and strength $\omega_y = \omega_W$, $\omega_x = \omega_z = 0$. The passage of the three-dimensional waves described by (4.31) displaces fluid elements a distance

$$X_1 \sim N\left(1+\frac{1}{2}\sin^2(k_y y)\right)a^2\frac{(k_x^2+k_y^2)}{k_x}e^{2\sqrt{k_x^2+k_y^2}z}, \tag{4.6}$$

where N is the number of waves which have passed $N \sim ctk_x/2\pi$. From Cauchy's result (4.28), vorticity generated by stretching the vortical elements is

$$\omega_x(t) = \frac{\partial X_1}{\partial y}\omega_y(0) = \frac{1}{2}Nk_y\sin(2k_y y)\omega_W a^2\frac{(k_x^2+k_y^2)}{k_x}e^{2\sqrt{k_x^2+k_y^2}z} \tag{4.7}$$

for $0 < -z < l_v$ and zero otherwise. Thus, the passage of a few waves creates alternate rolls of positive and negative vortices aligned with the wind direction. The action of the wind stress τ increases the strength of the vorticity ω_W linearly with time, $\omega_W \sim \tau t/\rho c$. Thus the induced flow created by the vortices increases as $l_v\omega_x \sim (\nu t)^{\frac{1}{2}}(tk_x)(\tau t/\rho) \sim (k_x\tau/\rho)\nu^{\frac{1}{2}}t^{\frac{5}{2}}$ much faster than the flow induced by the wind stress, which was shown by Vernon & Melville (2001) to grow as $\omega_w l_v \sim \tau\nu^{\frac{1}{2}}t^{\frac{3}{2}}/\rho$. The consequence of these calculations is the production of alternately signed vortices aligned with the wind direction (see figure 3(b)). The mechanism described above is essentially how small scale 'Langmuir cells' are created by wind driven waves and observed by Vernon & Melville (2002). The key result is that the vortices are created more rapidly than by the conventional formulation due to Craik & Leibovich (1976). Note that the sign of the longitudinal vorticity generated is independent of whether the free surface deformation has a positive or negative curvature, since both give rise to a drift. This interaction of individual waves may be important for mixing because the mechanism proposed by Craik & Lebovich (1976) is based on a slower average amplification of vorticity over many wavelengths by the Stokes drift straining.

5. Concluding remarks

This paper has shown how some of the critical problems in the dynamics of water waves are related to other fundamental problems in fluid mechanics, so that progress in the latter should be able to help with solutions of the former (Hunt 2002). Thus studies of the propagation of vorticity into irrotational

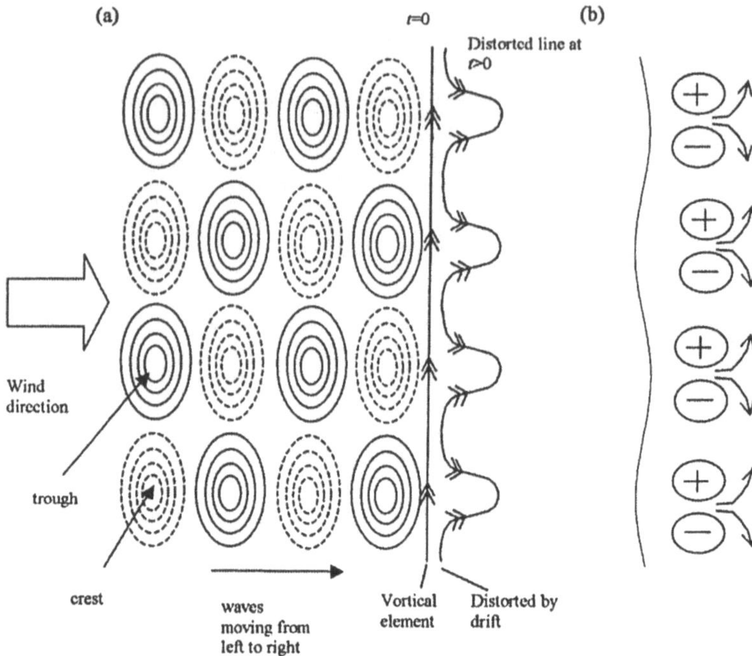

FIGURE 3. (a) Schematic plan view of the distortion of wind generated shear by three-dimensional waves and the generation of alternately signed vortices parallel to the wind. (b) A cross section through the water depth is shown, illustrating the position of the roll vortices related to the wave crests and troughs.

straining flows have indicated in §2 why in non-breaking waves, even of finite amplitude, the flow is irrotational except very close to the surface. Clearly, it would be interesting to extend this study to situations when breaking of capillary waves begins and vorticity is transported more rapidly into the interior (Melville 1996).

When waves break as plunging breakers, vorticity predominatly of one sign is generated over a wide region, so that straining effects cannot cancel out the vorticity. When they break as spilling breakers, the vorticity generated near the surface is subject to some cancellation by cross diffusion at the downstream saddle point S_2. Detailed study of the flow in this region in the three phases, water, air and foam is needed to provide more accurate estimates. The approximate and incomplete model for the three flow regions in the quasi-steady wind driven spilling breaker and separated air flow needs detailed computational modelling and further analysis. But the new approach developed here does appear to provide a consistent explanation and order of magnitude esimate for more of the features of this highly complex flow, than previous models. In particular, for the mechanisms controlling the form of the two-phase flow gravity current at its downwind end, and the interaction between the vortical flows associated with the breaking waves and the wind-drift current. When a model for these flows is more clearly established it will lead to estimates with greater accuracy of the distribution of breaking waves over the oceans and their contribution to momentum, mass and energy transfer.

Finally the paper has explored the equally difficult problem of describing quantitatively the individual eddying events caused by interactions between the vorticity and turbulence of wind driven currents and three-dimensional waves. We have argued that the Lagrangian (or Darwin) drift concept for vorticity distortion provides a framework for analysing both this short timescale mechanism and also the longer term interactions, hitherto examined using Stokes Drift. This equivalence had been hinted at by Belcher & Hunt (1998) but was only formally established by Eames & McIntyre (1999). If this effect is significant, it implies that there should be a marked change in mass and heat transfer at higher wind speeds or in complex waves where waves become three-dimensional. Some laboratory data supports this hypothesis.

ACKNOWLEDGEMENT We are grateful for support at the INI Summer Program on 'Mathematics of Surface Waves' and for many instructive conversations with K. Melville, M.E. McIntyre, H. Peregrine, M.S. Longuet-Higgins, H. Banner and J. Battjes. J.C.R. Hunt acknowledges support from NERC through the grant to CPOM (Centre for Polar Observation and Modelling) at UCL. I. Eames acknowledges support for an EPRSC Advanced Research Fellowship, held at UCL.

REFERENCES

BANNER, M.L. & PEREGRINE, D.H. 1993 Wave breaking in deep water. *Ann. Rev. Fluid Mech.*, **25**, 373–397.

BANNER, M.L. & MELVILLE, W.K. 1976 On the separation of air flow over water waves. *J. Fluid Mech.*, **77**, 825–842.

BATCHELOR, G.K. 1967 An Introduction to Fluid Dynamics. *Cambridge University Press.*

BELCHER, S.E. & HUNT, J.C.R. 1998 Turbulent air flow over hills and waves, *Ann. Rev. Fluid Mech.*, **30**, 507–538.

BELCHER, S.E., HARRIS, J.A. & STREET, R.L. 1994 Linear dynamics of wind waves in coupled turbulent air-water flow. Part 1. Theory. *J. Fluid Mech.*, **271**, 119–151.

CRAIK, A.D.D. & LEIBOVICH, S. 1976 A rational model for Langmuir circulations. *J. Fluid Mech.*, **73**, 401–426.

DARWIN, C.G. 1953 A note on hydrodynamics. *Trans. Camb. Phil. Soc.*, **49**, 342–354.

EAMES, I., BELCHER, S.E. & HUNT, J.C.R. 1994 Drift, partial drift and Darwin's proposition. *J. Fluid Mech.*, **275**, 201–223.

EAMES, I. & MCINTYRE, M.E. 1999 On the connection between Stokes and Darwin drift. *Proc. Cam. Phil. Soc.*, **126**, 171–174.

GRILLI, S.T., GUYENNE, P. & DIAS, F. 2001 A fully non-linear model for three-dimensional overturning waves over an arbitrary bottom. *Int. J. Numerical Meth. Fluids*, **35**, 829–867.

HARRIS, J.A., BELCHER, S.E. & STREET, R.L. 1996 Linear dynamics of wind waves in coupled turbulent air-water flow. Part 2. Numerical model. *J. Fluid Mech.*, **308**, 219–254.

HAWTHORNE, W.R. & MARTIN, M.E. 1955 The generation of secondary vorticity in the flow over a hemisphere due to a density gradient and shear. *Proc. R. Soc. Lond. A*, **232**, 184–195.

HUNT, J.C.R. & RICHARDS, K.J. 1985 Stratified air flow over one or two hills. *Boundary-layer Meteor.*, **30**, 223–259.

HUNT, J.C.R. & MORRISON, J.F. 2002 Eddy structure in turbulent boundary layers. *Euro. J. Mech. B-Fluids*, **19**, 673–694.

HUNT, J.C.R. 2002 I.M.A. Conference 'Wind over Waves'. *Mathematics Today.*

HUNT, J.C.R. & EAMES, I. 2002 Wake disappearance in complex flows. *J. Fluid Mech.*, **457**, 111-132.

KOMORI, S., NAGAOSA, & MURAKAMI, Y. 1993 Turbulence structure and mass transfer across a sheared air-water interface in wind-driven turbulence. *J. Fluid Mech.*, **249**, 161–183.

LAMB, H. 1932 Hydrodynamics. *Dover.*

LOMBARDI, P., DE ANGELI, V. & BANNERJEE, S. 1996 Direct numerical simulations of near-interface turbulence in complex gas-liquid flow, *Phys. Fluids*, **8**, 1643–1665.

LONGUET-HIGGINS, M.S. & TURNER, J.S. 1974 An "entraining plume" model of a spilling breaker. *J. Fluid Mech.*, **63**, 1–20.

LONGUET-HIGGINS, M.S. 1992 Capillary rollers and bores. *J. Fluid Mech.*, **240**, 659–679.

MAKIN, V. & KUDRYAVSTEX, V.N. 1999 Coupled sea surface-atmosphere model 1. Wind over waves coupling. *J. Geophys. Res.*, **104**, 7613–7623.

MAXWORTHY, T. 1999 The dynamics of sedimenting surface gravity currents. *J. Fluid Mech.*, **392**, 27–44.

MELVILLE, W.K., VERON, F. & WHITE, C.J. 2001 The velocity field under breaking waves: coherent structures and turbulence. *J. Fluid Mech.*, **454**, 203–233.

MELVILLE, W.K. 1996 The role of surface-wave breaking on air-sea interactions. *Ann. Rev. Fluid Mech.*, **28**, 279–321.

PARKINSON, G.V. & JANDALI, T. 1970 A wake source model for bluff body potential flow. *J. Fluid Mech.*, **40**, 577-594.

PHILLIPS, O.M. 1977 The dynamics of the upper ocean. *Cambridge University Press.* Second Edition.

PHILLIPS, O.M. 1957 On the generation of waves by turbulent wind. *J. Fluid Mech.*, **2**, 417–445.

SAJJADI, S.G. 2002 Vorticity generated by pure capillary waves. *J. Fluid Mech.*, **459**, 277–288.

SIMPSON, J.E. & BRITTER, R.E. 1979 The dynamics of the head of a gravity current advancing over a horizontal surface. *J. Fluid Mech.*, **94**, 477–495.

SMITH, F.T. 1973 Laminar flow over a small hump on a flat plate. *J. Fluid Mech.*, **57**, 803–824.

STOKES, G. G. 1847 On the theory of oscillating waves. *Trans. Camb. Phil. Soc.* , **8**, 441–455.

TEIXEIRA, M.A.C. & BELCHER, S.E. 2002 On the distortion of turbulence by a progressive surface wave. *J. Fluid Mech.*, **458**, 229–267.

THOMAS, N.H., AUTON, T.R., SENE, K. & HUNT, J.C.R. 1983 Entrapment and transport of bubbles by transient large eddies in multiphase turbulent shear flows. *Int. Conf. Physical Modelling of Multiphase Flow*, Coventry, April 1983.

THORPE, S.A. 1984 The effect of Langmuir circulation on the distribution of submerged bubbles caused by breaking wind waves, *J. Fluid Mech.*, **142**, 151–169.

VERON, F. & MELVILLE, W.K. 2001 Experiments on the stability and transition of wind-driven water surfaces. *J. Fluid Mech.*, **446**, 25–65.

ZHU, Y., OGUZ, H.N. & PROSPERETTI, A. 2000 On the mechanism of air entrainment by liquid jets at a free surface. *J. Fluid Mech.*, **404**, 151–177.

Langmuir Circulations

W.R.C. Phillips
Department of Theoretical & Applied Mechanics, University of Illinois at Urbana-Champaign, Urbana, IL 61801-2935, USA

ABSTRACT

Langmuir circulations in the open ocean typically form in an environment of surface waves and a shear current, and a mechanism which exploits these features and leads to shear-aligned longitudinal vortices much like Langmuir circulations is Craik-Leibovich instability theory. This theory is discussed in detail. The theory is first constructed in a form which accounts for both rotational and irrotational waves in all levels of shear. This is done from the generalized Lagrangian mean equations of Andrews & McIntyre which describe an exact theory of nonlinear waves on a Lagrangian mean flow. The instability theory is then discussed at length for both weak and strong shear, as both arise in the ocean. In weak shear the instability is centrifugal and catalysed by the Stokes drift without wave modulation, while in strong shear it is calalysed by the pseudomomentum with wave modulation and is not centrifugal. Accordingly the criteria for instability in weak and strong shear are different; both criteria are given. To accentuate the differences, the instability mechanism in weak shear is denoted CL2, while in strong shear it is denoted CLg, for generalized Craik-Leibovich. Recent studies of the first bifurcation to both CL2 and CLg are outlined.

1. INTRODUCTION

Mariners have for eons observed long froth-like rows on the ocean surface that more or less align with the wind, but their cause remained a mystery until Langmuir (1938) realized they are surface manifestations of counterrotating rolls in the water beneath. Now known as Langmuir circulations, or LCs, their spacings range from millimeters to several hundred meters, and rule of thumb observations suggest they form in the presence of surface waves tens of minutes after the onset of winds above 3 m s^{-1}, often, although not always, in conjunction with wave breaking. Langmuir's observations further led him to believe that the rolls are largely responsible for the formation of thermoclines and the maintenance of mixed layers in lakes and oceans, and contemporary observations concur.

Wind aligned rolls also form in the atmosphere with spacings as large as several kilometers, and we refer to these as atmospheric Langmuir circulations (ALCs). LCs and ALCs are together thought responsible for the interchange of heat, mass and momentum between the atmosphere and ocean. From the viewpoint of next generation global change models (*i.e.* computer simulations of long range planetary fluid mechanics), therefore, it is important not only to include both LCs and ALCs, but to model them credibly, ideally using the mechanism(s) responsible for each. To this end the present article is concerned with the prevailing theory for LCs, viz Craik-Leibovich instability theory (Craik

1977, Leibovich 1977), and our efforts to deduce whether or not it is physically realizable.

Craik-Leibovich instability theory assumes a wave mean-flow interaction, in which the waves are spanwise independent and of small slope ϵ, while the mean shear is measured by the ratio of the surface velocity to the wave speed, as $O(\epsilon^s)$, where $s \geqslant 0$ (see §2). Ocean LCs are thought to arise predominantly in circumstances in which $s = 2$, so that the waves are essentially irrotational. But observations by Smith (1992) in which LCs are noticeably absent in a consistent breeze of 8 m s^{-1}, but form within fifteen minutes of a sudden onset of wind to 13 m s^{-1}, fuel the notion (Phillips, Wu & Jahnke 1999) that ocean LCs may also originate in the strong shear ($s = 0$) regime. The LCs then grow in scale as they cascade through medium ($s = 1$) to ultimately weak ($s = 2$) shear, only to be sustained by the dominant waves (see also Phillips 2001b).

Of course Craik-Leibovich instability theory continues to apply in stronger shear (Craik 1982), although the requirements for instability are somewhat different. Different too is the character of the instability, which for $s = 2$ is centrifugal without wave modulation, while for $s = 0$ is not centrifugal with wave modulation, although in both cases the instability is inviscid. The $s = 2$ instability is commonly denoted CL2, while we refer to the $s = 0$ case as the generalized, or CLg instability; see §3.2.

Our object here is to discuss recent work pertaining to the first bifurcation to CL2 and CLg, and we do so in §§3, 4, 5, while various nonlinear consequences of CLg are outlined in Phillips (1998b) and Phillips *et al.* (1999). We begin in §2 by writing the momentum equations which describe the generalized Craik-Leibovich instability. These were derived from the generalized Lagrangian-mean (GLM) equations of Andrews & McIntyre (1978), a formulation which describes an exact theory of nonlinear waves on a Lagrangian mean flow. Of specific relevance here is that GLM describes mean vorticity kinematics in the same way instantaneous vorticity kinematics are described, rendering it canonical as an avenue to elucidate structures like LCs (Phillips 2001a).

2. Generalized Craik Leibovich instability

Following Craik (1982), who restricted attention to inviscid flows, and Phillips (1998a), who allowed for viscous effects, we apply the GLM formulation to a class of unidirectional shear flows that have imposed on them, or are unstable to, small amplitude (rotational or irrotational) waves that are independent of spanwise direction. Of particular interest is the instability of the ensuing wave-mean interaction to longitudinal vortices. Our intent is to restrict only the slope of the waves but remain general in regard to the level of the imposed shear. In consequence the ensuing equations are relevant to a range of bounded and unbounded flows, but of particular interest are LCs, and the instabilities that give rise to them, when they form beneath growing wind driven surface gravity waves.

Consider then the interaction between a unidirectional shear flow with characteristic velocity $\mathcal{V}$ and two-dimensional straightcrested waves of wavelength λ that propagate in (or opposite to) the direction of the basic flow. The amplitude of the waves is assumed to grow from infinitesimal to finite, but we require their slope ϵ to satisfy $\epsilon < O(1)$ at all times. Orbital velocities are thus characterized by $\epsilon\mathcal{C}$, where $\mathcal{C}$ is a typical phase speed. We next suppose that the characteristic thickness of the shear layer is $\mathcal{L}$ and make variables dimensionless with respect to $\mathcal{L}$ and $\mathcal{C}$. Finally we write $\mathcal{V}/\mathcal{C} = O(\epsilon^s)$ and $\mathcal{L}/\lambda = O(\epsilon^\beta)$, where $s \geqslant 0$ while β is real and of either sign. Then the level of shear is also $O(\epsilon^s)$ and in the event viscosity plays a role, the Reynolds number $R \equiv \mathcal{LC}/\nu$. Finally we invoke space coordinates (x, y, z) and choose a reference frame that moves in the x-direction with the phase speed of the waves c.

We use uppercase letters to denote quantities pertaining to the primary flow, which by design is devoid of spanwise (y) dependence, and lower case letters otherwise, while an overbar on the unscaled dimensionless variable denotes a streamwise average. Our unperturbed Eulerian shear flow in $[z_1, z_2]$ is then $\bar{\mathbf{U}}(z,t) + \mathbf{i}c = \epsilon^s[U, 0, 0]$.

Envisage now an $O(\epsilon)$ wave field $\breve{\mathbf{U}}$ that interacts with itself and the primary shear flow to excite streamwise averaged spanwise varying Eulerian velocity perturbations $\tilde{\mathbf{u}}$, whose strength relative to the primary shear flow is measured by the parameter Δ, and express the resulting flow field in GLM variables. The outcome is the velocity associated vector field $\bar{\mathbf{q}} = \bar{\mathbf{Q}} + \tilde{\mathbf{q}}$, which we expand as

$$\bar{\mathbf{q}}(y, z, t) = \epsilon^s\{[Q_1, 0, \epsilon^{2-s}Q_3] + \Delta[q_1, \epsilon^n q_2, \epsilon^n q_3] + ...\} \quad (n \geqslant 0), \tag{2.1}$$

along with an affiliated scalar field Π as $\epsilon^s[\mathcal{P}(x, z, t) + \Delta\wp(x, y, z, t) + ...]$. Note that Π includes the pressure and that the power n can have values other than zero, as we shall see in §2.1.

The waves interact with themselves and the shear flow to produce $O(\epsilon^2)$ primary fields of pseudomomentum $\bar{\mathbf{P}}$ and Stokes drift $\bar{\mathbf{D}}$ (for details see Phillips 2000, 2001a). Since the Eulerian and Lagrangian mean velocity fields are related through $\bar{\mathbf{q}} = \bar{\mathbf{u}} + \bar{\mathbf{d}} - \bar{\mathbf{p}}$, we see that $Q_3 = D_3 - P_3$, which explains the extra primary mean field component in (2.1) (in contrast to the primary Eulerian flow which by design has only one component). Moreover the $O(\epsilon^s\Delta)$ axial velocity perturbation (owing to the interaction between the waves and mean flow) may in turn act to modulate the wave field and produce an $O(\epsilon^{s+2}\Delta)$ spanwise varying component of pseudomomentum Craik (1982). So with no loss of generality we write $\bar{\mathbf{p}}$ or $\bar{\mathbf{d}}$ as $\bar{\mathbf{P}} + \tilde{\mathbf{p}}$, expand as

$$\bar{\mathbf{p}}(y, z, t) = \epsilon^2\{[P_1, 0, P_3] + \epsilon^s\Delta[p_1, \epsilon^n p_2, \epsilon^n p_3 + ...]\}, \tag{2.2}$$

and with (2.1) substitute into the GLM equations.

2.1. *Governing equations*

Of interest are secondary instabilities which lead to the growth of q_1 and $\mho_1$ (defined below) with time and we begin with the premise that likely instabilities occur when the component GLM equations are coupled. Of particular interest is the case for which the wave-wave nonlinearities $\bar{\mathbf{P}}$ and $\bar{\mathbf{D}}$ have nonzero gradients in z and are spanwise independent.

In this instance the component equations ((3.9) and (3.10) in Phillips (1998a)) are coupled via $q_3\partial Q_1/\partial z$ and $\partial q_1/\partial y \partial P_1/\partial z$. To explore such coupling we require $n = (2-s)/2$ and rescale time as $\tau = \epsilon^{(s+2)/2}t$. Then provided $\beta \geqslant -m/2$, where $m = [0, 1]$ for rotational and irrotational waves respectively, Phillips (1998a) obtains

$$\frac{\partial q_1}{\partial \tau} + \Delta \left(q_2 \frac{\partial q_1}{\partial y} + q_3 \frac{\partial q_1}{\partial z} \right) + \epsilon^{(2-s)/2} D_3 \frac{\partial q_1}{\partial z} + q_3 \frac{\partial Q_1}{\partial z}$$
$$= \epsilon^{-(s+2)/2} R^{-1} \nabla^2 q_1 + O(\epsilon^{(2-s)/2} R^{-1}) \qquad (2.3a)$$

and

$$\frac{\partial \mho_1}{\partial \tau} + \Delta \left(\frac{\partial \mho_1 q_2}{\partial y} + \frac{\partial \mho_1 q_3}{\partial z} \right) + \epsilon^{(2-s)/2} \frac{\partial}{\partial z} (\mho_1 D_3) + \frac{\partial q_1}{\partial y} \frac{\partial P_1}{\partial z} - \epsilon^s \frac{\partial Q_1}{\partial z} \frac{\partial p_1}{\partial y}$$
$$+ \epsilon^s \Delta \{ \frac{\partial q_1}{\partial y} \frac{\partial p_1}{\partial z} - \frac{\partial q_1}{\partial z} \frac{\partial p_1}{\partial y} \} = \epsilon^{-(s+2)/2} R^{-1} \nabla^2 \mho_1 + O(\epsilon^{(2-s)/2} R^{-1}) \qquad (2.3b)$$

where

$$\mho_1 = \frac{\partial q_3}{\partial y} - \frac{\partial q_2}{\partial z}.$$

Note that because n varies with s it is evident from (2.1) that transverse and axial velocity perturbations may differ in order. Accordingly, while modulation of the primary wave field by the secondary flow may be ignored for shear of $O(\epsilon)$ or less, it plays an important role through p_1 at $O(1)$, where a further equation must enter to complete the set (Craik 1982). That notwithstanding, wave distortion in the y and z directions is $O(\epsilon^{3+s/2}\Delta)$ and may be neglected for all $s \geqslant 0$, allowing us to write

$$\frac{\partial q_2}{\partial y} + \frac{\partial q_3}{\partial z} = 0 \qquad \text{and thus} \qquad \mho_1 = -\nabla^2 \psi. \qquad (2.3c)$$

Of course in $s = 2$ shear with irrotational neutral waves, P_1 reduces to D_1 (see §6, Andrews & McIntyre 1978; Craik 1982), $D_3 = 0$ and $[q_1, q_2, q_3] \to [u, v, w]$ so that (2.3) reduce to the CL equations of Craik & Leibovich (1976). But (2.3) and the instability take on a different character at higher levels of shear, which we shall now discuss in detail.

3. The role of shear

Craik (1977) further realized that the linear form of (2.3) is separable and thus that exponentially growing solutions are admissible; he then proved they occur (when $s = 2$) for both homogeneous Dirichlet and Neumann boundary conditions. The resulting $s = 2$ instability has come to be known as the Craik-Leibovich type 2 or CL2 instability (Leibovich 1983). CL2 is an inviscid instability, although the $O(\epsilon^{(s+2)/2})$ growth rate predicted by inviscid theory will be annihilated by viscous damping unless $R \geqslant O(\epsilon^{-(s+2)/2})$.

3.1. *Change in character of the instability with the level of shear*

Craik (1982) later determined that the $s = 2$ instability is centrifugal and akin to the Taylor-Görtler instability, albeit in an averaged sense for an $O(\epsilon^2)$ mean curvature. But Phillips (1998a) shows that is not the case at higher levels of shear. To see why consider the root mean square kinetic energy K of each of the Fourier modes in the expansion of the perturbation velocity $\mathbf{u}$ as

$$\mathbf{u}(\mathbf{x},t) = \sum_{|m|\leqslant M}\sum_{|n|\leqslant N} \mathrm{e}^{i(m\alpha^* x+nl^* y)}\mathbf{U}_{m,n}(m\alpha^*,nl^*,z,t)$$

so that

$$K(m\alpha^*,nl^*) = \left\{\int_{z_1}^{z_2}[U_{m,n}^2+V_{m,n}^2+W_{m,n}^2]dz\right\}^{\frac{1}{2}}, \tag{3.1}$$

where α^* and l^* are the fundamental wavenumbers in the x and y directions respectively. Then in view of the relative magnitudes of the component velocities (see §5, Phillips 1998a) it follows that because, for example $\mathbf{U}_{0,1} = O(\epsilon^s\Delta, \epsilon^{s+1}\Delta)$, then

$$K(0,l^*) = O(\epsilon^s\Delta,\epsilon^{s+1}\Delta), \quad K(\alpha^*,0) = O(\epsilon) \quad \text{and} \quad K(\alpha^*,l^*) = O(\epsilon^{s+1}\Delta).$$

Thus, in $O(\epsilon^2)$ shear, the greatest kinetic energy is to be found in the α^* and higher order modes, as would be expected with a centrifugal instability. But because the extent of the mean flow modification is bound by the level of shear and not by the strength of the waves, the measure Δ may exceed ϵ; indeed a useful estimate in the fully nonlinear state is that $\Delta \equiv \epsilon^{s+\frac{1}{2}}$. In consequence $K(0,l^*)$ dominates when $s=0$; and because $K(0,l^*)$ comprises modes that are streamwise independent, this form of the instability is not centrifugal.

3.2. *Different notation at different levels of shear*

To summarize, while CL2 is a centrifugal instability catalysed by the Stokes drift without wave modulation in $s = 2$ shear, it is catalysed by the pseudomomentum with wave modulation and is not centrifugal in $s = 0$ shear. Necessary at all levels of shear, however, are spanwise independent gradients of the pseudomomentum and Stokes drift which act to tilt vertical vorticity streamwise. Physically this is doubtless done by the irrotational portion of the pseudomomentum (*i.e.* the Stokes drift) in $s = 0$ shear, but as we shall see in §5, wave modulation can severely suppress and in some instances curtail the $s = 0$ instability, particularly when $\alpha \leqslant O(1)$. In consequence not only does the character of the instability change with s as we saw in §3.1, but so too does the criteria for instability (see §4.1, 5.1). For this reason Phillips, Wu & Lumley (1996) introduced the notation CL2-$O(\epsilon^s)$ to distinguish the classes of instability, and this has to some degree been reflected in the literature. But such notation is cumbersome and for simplicity and historical reasons we shall henceforth refer to the $s = 2$ class as CL2, and all stronger shear classes as the generalized Craik-Leibovich, or CLg instability.

4. Recent calculations of CL2

In the open ocean, at least with regard to the dominant waves, shear is usually in the $s = 2$ regime, while its counterpart in the laboratory is more likely $s = 0$. That is not to say the open ocean is devoid of $s = 0$ shear, in fact the range $s \in [0, 2]$ is common (Melville, Shear & Veron 1998) and, as noted in §1, it is not implausible that ocean LCs originate in the strong shear ($s = 0$) regime and then grow in scale to $s = 2$ shear. But their aetiology notwithstanding, Smith's lament (1992), and Veron & Melville's (2001) too, is the lack of calculations describing the first bifurcation to CL2 and CLg, and we should now like to outline our efforts in that direction. Of particular interest to observationists *vis a vis* CL2, are the role of Prandtl and Richardson numbers and the effect of growing waves; these are outlined below following a discussion of the instability criterion for CL2.

4.1. *Criterion for instability*

To occur, the CL2 instability requires the presence of a neutral wavy disturbance giving rise to a sheared spanwise independent Stokes drift, together with pre-existing vorticity imparting an Eulerian-mean shear in the same sense as the Stokes drift. Kinematically (an inviscid flow subject to) the CL2 instability occurs because the Stokes drift gradient causes vortex lines (which move with the fluid) to tilt streamwise wherever the Eulerian-mean shear is laterally distorted, giving rise to a longitudinal component of vorticity and ultimately vortices. Mathematically CL2 arises because of the presence of the force term $\bar{\mathbf{p}} \wedge \nabla \wedge \bar{\mathbf{q}}$ in the GLM equations; this term reduces, if the wave field is irrotational, to $\bar{\mathbf{d}} \wedge \nabla \wedge \bar{\mathbf{u}}$, which Leibovich (private communication 2001) terms the Langmuir force.

4.1.1. *The role of Prandtl and Richardson numbers*

Phillips (2001b) considers the instability of a weakly sheared density stratified two dimensional wavy flow to LCs. Of interest is the influence to CL2 of Prandtl Pr and Richardson Ri numbers according to linear theory. The basis for the study is an initial value problem posed by Leibovich & Paolucci (1981) in which the liquid substrate is of semi-infinite extent and the wind driven current is permitted to grow in the presence of neutral waves. The Pr number is varied from zero to infinity, and both stabilizing and destabilizing Ri numbers are considered; so too are monochromatic and measured wave fields, and laminar and turbulent velocity profiles. Only the $Ri = 0$ results recover the numerics of Leibovich & Paolucci, but their conclusions are otherwise correct. For stabilizing Ri numbers, it is found in general that diminishing Pr numbers are destabilizing to LCs, and thus that LCs can be present or absent at the same Langmuir number La provided $Ri \neq 0$. It is further found that two branches of neutral curves occur for some combinations of Pr and Ri, and that minor changes in either parameter permit the preferred spacing to switch from one branch to the other. In consequence the preferred spanwise spacing may change from smaller than the wavelength of the dominant waves to larger than it. Furthermore, although LCs will not form at inverse La numbers below a global lower bound given by an energy stability analysis, the actual value of La at onset is found to depend greatly upon local details of the wave and shear fields. Interestingly although this global lowerbound is independent of Ri and Pr for $Ri \geqslant 0$, that is not the case for $Ri < 0$, where it approaches zero as $Ri \longrightarrow -\infty$.

4.1.2. *The role of growing waves*

Craik & Leibovich do not study growing waves, but as is evident from (2.3) such waves have the greatest influence on the instability in $s = 2$ shear; of course CL2 remains the underlying instability mechanism, but growing waves may influence the criterion for instability given in §4.1. Furthermore, because D_3 is temporal, the growth rate of the secondary flow will not be exponential, unless of course the growth rates of D_3 and the evolving LCs are disparate, which is the case studied.

With this in mind Phillips (2001c) considers the instability to longitudinal vortices of a temporally evolving two dimensional weakly sheared density stratified wavy flow. Waves aligned both with the wind and counter to it are considered, as is the role of stratification. As before the basis for the study is an initial value problem posed by Leibovich & Paolucci, but here it is generalized and the current has two components: a wind driven portion owing to the wind stress applied at the free surface and a second due to the diffusion of momentum owing to the wave amplitude squared free surface stress condition (Longuet-Higgins 1953). The waves may grow or decay. Using the case for neutral waves in nonstratified uniform shear for reference, it is found, in general, that growing waves are stabilizing while decaying waves are destabilizing to the formation of LCs, although the latter applies only for sufficiently large spanwise spacings and is subject to a globally stable lower bound. Decaying waves in the absence of wind can also be destabilizing to LCs. When the wind is counter to the waves, a scenario stable to CL2 in neutral waves, we find that decaying waves are unstable to LCs. Furthermore, while growing waves are stable to the formation of LCs in the presence of stable stratification, decaying waves are unstable in both aligned and opposed windwave conditions. Unstable stratification on the other hand, is destabilizing to LCs for all temporal waves in both aligned and opposed wind-wave conditions.

4.2. *Comments on $s = 1$ shear*

No wave modulation occurs in $s = 1$ shear, although as Craik (1982) notes and we see from §2.1, n here equals 1/2 and the velocity components are $\epsilon\Delta[q_1, \epsilon^{\frac{1}{2}} q_2, \epsilon^{\frac{1}{2}} q_3]$. This level of shear has not to this point been studied because there is no wave modulation and the form (2.3) is unchanged from the $s = 2$ case, at least for neutral waves. However there are physical differences, because here the orbital velocities of the wave are of the same order as the surface shear flow, which may render the waves weakly rotational. We note too that in contrast to $s = 2$ scaling, $s = 1$ scaling more closely concurs with Weller & Price's (1988) ocean observations, in particular their discovery that peak downwelling velocity can vary from an order of magnitude less than the peak windward current anomaly to be almost equal to it.

5. Recent calculations of CLg

Laboratory LCs form in the wind driven wave experiments of Melville *et al.* (1998) and Veron & Melville (2001). Here the shear level is $s = 0$ and an obvious question is whether the instability mechanism giving rise to them is CLg? We note that to this point there is no concrete evidence that either CL2 or CLg are physically realizable in the free surface wave context, although there is evidence that CLg can occur in the presence of rigid boundaries (see §5.2). More to the point, if CLg is occurring in the above experiments we are to some degree justified in using CLg and CL2 as a metaphor to describe ocean LCs in global change models. But if the instability mechanism is not CLg, the question becomes: what mechanism is it? The following sections outline our efforts in the CLg area.

5.1. *Criterion for instability*

Craik (1982) took the first step toward determining an instability criterion for CLg by studying uniform inviscid shear flow between rigid wave walls in the limit $l^2 \gg \alpha^2$, $\alpha = O(1)$, a parameter range Phillips & Wu (1994) later showed numerically to be the least stable. Phillips & Shen (1996) then considered, analytically, the instability via CLg of a more general class of flows $U = \pm|z|^\kappa$, where κ is a constant, subjected to a range of neutral rotational or irrotational Rayleigh waves whose amplitudes either diminish or diverge with αz. This led to the Craik-Phillips-Shen criterion for instability via CLg: viz that from the reference frame of the wave, and in the direction of increasing mean flow, the gradient of the mean flow (normalized by the mean flow) must exceed the gradient of the wave amplitude (normalized by the wave amplitude).

Note that the gradients of the Stokes drift and shear can be in the same sense in this criterion, but it is not a requirement (*cf* §4.1). Thus while $s = 2$ shear flows with counterflowing neutral waves are stable to CL2, counterflowing $s = 0$ shear flows may be unstable to CLg. This later situation may explain Nepf & Monismith's (1991) laboratory observation of LCs, which form in their $s = 0$ counterflow.

There are two features which lead to different instability criteria: the first is the presence (absence) of wave modulation in $s = 0$ ($s = 2$) flows; the second is that the instability is catalysed by the pseudomomentum in rotational wave fields and by the generalized Stokes drift in irrotational fields. To gain further insight into the role of wave modulation, Phillips & Wu considered the presence and (by switching it off numerically) absence of wave modulation with irrotational waves in $s = 0$ flow. They found that while modulation is substantial and acts to suppress the instability for all l when $\alpha = O(1)$, it diminishes with increasing α and is negligible once $\alpha \gg 1$. Since wave modulation is caused by a sufficiently strong axial velocity perturbation, a lack of modulation means a diminution in the axial perturbation. As is evident from (3.1), this means a change in character of the instability from non-centrifugal to centrifugal. In essence, it means that the CLg instability encompasses, and in appropriate circumstances contracts to, the more restrictive CL2, at least when the wave field is irrotational and $\alpha \gg 1$.

5.1.1. *Is CLg a new instability mechanism?*

To this point we have determined that CLg is an inviscid noncentrifugal instability which, in the presence of neutral waves, initially grows exponentially fast. Moreover Phillips & Shen (1996) have shown that CLg is ubiquitous to a wide range of physically occurring bounded and unbounded flows and, by comparison with the data of Gong, Taylor & Dörnbrack (1996), Phillips *et al.* (1996) determined that CLg is physically realizable. Such findings beg the question whether the CLg instability has a role in the well known secondary instability observed by Orszag & Patera (1983). Both are catalyzed by two-dimensional finite amplitude waves; both require concurrent stretching and tilting of vortex lines that lead to longitudinal vortices which grow exponentially fast on a convective scale; and both are ubiquitous. Indeed, Phillips (1998a) questions whether CLg and Orszag-Patera are identical instabilities viewed from different reference frames, but was able to conclude only that they have much in common.

5.2. *Studies of $O(1)$ shear flows between rigid wavy walls*

To model turbulent boundary layer flow over small amplitude rigid wavy terrain, Phillips *et al.* (1996) used parallel inviscid $O(1)$ shear interacting with $O(\epsilon)$ spanwise independent neutral rotational Rayleigh waves. Of specific interest was the instability of the flow to spanwise-periodic initially exponentially growing longitudinal vortex modes via CLg and whether it is this instability mechanism that gives rise to longitudinal vortices evident in the experiments of Gong *et al.* (1996). In modelling the flow, wave and turbulence length scales were assumed sufficiently disparate to cause minimal interaction. This allowed the primary mean velocity profile to be specified. Two profiles were chosen: a power law and the logarithmic law-of-the-wall. Both primary mean flows were found to be unstable to longitudinal vortex form in the presence of Rayleigh waves whose amplitudes diminish with altitude. Moreover the interaction is most unstable for streamwise wavenumbers $\alpha = O(1)$, the growth rate increasing with increased spanwise wavenumber.

In comparing the results with experiment, Phillips *et al.* first show that spanwise independent waves excited in Gong *et al.*'s experiment depict velocity fluctuations whose amplitudes diminish with altitude in accord with those for appropriate Rayleigh waves. Concordantly, they next show the longitudinal vortices depict transverse velocity components that are weaker by a factor of ϵ than the axial perturbation and are observed to grow at rate consistent with exponential growth. All are key features of CLg, although the observed growth rate is not in accord with the maximal suggested by inviscid instability theory: Rather it appears that the spanwise wavenumber takes a value at which energy is extracted from the mean motion in an optimal volume averaged sense while minimizing energy loss to both viscous dissipation and small scale turbulence. Finally, they conclude that the CLg instability mechanism is physically realizable and that the data of Gong *et al.* represent the first documented observations thereof.

5.3. *Studies of flow fields beneath surface gravity waves*

The inviscid instability of $O(\epsilon)$ two-dimensional periodic flows to spanwise-periodic longitudinal vortex modes in parallel $O(1)$ shear flows beneath surface gravity waves was first considered by Phillips & Wu (1994). However they exclude velocity anomalies at the surface and restrict attention to a parameter range of α and l that renders the surface rigid. Of specific interest is whether the spanwise distortion of the wave field feeds back to enhance or inhibit instability to longitudinal vortex form, and whether the free surface boundary conditions for wave distortion, which are spanwise dependent, act to give rise to a preferred spanwise spacing of the vortices. Recent unpublished calculations, which treat the free surface boundary conditions fully for all α and l, suggest that wave modulation acts to enhance the instability and that a preferred spanwise spacing can occur, although it is a factor of two smaller than that observed in Melville *et al.* 's (1998) experiments.

6. Discussion

Craik-Leibovich instability theory, particularly when derived from GLM, provides an exact description of the formation of longitudinal vortices in a wave-mean flow environment, but we do not as yet know whether this theory is realizable physically or whether in practise it is primarily responsible for LCs, at least in the free surface context. Nevertheless, it continues to be a useful metaphor by which to describe LCs, but with limitations: for example LCs commonly occur concurrently with breaking waves of slope significantly larger than allowed for in the theory; and the aftermath of wave breaking infuses the substrate with energy that likely influences the development and possibly inception of the LCs. Also, while mean shear and surface waves are ubiquitous features of all observations of LCs, LCs are by no means ubiquitous features of all mean shear wave interactions: Indeed, LCs form in Melville *et al.* 's (1998) temporal wind wave experiments, but are not evident in the spatial wind wave experiments of Caulliez, Ricci & Dupont (1998). Why? Is wave modulation sufficently fierce in the latter to annihilate any structure that is formed, or are there ingredients theorists have so far missed? Again, future work should address not just the temporal problem but the spatial problem as well. Finally in analyzing LCs, we should always bear in mind a comment from the great oceanographer Henri Stommel, who mused: "it would be naive to think there is only one mechanism responsible for all observed Langmuir circulations".

Acknowledgement: My thanks to Stephen Belcher, Alex Craik, Tetsu Hara, Julian Hunt, Ken Melville, Michael McIntyre and Miguel Teixeira for their interest, questions and comments. The work was supported by National Science Foundation grants OCE-9696161, 9818092 and 0116921.

REFERENCES

Andrews, D. G. & McIntyre, M. E. 1978 An exact theory of nonlinear waves on a Lagrangian-mean flow. *J. Fluid Mech.* **89**, 609–646.

Caulliez, G., Ricci, N. & Dupont, R. 1998 The generation of the first visible wind waves. *Phys. Fluids* **10**, 757–759.

Craik, A. D. D. 1977 The generation of Langmuir circulations by an instability mechanism. *J. Fluid Mech.* **81**, 209–223.

CRAIK, A. D. D. 1982 Wave-induced longitudinal-vortex instability in shear layers. *J. Fluid Mech.* **125**, 37–52.

CRAIK, A. D. D., & LEIBOVICH, S. 1976 A rational model for Langmuir circulations. *J. Fluid Mech.* **73**, 401–426.

GONG, W., TAYLOR, P. A. & DÖRNBRACK, A. 1996 Turbulent boundary-layer flow over fixed, aerodynamically rough two-dimensional sinusoidal waves. *J. Fluid Mech.* **312**, 1–37.

LANGMUIR, I. 1938 Surface motion of water induced by wind. *Science*, **87**, 119–123.

LEIBOVICH, S. 1977 Convective instability of stably stratified water in the ocean. *J. Fluid Mech.* **82**, 561–585.

LEIBOVICH, S, 1983 The form and dynamics of Langmuir circulations. *Ann. Rev. Fluid Mech.*, **15**, 391–427.

LEIBOVICH, S & PAOLUCCI, S. 1981 The instability of the ocean to Langmuir circulations. *J. Fluid Mech.* **102**, 141–167.

LONGUET-HIGGINS, M. S., 1953 Mass transport in water waves. *Phil. Trans. R. Soc. Lond.* A, **245**, 535–581.

MELVILLE, W. K., SHEAR, R. & VERON, F. 1998 Laboratory measurements of the generation and evolution of Langmuir circulations. *J. Fluid Mech.* **364**, 31–58.

NEPF, H. M. & MONISMITH, S. G. 1991 Experimental study of wave-induced longitudinal vortices. *J. Hydr. Engg.* **117**, 1639–1649.

ORSZAG, S. A. AND PATERA, A. 1983 Secondary instability of wall bounded shear flows. *J. Fluid Mech.* **128**, 347–385.

PHILLIPS, W. R. C. 1998a Finite amplitude rotational waves in viscous shear flows. *Stud. Appl. Math*, **101**, 23–47.

PHILLIPS, W. R. C. 1998b On the nonlinear instability of strong wavy shear to longitudinal vortices. In *Nonlinear Instability, Chaos and Turbulence* (ed Debnath, L. & Riahi, D.N.), Comp. Mech. Publns, UK., **1**, 251–273.

PHILLIPS, W. R. C. 2000 Eulerian space-time correlations in turbulent shear flows. *Phys. Fluids*, **12**, 2056–2064.

PHILLIPS, W. R. C. 2001a On the pseudomomentum and generalized Stokes drift in a spectrum of rotational waves. *J. Fluid Mech.* **430**, 209–220.

PHILLIPS, W. R. C. 2001b On an instability to Langmuir circulations and the role of Prandl and Richardson numbers. *J. Fluid Mech.* **442**, 335–358.

PHILLIPS, W. R. C. 2001c Langmuir circulations beneath growing or decaying surface waves. *J. Fluid Mech.* **469**, 317-342.

PHILLIPS, W. R. C. & WU. Z. 1994 On the instability of wave-catalysed longitudinal vortices in strong shear. *J. Fluid Mech.* **272**, 235–254.

PHILLIPS, W. R. C. & SHEN, Q. 1996 On a family of wave-mean shear interactions and their instability to longitudinal vortex form. *Stud. Appl. Math* **96**, 143–161.

PHILLIPS, W. R. C., WU, Z. & JAHNKE, C. C. 1999 Longitudinal vortices in wavy boundary layers. In *Wind-Over-Wave Couplings: Perspectives and Prospects*, (ed Sajjadi, S.G., Thomas, N.H. & Hunt, J.C.R.), Oxford Univ Press, pp 41-47.

PHILLIPS, W. R. C., WU. Z. & LUMLEY, J. L. 1996 On the formation of longitudinal vortices in a turbulent boundary layer over wavy terrain. *J. Fluid Mech.* **326**, 321–341.

SMITH, J. A. 1992 Observed growth of Langmuir circulation. *J. Geophys. Res.* **97**, 5651–5664.

VERON, F. & MELVILLE, W. K. 2001 Experiments on the stability and transition of wind-driven water surfaces. *J. Fluid Mech.* **446**, 25–65

WELLER, R. A. & PRICE, J. F. 1988 Langmuir circulation within the oceanic mixed layer. *Deep-Sea Res.* **35**, 711–747.

Dispersive Dynamics of Waves in Euler Systems

K. Kirchgässner

Mathematisches Institut A, Universität Stuttgart, Germany

Abstract

This contribution analyses the time evolution of a gravity-driven potential flow with a free upper surface. A new method – based on the Eulerian formulation – is developed using the complex dispersion relation, the Paley-Wiener theorem and a new Dirichlet-Neumann procedure. We prove, for small initial datas, global existence in time and stability of the quiescent state for Froude-numbers greater than one. Also a reduction method is indicated and instability is shown to occur when the Froude-number is less than one.

1. Introduction

Time-dependent, gravity-driven waves of an incompressible, irrotational fluid are considered in a layer of finite depth h with a free upper surface. The observer is travelling with a finite speed c relative to the quiescent state. A new method is described for solving the initial-value problem of the $2d$-Euler-system. Existence and uniqueness is shown for small initial datas globally in time when the Froude-number $c/(gh)^{1/2}$ is greater than one; moreover stability of the quiescent state follows and a reduction scheme as well. Several possible applications are indicated. The main features of the method are: an analysis of the complex dispersion relation, a construction of a nonlinear Dirichlet-Neumann (DN)-operator, and the use of the Paley-Wiener-theorem.

The method is a generalisation of a theory to construct travelling-wave solutions of systems as the one described here (Kirchgässner (1988), Iooss & Kirchgässner (1992)) by using a version parameterised by the complex Laplace-variable s.

Due to the page-restriction of this note, it will not be possible to be generous with explanations. Instead, the procedures and results are clearly stated, but proofs are replaced by examples and hints; thus, providing the reader with enough information to do the rigour with vigour.

The existing literature concerning the given problem is overseeable. Let me mention two more recent contributions, one by Craig (1985) and one by Wu (1997), where the older literature is quoted. Both papers use Lagrange-formulation. In Craig (1985) the longtime existence is shown in the long wave-limit for small initial data, whereas in Wu (1997) global existence for nonself-intersecting surfaces is proved locally in time for layers of infinite depth.

2. Basic equations

In a frame where depth, speed and time are measured relative to h, c and h/c, the fluid domain, written in Cartesian coordinates (x, y) is given by $\mathbb{D}_{\text{free}} := \{(x, y)/x \in \mathbb{R}, 0 \le y \le 1 + z(t, x)\}$, where $y = 1 + z(t, x)$ describes the free surface and $(t, x, y) \in [0, \infty] \times \mathbb{D}_{\text{free}}$. We flatten the free surface via the transform $y \mapsto y/(1 + z(t, x))$, $z > -1$ being a permanent condition. Then we obtain the following system in $\mathbb{R}^+ \times \mathbb{R} \times [0, 1] =: \Omega$ (Whitham (1974), Chapter 13):

$$\begin{aligned} D\underline{u}_+ + \underline{u}_x - A(\partial_y; \mu^2)\underline{u} &= \underline{f}(\underline{u}) \\ U_y|_{y=0} = 0 \quad , \quad U|_{y=1} &= u_1 \\ D\underline{u}|_{t=0} &= D\underline{u}^0 \quad . \end{aligned} \tag{2.1}$$

Here, $\mu^2 = gh/c^2$, $\underline{u} = (u_0, u_1, U, U_1)^T$, $u_j = u_j(t, x)$, $U_j = U_j(t, x, y)$, $j = 0, 1$, $U_0 = U$:

$$\begin{aligned} D\underline{u} &= (u_0, u_1, 0, 0)^T \\ A\underline{u} &= (U_y|_{y=1}, -\mu^2 u_0, U_1, -U_{yy})^T \\ \underline{f} &= (f_0, f_1, F, F_1)^T \\ f_0 &= -u_{0x}u_{1x} - \frac{u_0 - u_{0x}^2}{1+u_0} U_y^1 \\ f_1 &= \frac{1}{1+u_0} U_y^1 (U_y^1 + f_0) - \frac{1}{2(1+u_0)^2} (U_y^1)^2 - \frac{1}{2}\left(u_{1x} - \frac{u_{0x}}{1+u_0} U_y^1\right)^2 \\ F &\equiv 0 \\ F_1 &= a_{01}U_y + a_{11}U_{xy} + a_{02}U_{yy} \end{aligned} \tag{2.2}$$

where

$$a_{01} = -y\left(\frac{u_{0x}}{(1+u_0)^2}\right)_x \quad , \quad a_{11} = -2y\frac{u_{0x}}{1+u_0} \quad ,$$

$$a_{02} = -y^2\left(\frac{u_{0x}}{1+u_0}\right)^2 - \frac{1}{(1+u_0)^2} + 1 \quad .$$

It is easy to see that (2.1) is *reversible* in the following sense: if $\underline{u}(t, x, y)$ solves (2.1), and if $D\underline{u}^0 = (u_0^0, -u_1^0, 0, 0)^T(-x, y)$, then $S\underline{u}(-t, -x, y)$ solves the system (2.1) with $D\underline{u}^0$ as initial condition, when $S\underline{u} = (u_0, -u_1, -U, U_1)^T$ holds. Observe that for $t = 0$, only u_0 and u_1 can be chosen. This reversibility explains the symmetries in the dispersion relation discussed in the following section.

3. The dispersion relation

Consider the linearised version of (2.1), i.e. $\underline{f} = \underline{0}$, and replace ∂_t by s (Laplace-variable, and x by σ (Fourier-variable resp. eigenvalue of A), then solvability yields the dispersion relation

$$(s + \sigma)^2 = \mu^2 \sigma \tan \sigma \quad , \quad (s, \sigma) \in \mathbb{C}^2 \quad . \tag{3.1}$$

For given $s \in \mathbb{C}$, we denote by $DR^{-1}(s)$ (pre-image of s), the set of all σ satisfying (3.1). The structure of the pre-image of $i\beta = s$ resp. $-\alpha + i\beta = s$, $\alpha > 0$ fixed, $\beta \in \mathbb{R}$, in the complex σ-plane will be essential for our analysis.

Relation (3.1) can be rewritten as

$$s = -\sigma \pm \mu\sigma d(\sigma) \quad , \quad d(\sigma) = \left(\frac{\tan\sigma}{\sigma}\right)^{1/2} \tag{3.2}$$

d has branch-points of order 2 for each $\sigma = k\pi$, $k \in \mathbb{Z}/\{0\}$, but is holomorphic near $\sigma = 0$. For $s = 0$, (3.1) reduces to $\tan\sigma/\sigma = \mu^{-2}$ which holds, for $\mu < 1$, exactly for a real sequence $(\pm p_0, \pm p_1, \ldots)$, $0 < p_0(\mu) < p_1(\mu) < \ldots$. At $\sigma = p_0(1) = 0$, a solitary travelling wave of elevation branches to the left, and Stokes' waves to the right. For $\mu > 1$, $\sigma_0 = \pm i q_0(\mu)$ holds, and the other $\sigma_j = p_j$ remain real (c.f. Kirchgässner (1988)).

When $s = i\beta$ and β becomes nonzero, a continuum of solutions of (3.1) emanates at each p_j: the "Laplace-haloes". To show that $DR^{-1}(i\beta)$ consists solely of the Laplace-haloes and a pair of lines emanating from 0 for $\beta = 0$, one can use Poincaré-Bendixon-theory in any simply-connected domain of the Riemann-surface to (3.1). To do this, it is useful to know that, whenever $\sigma(s)$ is a simple root of (3.1), the following differential equation holds:

$$\frac{d\sigma}{ds} = \frac{2d(\sigma)}{\mu(1+d^2+\sigma^2 d^4) - 2d} \quad . \tag{3.3}$$

Set $s = \alpha + i\beta$, $\sigma = p + iq$, then (3.1) reads explicitly

$$\begin{aligned} (\alpha+p)^2 - (\beta+q)^2 &= \frac{\mu^2}{N_0}(p \sin p \cos p - q \operatorname{sh} q \operatorname{ch} q) \\ 2(\alpha+p)(\beta+q) &= \frac{\mu^2}{N_0}(q \sin p \cos p + p \operatorname{sh} q \operatorname{ch} q) \end{aligned} \tag{3.4}$$

with $N_0 = \operatorname{ch}^2 q - \sin^2 p$. To reach the results of Proposition 1, it is useful to study the asymptotic behaviour of q and p as β tends to ∞. In the following, the suffix 'c' resp. 'h' indicate the central resp. hyperbolic part of DR^{-1}, i.e. $DR^{-1}(\alpha+i\beta)_c = DR^{-1} \cap i\mathbb{R}$ resp. its complement $DR^{-1}(\alpha+i\beta)_h$ in $\mathbb{C}$.

PROPOSITION 1.

I Consider the case $\mu < 1$.

(i) Then the central part $DR_c^{-1}(i\beta)$ consists of two straight lines $\sigma_0^{\pm}(s)$, given by the equations

$$\beta = -q \pm \mu(q \operatorname{th} q)^{1/2} \quad , \quad p = 0 \tag{3.5}$$

$\sigma_0^{\pm}(s)$ are holomorphic in s.

(ii) The hyperbolic part $DR_h^{-1}(i\beta)$ consists of countably many components C_j, $j \in \mathbb{Z}$, which are compact and satisfy

$$\sigma_j(0) = p_j \quad , \quad \sigma_j(\pm\infty) = \frac{\pi}{2} \bmod \pi$$

where $\tan p_j/p_j = \mu^{-2}$ holds.

(iii) There exist positive constants c and ρ_0, such that

$$DR_h^{-1}(i\beta) \subset \{|p| > c\,\mathrm{ch}\, q\} \cap \{|p| \geq \rho_0\} \quad .$$

(iv) $DR^{-1}(-\alpha+i\beta)$, $\alpha > 0$ and small, consists of two components $\sigma_\alpha^\pm(s)$ satisfying $\sigma_\alpha^\pm(s) \cap i\mathbb{R} = \emptyset$ and

$$\sigma_\alpha^\pm(-\alpha \pm i\infty) = \alpha \quad , \quad \sigma_\alpha^\pm \notin i\mathbb{R} \quad ,$$

$\sigma_\alpha^\pm(s)$ are holomorphic. Moreover, $DR^{-1}(-\alpha+i\beta)$ without $\sigma_\alpha^\pm(s)$, has properties analogous to I(ii) and I(iii).

II In the case $\mu > 1$, the central part $DR_c^{-1}(i\beta)$ is still of the form (3.5), $\sigma_0^\pm(s) = iq$. However, only $\sigma_0^-(s)$ is holomorphic in s; $\sigma^+(s)$ has two branch points in $\pm i\beta^*(\mu)$ corresponding to $\pm iq^*(\beta^*)$. The statements in I for the hyperbolic part are still valid; only $\pm p_0(\mu)$ has moved to $\pm iq^*(\beta^*)$.

4. A NONLINEAR DN-OPERATOR

First we define the Hilbert-spaces over Ω used in this analysis. Let $l, m, n \in \mathbb{N}_0$ and ∂_t etc. denote weak derivatives.

$$H^{l,m,n} = \{U : \Omega \to \mathbb{C} / \text{ measurable} \quad ,$$
$$\sum_{\substack{0 \leq \lambda \leq l \\ 0 \leq \mu \leq m \\ 0 \leq \nu \leq n}} \int_\Omega |\partial_t^\lambda \partial_x^\mu \partial_y^\nu U|^2 d\omega =: \|U\|^2_{H^{l,m,n}} < \infty\} \quad .$$

When Laplace resp. Fourier-transform is used, the corresponding Nicholski-spaces are denoted by $\widetilde{H}^{l,m,n}$ resp. $\widehat{H}^{l,m,n}$. The spaces $H^{l,m}$ are subspaces of $H^{l,m,n}$, when y is treated as a parameter. Moreover $\|u\|_\infty := \sup_\Omega \{u(t,x,y)\}$.

For given $u_0, u_1 \in H^{0,4} \cap H^{1,3} =: X$ the following boundary-value problem will be solved

$$\begin{aligned} U_{xx} + U_{yy} &= a_{01} U_y + a_{11} U_{xy} + a_{02} U_{yy} \\ U_y|_{y=0} = 0 \quad &, \quad U|_{y=1} = u_1 \end{aligned} \tag{4.1}$$

the a_{ij} being dependent of u_0, u_1 as given in (2.2); u_0, u_1 are assumed to be small in X. Set $\alpha_{ij} = a_{ij}/1{-}a_{02}$ and $U = u_1 + V$, then

$$\begin{aligned} V_{xx} + V_{yy} &= -u_{1xx} - \alpha_{02}(u_{1xx}+V_{xx}) + \alpha_{11} V_{xy} + \alpha_{01} V_y \\ V_y|_{y=0} = 0 \quad &, \quad V|_{y=1} = 0 \quad . \end{aligned} \tag{4.2}$$

Take Fourier-transform '$\widehat{\ }$' or '$\mathcal{F}$' in x and set $\eta = ky$, for positive k, $\widehat{V}(t,k,y) =: \widehat{W}_0$, $\underline{\widehat{W}} = (\widehat{W}_0, \widehat{W}_1)$, $B = \left(\begin{smallmatrix} 0 & 1 \\ 1 & 0 \end{smallmatrix}\right)$, then (4.2) yields

$$\begin{aligned} \underline{\widehat{W}}'(\eta) &= B\underline{\widehat{W}} + \begin{pmatrix} 0 \\ -\widehat{u}_1 + \frac{1}{k^2}\widehat{F}_1 \end{pmatrix} \\ \widehat{W}_0(k) &= 0 \quad , \quad \widehat{W}_1(0) = 0 \end{aligned} \tag{4.3}$$

where

$$\widehat{F}_1 = -\mathcal{F}(\alpha_{02}(u_{1xx} + W_{0,xx})) + \mathcal{F}(\alpha_{11} W_{0,x\eta}) + \mathcal{F}(\alpha_{01} W_{0\eta}) \quad .$$

Exploiting the regularity of u_0, u_1 and $\widehat{W}_0 \in \widehat{X}$, we have e.g.

$$\mathcal{F}(\alpha_{02}(u_{1xx}+w_{0,xx})) = \frac{1}{k^2}\mathcal{F}\left([\alpha_{02}(u_{1xx}+W_{0,xx})]_{xx}\right) \quad .$$

Where the term in [ ]-bracket can be bounded in the $L^2(\mathbb{R})$-norm (use boundedness of u_{0xx} and $W_{0,xx}$ by the $H^{1,3}$-norm). Similarly, one can proceed for the xt-derivative (one partial integration is needed for $(u_{0x}^2)_t W_{0xxx}$ e.g.).

In this way one concludes that (4.3) is well-posed for $\underline{W} : [0,k] \to H^{0,4} \cap H^{1,3}$. Its unique solution is found by the representation

$$\begin{aligned} \widehat{W}_+(\eta) &= e^{B_+(\eta-k)}\widehat{W}_+(k) - \int_\eta^k e^{B_+(\eta-\zeta)}\widehat{g}_+(\zeta)dy \\ \widehat{W}_-(\eta) &= e^{B_-\eta}\widehat{W}_-(0) + \int_0^\eta e^{B_-(\eta-\zeta)}\widehat{\underline{g}}_-(\zeta)d\zeta \end{aligned} \tag{4.4}$$

B_+ resp. B_- act in the unstable resp. stable subspace of B and $\widehat{g} := \begin{pmatrix} 0 \\ \hat{u}_1+\frac{1}{k^2}\widehat{F}_1 \end{pmatrix}$. $\underline{\widehat{W}}_+(k)$ and $\underline{\widehat{W}}_-(0)$ can be calculated using the boundary conditions.

The contraction-mapping argument to solve (4.4) is straight-forward for small datas $u_0, u_1 \in H^{0,4} \cap H^{1,3}$ and $W_j : [0,k] \to H^{0,4} \cap H^{1,3}$. The dependence of $\underline{W}$ on u_0, u_1 is analytic. The consequence for $U(y)$ is as follows: $U \in \bigcap_{j=0}^2 H^{0,4-j,j}$, since $\widehat{U}_y = kW_\eta$.

A similar analysis works for $k < 0$. And the case $[-k_0, k_0]$, $k_0 > 0$ fixed, can be done without stretching y to η.

5. Inversion of the hyperbolic part of DISP

Consider (2.1) as a fixed point-problem of the map

$$\begin{aligned} \text{DISP} : \underline{u} &\longmapsto \underline{f}(\underline{u}) =: \underline{f} \overset{\mathcal{L}}{\longmapsto} \underline{\widetilde{f}} \longmapsto \text{disp}(\underline{\widetilde{f}}+D\underline{u}^0) := \\ &= (\partial_x - A(s))^{-1}(\underline{\widetilde{f}}+D\underline{u}^0) = \underline{\widetilde{U}} \overset{\mathcal{L}^{-1}}{\longmapsto} \underline{U} =: \text{DISP}\,\underline{u} \end{aligned} \tag{5.1}$$

where $A(s) = A(\partial y; \mu^2) - sD$, $\mathcal{L}$ resp. $\mathcal{L}^{-1}$ denote Laplace-transform in t with variable s. We work in the space $(u_0, u_1, U) \in X^2 \times \mathbb{X}$, where $X = H^{0,4} \times H^{1,3}$, $\mathbb{X} = \bigcap_{j=0}^2 H^{0,4-j,j} \cap \bigcap_{k=0}^1 H^{1,3-k,k}$; note $U_1 = U_x$. One can show, by imitating the arguments of the previous section and the results of DN, that f_0, f_1 belong to $H^{0,3}$. The Paley-Wiener-theorem (Yosida, p. 161 f.) yields that $\widetilde{f}$ belongs to a corresponding Hardy-Lebesgue space which we denote by $\widetilde{H^{03}}$. Roughly speaking, these are functions $s \mapsto \widetilde{f}(s) \in \widetilde{X}^2 \times \widetilde{\mathbb{X}}$, holomorphic for $\operatorname{Re} s > 0$, L_2-extendible to $\operatorname{Re} s = 0$ and L_2-integrable along each $s = \alpha + i\beta$, $0 < \alpha$, fixed, $\beta \in \mathbb{R}$.

The necessary gain of regularity, matching the loss from X to $H^{0,3}$, rests on $(\partial_x - A(s))^{-1} = \text{disp}$. For $\mathcal{L}^{-1}$ we need $s = i\beta$, and therefore we have to discuss the DR (3.1) in this case.

We have $\sum_c A(i\beta)$ given by $\alpha = 0$, $p = 0$ and

$$\beta = -q \pm \mu(q\,\text{th}\,q)^{1/2} \quad , \quad \sigma_\pm(\beta) = iq^\pm \tag{5.2}$$

where $q^{\pm}$ are the resolutions of (5.2) for $\mu < 1$. For $\mu > 1$ the situation is more complicated and shall be illuminated at the end of the next section. However, the analysis of the hyperbolic part is valid with minor corrections also for $\mu > 1$.

We assume now that $\mu < 1$ holds. The eigenfunctions and their adjoints to $\sigma^{\pm}$ are as follows

$$\begin{aligned} \underline{\varphi}_{\pm} &= \left(-\frac{\sigma \sin \sigma}{\sigma + s}, \cos \sigma, \cos \sigma y, \sigma \cos \sigma y\right)^T \\ \underline{\psi}^{\pm} &= \left(\cos \overline{\sigma}, -\frac{1}{\overline{s}+\overline{\sigma}} \overline{\sigma} \sin \overline{\sigma}, \overline{\sigma} \cos \overline{\sigma} y, \cos \overline{\sigma} y\right) \end{aligned} \tag{5.3}$$

with scalar product ($\sigma = \sigma^{\pm}$)

$$[\varphi_{\pm}, \underline{\psi}^{\pm}] = \sigma + \frac{1}{2} \sin(2\sigma) - \frac{2}{\mu^2}(\sigma + s) \cos^2 \sigma \quad .$$

Determining $(\partial_x - A_h(i\beta))^{-1}$ we take Fourier-transform but deform the path of integration to

$$\zeta(k) = \begin{cases} \rho_0(1-\tau) + k(\tau + i), & k \geq \rho_0 \\ \rho_0 + ik, & -\rho_0 \leq k \leq \rho_0 \\ -\rho_0(1-\tau) + k(\tau + i), & k \leq -\rho_0 \end{cases} \tag{5.4}$$

for some $\tau \in (0, 1)$.

We have to solve

$$\begin{aligned} (\zeta + s)\widehat{\widehat{u}}_0 - \widehat{\widehat{U}}^1_y &= \widehat{\widehat{f}}_0 + \widehat{u}^0_0 \\ (\zeta + s)\widehat{\widehat{u}}_1 + \mu^2 \widehat{\widehat{u}}_0 &= \widehat{\widehat{f}}_1 + \widehat{u}^0_1 \\ \widehat{\widehat{U}}_{yy} + \zeta^2 \widehat{\widehat{U}} &= \widehat{\widehat{F}}_1 \\ \widehat{\widehat{U}}_y|_{y=0} = 0 \quad , &\quad \widehat{\widehat{U}}|_{y=1} = \widehat{\widehat{u}}_1 \end{aligned} \tag{5.5}$$

under the conditions $[\widehat{\underline{f}}, \underline{\psi}^{\pm}] = 0$. The latter shows that $(D\underline{u}^0)_h = 0$. The function $\widehat{U}$ is the analytic continuation of the Laplace-transform of U via DN. Its linear part reads $\widehat{\widehat{U}} = \widehat{\widehat{u}}_1 \cos \zeta y / \cos \zeta$. Its nonlinear part is added to the right side of (5.5). Then we have

$$\begin{aligned} \widehat{\widehat{u}}_0 &= -\frac{\zeta}{\zeta + s} \tan \zeta \cdot \widehat{\widehat{u}}_1 + \frac{1}{\zeta + s} \widehat{\widehat{f}}_0 \\ \widehat{\widehat{u}}_1 \left(1 - \frac{\mu^2 \zeta \tan \zeta}{(\zeta + s)^2}\right) &= -\frac{\mu^2 \widehat{\widehat{f}}_0}{(\zeta + s)^2} + \frac{\widehat{\widehat{f}}_1}{\zeta + s} \quad . \end{aligned} \tag{5.6}$$

Since $s = i\beta$, $\zeta \approx ik$ for large k, one concludes that both, u_0 and u_1, alternatively gain a t resp. x-derivative. Thus, disp_h maps $X^2 \times \mathbb{X}$ into itself.

To prove that $\widehat{U}$ in (5.5) is the analytic continuation of the one in Section 4, we use (5.4), set $\rho_0 = \tau$ and write $\zeta(k; \tau)$. For $\tau \to 0+$ we obtain (4.1) by which U is uniquely determined. Write $\widehat{U} = \widehat{u}_1 + \widehat{V}$ in (5.5) and $\eta = ky$ for

$k > 0$; moreover $\underline{V} = (V_r, V_i)$ and $A(\tau) = I - 2\tau J$, where $J = \left(\begin{smallmatrix} 0 & -1 \\ 1 & 0 \end{smallmatrix}\right)$. Then, $\widehat{U}_{yy} + \zeta^2 \widehat{U} = \widehat{F}_1$ yields – up to terms of the order $O(1/k)$ –

$$\begin{aligned} \underline{V}_\eta &= \underline{W} \\ \underline{W}_\eta &= A(\tau)\underline{V} - \frac{1}{k^2}\underline{F}_1 + A(\tau)\underline{u}_1 \\ \underline{W}(0) &= 0 \quad , \quad \underline{V}(k) = 0 \quad . \end{aligned} \tag{5.7}$$

The matrix $\mathcal{A}(\tau) = \left(\begin{smallmatrix} 0 & I \\ A(\tau) & 0 \end{smallmatrix}\right)$ has two eigenvalues in $\mathbb{C}^+$ and 2 eigenvalues in $\mathbb{C}^-$. Then, (5.7) can be solved in a similar spirit as we did in Section 4.

$$\underline{W}_+(\eta) = e^{\mathcal{A}_+(\eta-k)}\underline{W}_+(k) - \int_\eta^k e^{\mathcal{A}_+(\eta-y)}G_+(y)dy$$

$$\underline{W}_-(\eta) = e^{\mathcal{A}_-\eta}\underline{W}_-(0) + \int_\sigma^\eta e^{\mathcal{A}_-(\eta-y)}G_-(y)dy$$

where $G = (0, -\frac{1}{k^2}F_1 + A(\tau)u_1)^T$; $\underline{W}_+(k)$ and $\underline{W}_-(0)$ are uniquely determined by the boundary conditions in (5.7). We conclude

PROPOSITION 2. *Consider the mapping* $\mathrm{disp}_h = (\partial_x - A(i\beta))^{-1}$ *restricted to the hyperbolic part of* $A(i\beta)$, $\beta \in \mathbb{R}$. *It acts on* $\widetilde{f}_h + (D\underline{u}^0)_h$. *We have, for* $\underline{u} \in X^2 \times \mathbb{X}$ *small,*

(i) $(D\underline{u}^0)_h = 0$

(ii) $\underline{f}_h : X^2 \times \mathbb{X} \longrightarrow \widetilde{X}_1^2 \times \widetilde{X}_1$

(iii) $\mathrm{disp}_h : \widetilde{X}_1^2 \times \widetilde{\mathbb{X}}_1 \longrightarrow \widetilde{X}^2 \times \widetilde{\mathbb{X}}$

and $\mathrm{disp}_h \circ \widetilde{f}_h$ *maps smoothly* $X^2 \times \mathbb{X}$ *into* $\widetilde{X}^2 \times \widetilde{\mathbb{X}}$.

Here use the following notations:

$$X = H^{0,4} \cap H^{1,3} \quad , \quad X_1 = H^{0,3}$$

$$\mathbb{X} = \bigcap_{j=0}^{2} H^{0,4-j,j} \cap H^{1,3-j,j}$$

$$\mathbb{X}_1 = H^{0,2,0} \cap H^{0,1,1} \cap H^{1,1,0} \cap H^{1,0,1}$$

6. THE CENTRAL PART OF disp AND RESULTS

A satisfactory analysis for the dynamics of solutions of (2.1) requires, for the moment, restriction to the parameter region $\mu < 1$, i.e., Froude-number greater than one. Existence of the initial-value-problem (2.1) can be shown globally in time when the initial datas are small. Liapunov stability of the quiescent state will follow. In fact, one can determine the leading amplitude-term of the solution explicitly by using the central part of disp alone.

For $\mu < 1$ – an assumption valid for this whole section – implies that $\sum_c A(i\beta) = \{\sigma^\pm = iq^\pm/\beta = -q \pm \mu(q \,\mathrm{th}\, q)^{1/2}\}$. The explicit form of the corresponding eigenvectors $\varphi_\pm$ and adjoint eigenvectors $\underline{\psi}^\pm$ were given in Section 5.

Observe that $\sigma^{\pm}(\beta)$ are simple roots of the DR, and that

$$q^{\pm}(\beta) = -\beta + \mu r_{\pm}(\beta) \tag{6.1}$$

holds with $r_{\pm}(\beta) = \pm(\beta \operatorname{th} \beta)^{1/2}$ in the highest order.

The central part of disp is determined by

$$\underline{\widetilde{u}}_{c,x} - A(i\beta)\underline{\widetilde{u}}_c = \underline{\widetilde{f}}_c + D\underline{u}^0 \quad . \tag{6.2}$$

Write $\widetilde{u}_c = \widetilde{u}_{c1} + \widetilde{u}_{c2}$, where

$$\widetilde{u}_{c2,x} - A(i\beta)\widetilde{u}_{c2} = D\underline{u}^0 \tag{6.3}$$

and $\widetilde{u}_{c1}$ solves the complementing equation with $\underline{\widetilde{f}}_c$ on the right side. Taking Fourier-transform it follows, since $\widetilde{f}_c \in \widetilde{H^{03}}$, $\underline{u}_{c1} \in X = H^{1,3} \cap H^{0,4}$. To solve (6.3) write $\widetilde{u}_{c2}$ and $D\underline{u}^0$ in the $\underline{\varphi}_{\pm}$-system with coordinates $a^{\pm}(s,x)$, $d_0^{\pm}(s,x)$. Then, (6.3) yields

$$\widehat{a}^{\pm} = \frac{\widehat{d_0^{\pm}}}{i(k-q^{\pm})} \quad . \tag{6.4}$$

$D\underline{u}^0 \in H^4(\mathbb{R})$ implies $d_0^{\pm}(i\beta,x) = \frac{[D\underline{u}^0,\underline{\psi}^{\pm}]}{[\underline{\varphi}_{\pm},\underline{\psi}^{\pm}]}\underline{\varphi}_{\pm}$ bounded in β and H^4 in x. Solving (6.3) via Fourier-transform yields

$$a^{\pm}(i\beta,x) = \frac{1}{2\pi i}\int_{\mathbb{R}} \frac{e^{ikx}d^{\pm}(k)}{k-q}dk = d^{\pm}(q) + \frac{1}{2\pi i}\int_{\mathbb{R}} e^{ikx}d_1(k,q)dk$$

where $d_1(k,q) = \frac{d^{\pm}(k+q)-d(q)}{k}$. Finally we obtain

$$\underline{U}^{\pm}(t-x,x,y) = \frac{1}{2\pi}\int_{\mathbb{R}} e^{i\beta(t-x)}\Delta(x,\beta)\underline{\varphi}_{\pm}(i\beta,y)d\beta \tag{6.5}$$

where

$$\Delta(x,\beta) = e^{-ir_{\pm}(\beta)x}\left(d^{\pm}(q)+\frac{1}{2\pi i}\int_{\mathbb{R}} e^{ikx}d_1(k,q)dk\right) \quad .$$

Hence

$$\underline{u}(t-x,x,y) = \mathcal{L}^{-1}\widetilde{u}_{c,2} = \underline{U}^{+}(t-x,x,y) + \underline{U}^{-}(t-x,x,y)$$

is the solution of (2.1) in lowest order of the amplitude.

We construct the image of $\underline{u} \in X^2 \times \mathbb{X}$ via DISP by solving

$$\widetilde{\underline{U}}_c = \operatorname{disp}_c(\underline{\widetilde{f}}_c + D\underline{u}^0)$$
$$\widetilde{U}_h = \operatorname{disp}_h \underline{\widetilde{f}}_h \quad .$$

All terms on the right side are at least quadratic in $\underline{u}$, except $\operatorname{disp}_c D\underline{u}_0$.

We have seen that DISP, acting in physical space via $\mathcal{L}$ and $\mathcal{L}^{-1}$, defines a smooth map in $X^2 \times \mathbb{X}$. Thus, a straightforward contraction argument for small $\underline{u}$ yields the existence, and the fact that the sum of $\underline{U}^{\pm}$ in (6.5) give the lowest order term.

PROPOSITION 3. *Given $\mu \in (0,1)$. For sufficiently small positive ρ_0, ρ_1, there exists, for every $D\underline{u}^0 \in B_{\rho_0}(0) \subset H^4(\mathbb{R})$ a unique $\underline{u} \in B_{\rho_1}(0) \subset X^2 \times \mathbb{X}$ which* *solves (2.1). In lowest order of amplitude, this solution is the superposition of* $U^{\pm}$ *given in (6.5).*

Some closing remarks: We have sketched a route to analyse an Euler-system on the basis of its dispersion properties. The current outcome is an existence result, a way to reduction and explicit formulae of leading terms of the dynamics. However, the expectation seems to be justified, that a number of long standing open problems may be solvable in this frame. The most accessible appears to be the stability of the solitary waves bifurcating from the quiescent state at $\mu = 1$ and existing for $\mu < 1$. Here, the reduction will be essential. Another area should be the analysis of external forcing under the concept of causality. Ship-distortions or the effect of surface-winds fall under this category.

Let us finally look at the instability of the quiescent state for $\mu > 1$ (Froude-number less that 1). It is intimately connected to the instability of the Stokes waves, which bifuracate at $\mu = 1$ and exist for $\mu > 1$. But let us concentrate on the dynamics at the state of rest for $\mu > 1$. The set $\mathrm{DR}^{-1}(i\beta)$ consists of a hyperbolic part looking qualitatively as for $\mu < 1$ except that $\pm p_0(\mu)$ has moved up the imaginary axis. Therefore, the center-part consists now of 4 Laplace haloes, namely $\sigma^-(\beta) = i(-\beta - \mu(\beta \,\mathrm{th}\, \beta)^{1/2})$ in highest order, and three branches $\sigma^+_+, \sigma^+_0, \sigma^+_-$ replacing σ^+. There are double roots of DR at $\sigma = \pm iq^*$, $\beta = \mp\beta^*$, at which these branches leave or enter the imaginary axis. All of them do not stay on $i\mathbb{R}$ for all $\beta \in \mathbb{R}$. Whenever its real parts are nonzero they have to approach a point $\frac{\pi}{2} \bmod \pi$ on the real axis. σ^+_+ and σ^+_- have one, σ^+_c has two of those limit-points.

Let us consider, for the case $\widehat{d}(k) = 1/\mathrm{ch}\, k$ the effect of the limit points. Assume $\frac{\pi}{2} \bmod \pi = \frac{\pi}{2}$. We have

$$\widetilde{d}(\beta, x) = \frac{1}{2\pi} \int_{\mathbb{R}} \frac{e^{ikx}}{\mathrm{ch}\, k} dk$$

and

$$\widetilde{a}(\beta, x) = \frac{1}{2\pi} \int \frac{e^{ikx}}{\mathrm{ch}\, k} \cdot \frac{1}{ik - \sigma(\beta)} dk \quad .$$

Observe that, up to a smooth, rapidly decaying function, we have for β close to $\pm\infty$ and $x < 0$

$$\widetilde{a}(\beta, x) = e^{(\frac{\pi}{2} + z)x} / \cos\left(\frac{\pi}{2} + z\right)$$

where $z = 1/\beta^2$. Therefore $a(\beta, x) \sim \beta^2$ for large β and thus, $a(t, x)$ will be a distribution allowing polynomial growth in t.

REFERENCES

CRAIG, W. 1985 An existence theory for water waves and the Boussinesq and Korteweg-de Vries scaling limits. *Comm. Partial Differential Equations* **10**, no. 8, 787–1003.

IOOSS, G. & KIRCHGÄSSNER, K. 1992 Water waves for small surface tension: an approach via normal form. *Proc. Roy. Soc. Edinburgh* Sect. **A 122**, no. 3-4, 267–299.

KIRCHGÄSSNER, K. 1988 Nonlinearly resonant surface waves and homoclinic bifurcation. *Advances in applied mechanics* **26**, 135–181.

WHITHAM, G.B. 1974 Linear and nonlinear waves. *Pure and Applied Mathematics.* Wiley-Interscience, John Wiley & Sons, New York-London-Sydney.

WU, S. 1997 Well-posedness in Sobolev spaces of the full water wave problem in 2-D. *Invent. Math.* **130**, no. 1, 39–72.

YOSIDA, K. 1968 Functional analysis. Second edition. Die Grundlehren der mathematischen Wissenschaften, Band 123, Springer-Verlag New York Inc., New York 1968.

Benjamin Memorial Lecture: Stability of Solitary Waves: Geometry, Symplecticity and Three-Dimensionality

T.J. Bridges

Department of Mathematics and Statistics, University of Surrey, Guildford, Surrey GU2 7XH, UK

ABSTRACT

This article contains an overview of the Benjamin Memorial Lecture given at the Wind-over-Waves conference in September 2001 on the stability of solitary waves. The starting point for the talk was the 1972 paper of Benjamin which introduced the "energy-momentum method" for the stability of solitary waves. The energy-momentum method uses Hamiltonian structure in a central way. In the talk, and in this paper, a range of recent theoretical developments on the stability of solitary waves using Hamiltonian structure are discussed. The backbone of the new developments is the multi-symplectic structures framework which gives a refined structure for Hamiltonian PDEs. New developments discussed include multi-symplectification of the energy momentum method, the symplectic Evans function for studying the spectral problem, and a geometric theory for three-dimensionalization of solitary waves through transverse instability. The motivation and examples discussed throughout are based on model PDEs for water waves and the full water-wave problem.

1. INTRODUCTION

The solitary wave, modelled by a solution of the KdV equation, has been a central paradigm of the localized shallow water wave for well over a century. In the 1960s it was discovered that the KdV equation is Hamiltonian and integrable. It was the Hamiltonian structure which was the starting point for the proof of Benjamin (1972) that the KdV solitary wave is nonlinearly stable. This proof introduced the *energy-momentum method* into the stability analysis of solitary waves. The KdV solitary wave is not a minimum of the energy, but it is a minimum when restricted to the constraint of constant momentum. With this observation as a starting point, there are still significant technicalities to showing that a combination of the energy and momentum serves as a Lyapunov functional in a suitable Hilbert space, cf. Benjamin (1972), Bona (1975).

The energy-momentum method has been considerably developed in the past thirty years. The list of references is too numerous to include here. A representative sample of important work in this direction includes Grillakis, Shatah & Strauss (1987), Grillakis, Shatah & Strauss (1990), Bona, Souganidis & Strauss (1987), Maddocks & Sachs (1993), Pelinovsky (2002).

The starting point for the energy-momentum method is that solitary waves can be characterized as critical points of the energy on level sets of the momentum. The Lagrange necessary condition associated with this variational principle is the governing equation for solitary waves. A key hypothesis in the energy-momentum method is that the second variation should have a finite number of negative eigenvalues. Although this hypothesis is satisfied in many cases, there is however a wide range of examples where it is not satisfied. In this case, the energy-momentum method is no longer conclusive.

There are two new directions in the case where the second variation is indefinite: study the spectral problem associated with the linearization about the solitary wave directly, and use multi-symplectic structure, a more refined Hamiltonian structure. Effectively, multi-symplecticity splits the usual Hamiltonian function into more than one part, providing more refinement of the structure. With this structure, stability results can be obtained which use the multi-symplectification of the energy-momentum method, but circumvent the second variation. This idea is developed in Sections 3 and 4.

The stability question considered by Benjamin was *longitudinal stability*; stability with respect to perturbations travelling in the same direction as the basic solitary wave. However, it is possible that a longitudinally stable wave can become unstable to transverse perturbations, which travel in a direction oblique to the basic solitary wave. This type of instability is interesting because it leads to a higher dimensional pattern, two dimensional for model equations and three-dimensional in the case of water waves. Here it is shown in Sections 5 and 6 that multi-symplecticity leads to a new geometric condition for transverse instability, and this result can be applied to the full water wave problem with intriguing consequences.

Curiously, numerical methods for computing the stability of solitary waves have been slow to develop. Here we comment on a new numerical framework based on exterior algebra, which is a generalization of the compound matrix method which has used in hydrodynamic stability to solve the Orr-Sommerfeld equation. This framework does not require Hamiltonian or symplectic structure, so it is only briefly discussed here, with pointers to the literature. Even slower in development has been numerical methods to compute the stability of solitary waves for the full water wave problem, and we comment on some recent work in this direction.

2. The energy-momentum method

The starting point of Benjamin's proof of stability of the KdV solitary wave is a constrained variational principle which uses the Hamiltonian structure of the KdV equation in a central way. The KdV equation, in the standard form $u_t + u_x + uu_x + u_{xxx} = 0$, can be formulated as a Hamiltonian PDE by taking

$$u_t = \mathbf{J}\nabla H(u)\,, \quad \mathbf{J} = -\frac{\partial}{\partial x}\,, \quad H(u) = \int_{\mathbf{R}} \left(\tfrac{1}{2}u^2 + \tfrac{1}{6}u^3 - \tfrac{1}{2}u_x^2\right)\, dx\,.$$

A solitary wave travelling at speed c satisfies $-(c-1)u_x + uu_x + u_{xxx} = 0$ or

$$\nabla H(u) = c\nabla I(u), \quad \text{where} \quad I(u) = \int_{\mathbf{R}} \tfrac{1}{2}u^2\, dx\,. \tag{2.1}$$

The equation (2.1) can be interpreted as the Lagrange necessary condition for a constrained variational principle: the KdV solitary wave corresponds to a critical point of the energy, H, restricted to level sets of the momentum, I, with c a Lagrange multiplier.

The constrained variational principle turns out to be important, because the KdV solitary wave is *not* a minimum of the energy H. It is however a minimum of H with I fixed. The energy and momentum can be combined to make a suitable Lyapunov functional leading to a proof of nonlinear stability. There are a number of substantial technical steps to make this argument rigorous, cf. Benjamin (1972), Bona (1975).

This approach to studying the stability of solitary waves is now known as the energy-momentum method. It is natural for PDEs with solitary wave solutions which are also Hamiltonian. If the PDE is translation invariant in time, the Hamiltonian (energy) will be conserved, and if it is translation invariant in space the momentum will be conserved. One can develop this idea in general, and it applied to a wide range of examples in fluid mechanics (cf. Benjamin, 1983). Formally, solitary wave states can then be characterized as critical points of the energy restricted to level sets of the momentum. Since 1972, this approach has been widely used to study the stability of solitary waves, and the list of references is too long to cite here. A cross-section of references can be found in Grillakis, Shatah & Strauss (1990) and Bona, Souganidis & Strauss (1987) and recent references in Pelinovsky (2002).

The energy-momentum argument has been used almost exclusively for model equations, either scalar-valued equations like KdV or the nonlinear Schrödinger equation and their generalizations, and some vector-valued systems. Benjamin hinted in his 1972 paper that it might be possible to use an energy-momentum argument to prove stability of the classical solitary wave state of the full water wave problem, since solitary water waves have an energy-momentum characterization. A proof of this has however remained elusive. In a recent preprint, Mielke (2002) shows that if surface tension is non-zero a proof based on an energy-momentum argument can be constructed to prove stability of a class of solitary waves for the full water wave problem. However, a proof of stability for gravity solitary wave states of the water-wave problem, which are of greatest interest in applications, remains elusive.

A principal hypothesis of the energy-momentum argument is that the second variation $\mathbf{L} = D^2H(\hat{u}) - cD^2I(\hat{u})$ – evaluated at a solitary wave $\hat{u}$ – has at most one negative eigenvalue. For the KdV equation, $\mathbf{L}$ has exactly one negative eigenvalue. In application of the energy-momentum argument it is this hypothesis on the second variation which is most difficult to verify. Indeed, in a wide range of examples this hypothesis is not satisfied, and in some cases the second variation may have a countable number of negative as well as positive eigenvalues. In this case the energy-momentum method will be inconclusive.

3. The spectral problem for the linearization about solitary waves

The set of PDEs which have an energy-momentum characterization of solitary waves is much larger than the set which satisfies the hypothesis on the second variation. An alternative approach to stability analysis is to use the geometry associated with the energy momentum method, but to apply this geometrical information to the spectral problem. For example, linearizing the KdV equation about a solitary wave state, leads to

$$u_t - (c-1)u_x + \hat{u}u_x + u\hat{u}_x + u_{xxx} = 0. \tag{3.1}$$

Let $u = \phi e^{\lambda t}$, then λ is an eigenvalue associated with the spectral problem

$$-\phi_{xxx} + (c-1)\phi_x - \hat{u}\phi_x - \hat{u}_x\phi = \lambda\phi. \tag{3.2}$$

The basic solitary wave is *spectrally stable* if there are no eigenvalues of (3.2) in the right-half plane.

Unfortunately, this spectral problem is very difficult to analyze. The operator is not symmetric, although it is the product of a symmetric and a skew-symmetric operator. On the other hand, the geometry associated with the energy-momentum method is still encoded in the operator.

A dynamical systems formulation for general spectral problems of the form (3.2) was introduced in Evans (1975) and Alexander, Gardner & Jones (1990) and this turns out to be a profitable way to proceed. For the KdV example, let $\mathbf{v} = (\phi, \phi_x, \phi_{xx})$, then the spectral problem (3.2) can be reformulated as $\mathbf{v}_x = \mathbf{A}(x,\lambda)\mathbf{v}$ with $\mathbf{v} \in \mathbb{C}^3$, and the matrix $\mathbf{A}(x,\lambda)$ has the property that it goes to a constant (λ–dependent) matrix in the limit as $x \to \pm\infty$. In general, the Evans function theory applies to λ–dependent linear systems arising in spectral theory of the form

$$\mathbf{v}_x = \mathbf{A}(x,\lambda)\mathbf{v}, \quad \mathbf{v} \in \mathbb{C}^n, \quad \text{with} \quad \lim_{x\to\pm\infty} \mathbf{A}(x,\lambda) = \mathbf{A}_\infty(\lambda). \tag{3.3}$$

The dimension of the subspace of solutions which is bounded as $x \to +\infty$ (respectively as $x \to -\infty$) is determined by the number of eigenvalues of $\mathbf{A}_\infty(\lambda)$ with negative real part (respectively postive real part). Let k be the number of eigenvalues of $\mathbf{A}_\infty(\lambda)$ with negative real part. Then it follows from standard theory of ODEs that there exists a k dimensional subspace of solutions of (3.3) which is bounded as $x \to +\infty$. The Evans function, $E(\lambda)$, is a complex analytic function which is zero precisely when the space of bounded solutions as $x \to +\infty$ intersects the space of bounded solutions as $x \to -\infty$: the zeros of $E(\lambda)$ are precisely the eigenvalues of the spectral problem.

There are a number a technical details to make this theory go through, and details can be found in the references. Here we sketch a geometric approach to analyzing the Evans function.

Pego & Weinstein (1992) first considered the idea of relating the geometry of the energy momentum method to the Evans function. They considered the stability of solitary waves of generalizations of the KdV equation, the Benjamin-Bona-Mahony equation and the regularized Boussinesq equation. For these

PDEs they proved that the Evans function satisfies

$$E(0) = 0\,, \quad E'(0) = 0\,, \quad \text{sign}(E''(0)) = \text{sign}\left(\frac{dI}{dc}\right)\,, \tag{3.4}$$

and $E(\lambda) \to 1$ as $\lambda \to \infty$ along the real axis. In (3.4) I is the momentum of the solitary wave, considered as a function of c. Note that if $\frac{dI}{dc} < 0$ one can conclude immediately that $E(\lambda) = 0$ for some λ with positive real part: the solitary wave in linearly (spectrally) unstable.

This result is interesting because information from the energy-momentum characterization of solitary waves is used, but without any information from the second variation. On the other hand the result of Pego & Weinstein required explicit information about the solitary waves, and quite special hypotheses on the system of PDEs.

The result (3.4) raises the general question: given any Hamiltonian PDE with solitary wave solutions, that have an energy-momentum characterization, how much can we say about the spectral problem using that geometry? The Hamiltonian structure is insufficient to give a general answer to this question. The natural setting for studying the linearized stability problem turns out to be the multi-symplectic structures formulation of Hamiltonian PDEs.

4. Multi-symplectic structures and geometric instability

In a classical Hamiltonian formulation of PDEs, there is a lot of information that is encoded in the Hamiltonian function that can be extracted. For illustration, consider the nonlinear Schrödinger equation $\mathrm{i}A_t + A_{xx} + V'(|A|^2)A = 0$, where $V(\cdot)$ can be taken to be any smooth potential. It has a classical Hamiltonian formulation

$$\mathbf{q}_t = \mathbf{J}_2 \nabla H(\mathbf{q})\,, \quad \mathbf{J}_2 = \begin{pmatrix} 0 & 1 \\ -1 & 0 \end{pmatrix}\,, \quad \mathbf{q} \in \mathbb{R}^2 \tag{4.1}$$

where $A = q_1 + \mathrm{i}\, q_2$ and $H(\mathbf{q}) = \frac{1}{2}\mathbf{q}_x \cdot \mathbf{q}_x - \frac{1}{2}V(\mathbf{q}\cdot\mathbf{q})$.

Further structure is obtained by taking the Legendre transform of H. This will separate H into two parts: an action for space, and a generalized Hamiltonian function which has no space derivatives. Carrying out these steps leads to the multi-symplectic formulation of NLS,

$$\mathbf{M}Z_t + \mathbf{K}Z_t = \nabla S(Z)\,, \quad Z \in \mathbb{R}^4 \tag{4.2}$$

where

$$\mathbf{M} = \begin{bmatrix} \mathbf{J}_2 & \mathbf{0} \\ \mathbf{0} & \mathbf{0} \end{bmatrix}\,, \quad \mathbf{K} = \begin{bmatrix} \mathbf{0} & -\mathbf{I}_2 \\ \mathbf{I}_2 & \mathbf{0} \end{bmatrix}\,, \quad Z = \begin{pmatrix} \mathbf{q} \\ \mathbf{p} \end{pmatrix}\,, \quad \mathbf{p} = \mathbf{q}_x\,, \tag{4.3}$$

and $S(Z) = \frac{1}{2}\mathbf{p}\cdot\mathbf{p} + \frac{1}{2}V(\mathbf{q}\cdot\mathbf{q})$.

It is the general abstract form of the equation (4.2) which will form the backbone of the theory: $\mathbf{M}$ and $\mathbf{K}$ can be any skew-symmetric matrices and $S(Z)$ is an algebraic function of Z, with $Z \in \mathbb{R}^n$ for some $n > 2$, cf. Bridges (1996), Bridges (1997).

The advantages of the above formulation are multi-fold. Each feature of the equation is identified with a geometric structure: $t-$derivatives and $x-$derivatives associated with independent symplectic structures and the right-hand side is the gradient of a function. It encodes the energy-momentum characterization of solitary waves in a fundamentally new way, and the linearized stability problem can be characterized geometrically. Here we will give a sketch of the theory and full details can be found in Bridges & Derks (1999,2001,2002a,2002b).

Consider any system in the form (4.2). A solitary wave $Z(x,t) = \widehat{Z}(x-ct)$ satisfies

$$\nabla S(\widehat{Z}) - \mathbf{K}\widehat{Z}_x = c(-\mathbf{M}\widehat{Z}_x), \tag{4.4}$$

This is in fact a multi-symplectification of the Lagrange necessary condition for the energy-momentum formulation. The right-hand side is c times the gradient of the momentum, and the left-hand side is the gradient of the energy. Linearizing the system (4.2) about the solitary wave state leads to the linear system

$$\mathbf{M}Z_t + \mathbf{J}_c Z_x = D^2 S(\widehat{Z})\, Z\,, \quad \text{where} \quad \mathbf{J}_c = \mathbf{K} - c\mathbf{M}\,, \tag{4.5}$$

and $D^2S(\widehat{Z})$ is the Hessian of S evaluated at the solitary wave. With the spectral ansatz: $Z = \mathrm{Re}(\mathrm{e}^{\lambda t}\mathbf{v})$, the linearized stability equation reduces to

$$\mathbf{v}_x = \mathbf{A}(x,\lambda)\mathbf{v}\,, \quad \mathbf{v} \in \mathbb{C}^n\,, \quad \mathbf{J}_c\,\mathbf{A}(x,\lambda) = D^2S(\widehat{Z}) - \lambda\mathbf{M}\,. \tag{4.6}$$

This system is in standard form for the Evans function theory: $\widehat{Z} \to 0$ as $x \to \infty$ implies that $\mathbf{A}(x,\lambda) \to \mathbf{A}_\infty(\lambda)$ as $x \to \pm\infty$. More important is the fact that $\mathbf{A}(x,\lambda)$ has a multi-symplectic decomposition: the $x-$dependence is associated with a symmetric matrix and λ multiplies a skew-symmetric matrix.

Application of the Evans function theory to (4.6) and using the above geometry leads to the following result $E(0) = E'(0)$ and

$$E''(0) = \chi_{00}\left[\frac{dI}{dc} - \tfrac{1}{2}\omega(Z_0^+, \partial_c Z_0)\right], \tag{4.7}$$

which is to be compared with (3.4). The coefficient χ_{00} is a geometric property of the basic solitary wave: see Bridges & Derks (2001) for a precise definition. Note that in general $\frac{dI}{dc} < 0$ does not imply instability; it depends on the sign of χ_{00} and other properties. In Bridges & Derks (1999) it is shown that in a single example $\frac{dI}{dc} < 0$ implies instability and for other parameter values the opposite sign implies instability (χ_{00} changes sign).

The second term inside the parentheses, $\frac{1}{2}\omega(\cdot,\cdot)$ is associated with the state at infinity. For classical solitary waves, which decay exponentially to zero as $x \to \pm\infty$ this term is zero. In Bridges (2001,2002a,2002b) examples are given where the solitary wave is non-trivial at infinity and this term is nonzero. $E''(0) < 0$ is sufficient for instability if the $E(\lambda)$ is positive for large real λ, and general results on this are given in Bridges & Derks (2001,2002b). This framework for the stability of solitary waves has been applied to a wide range of examples and details can be found in the references.

5. Three-dimensionality through transverse instability

So far the discussion of stability has concentrated on *longitudinal stability*, stability with respect to perturbations travelling in the same direction as the basic solitary wave. In any real oceanic or other physical environment, there is also the potential to be stable or unstable with respect to perturbations travelling in an oblique direction.

About the same time that Benjamin was proving that the KdV solitary wave was nonlinearly *longitudinally* stable, Kadomtsev & Petviashvili (1970) were raising the question of stability of the KdV solitary wave with respect to *transverse perturbations*. This question was their motivation for deriving the KP equation,

$$(u_t + u_x + uu_x + u_{xxx})_x + \sigma\, u_{yy} = 0\,, \quad \sigma = \pm 1\,, \tag{5.1}$$

and analysis of the KP equation linearized about the KdV solitary wave shows that it is longitudinal and transverse stable when $\sigma = +1$ and transverse unstable when $\sigma = -1$. For the case of water waves, $\sigma = +1$ corresponds to gravity waves, indicating that gravity solitary waves in shallow water are stable. However when the Bond number is greater than 1/3 the KP model for water waves changes sign, indicating the solitary water waves with large surface tension are unstable.

Much of what we know about transverse instability is based on model equations such as KP which can be analyzed in some detail; see Bridges (2000), Bridges (2001), Kivshar & Pelinovsky (2000) for a review and a list of references.

However, as we saw in the previous section, multi-symplectification encodes considerable geometry which then shows up in the linear stability problem. In two space dimensions and time, the canonical form for a multi-symplectic system is

$$\mathbf{M}Z_t + \mathbf{K}Z_x + \mathbf{L}Z_y = \nabla S(Z)\,, \quad Z \in \mathbb{R}^n\,, \tag{5.2}$$

where $\mathbf{M}$, $\mathbf{K}$ and $\mathbf{L}$ are skew-symmetric matrices and $S(Z)$ is an algebraic function, with $\nabla S(Z)$ a standard gradient on $\mathbb{R}^n$.

The energy-momentum characterization of solitary waves does not provide sufficient information for transverse instabilities. However, we embed the solitary wave in a two parameter family which turns out to contain geometric information about the transverse stability. Consider the rotated family of solitary waves $Z(x, y, t) = \widehat{Z}(x - ct + \ell y)$. Then $\widehat{Z}$ satisfies

$$\nabla S(\widehat{Z}) - \mathbf{K}Z_x = c\,(-\mathbf{M}Z_x) + \ell\,\mathbf{L}Z_x\,, \tag{5.3}$$

which can be interpreted as the Lagrange necessary condition for a constrained variational princple: critical points of a generalized Hamiltonian function restricted to level sets of the momentum and a second functional, which turns out to be the gradient of the transverse symplectic structure. Let $\mathcal{A}$ be the value of the momentum, depending on both the parameters c, ℓ, and let $\mathcal{B}$ be the value of the new functional associated with transverse symplecticity. Then it is proved in Bridges (2000), Bridges (2001) that if

$$\det \begin{bmatrix} \frac{\partial \mathcal{A}}{\partial c} & \frac{\partial \mathcal{A}}{\partial \ell} \\ \frac{\partial \mathcal{B}}{\partial c} & \frac{\partial \mathcal{B}}{\partial \ell} \end{bmatrix} > 0\,, \tag{5.4}$$

when evaluated on a solitary wave, it is transverse unstable. Note that this result is stronger than the geometric condition (4.7) in the sense that (4.7) requires knowledge of the Evans function for large λ to conclude, whereas the result (5.4) is independent of the spectral parameter, and indeed, it is independent of the Evans function.

One of the advantages of the condition (5.4) is that it applies to any Hamiltonian PDE which has a multi-symplectic structure, including the full water wave problem. The condition uses only information about the basic solitary wave, and is easy to verify given a basic solitary wave. If the condition (5.4) is satisfied it provides a mechanism for the creation of two dimension patterns, or in the case of the water-wave problem, three-dimensional – two horizontal, one vertical – of the water wave problem.

6. Stability of solitary water waves

Almost all work on the stability of solitary *water* waves has been in the context of model equations: KdV, KP, Boussinesq, NLS, Davey-Stewartson, Fifth-order KdV, for example. There has been very little work on stability analysis for the full water wave problem.

The first work in this direction is the computation of stability exponents by Tanaka (1986) for the full water wave problem linearized about the classical gravity solitary wave. An interesting result of that work was that the solitary wave is longitudinally unstable at very high amplitude, much like the superharmonic instability of large amplitude Stokes waves. Further numerical calculations by Longuet-Higgins & Tanaka (1997) showed that as the amplitude of the solitary wave increased, additional unstable eigenvalues were generated. The discovery of these "crest instabilities" led Tanaka, Dold, Lewy & Peregrine (1987) to compute the time evolution of a water wave, taking the unstable eigenfunction as an initial condition. There resulting direct numerical simulations show that these crest instabilities almost always lead to a form of wave breaking. All this work was based on strictly two-dimensional waves.

Application of the condition (5.4) to the full water wave problem in Bridges (2001) showed that a transverse instability occurs at precisely the same point in parameter space as the longitudinal instability found by Tanaka. Therefore, if three-dimensional perturbations based on the transverse unstable mode, in a direct numerical simulation, the crest instability may lead to a breaking wave with transverse modulation. The modulation wavelength is not predicted by (5.4) but can in principle be determined by numerical calculation. This provides a new theoretical mechanism for three-dimensionality in wave breaking. However, this is a very special type of wave breaking, associated with very large amplitude gravity solitary waves.

It is now known that in addition to the classical gravity solitary wave –

which limits on the KdV solitary wave at low amplitude – of the water wave problem, there is a wide range of other classes of solitary waves. The most well-known of which is the capillary-gravity solitary waves with oscillatory tails which bifurcate when the phase and group velocities are nearly equal. Recently, Calvo & Akylas (2002) have developed a numerical scheme to compute the stability of these waves.

7. Concluding remarks

Another direction of great interest is numerical schemes for computing the eigenvalues associated with the linearization about solitary waves, of both model PDEs and the full water wave problem.

For a typical PDE, the spectral problem associated with the linearization about a solitary wave state can be cast into the form $\mathbf{v}_x = \mathbf{A}(x,\lambda)\mathbf{v}$ with $\mathbf{v} \in \mathbb{C}^n$ as shown in §3. Generally, $\mathbf{A}_\infty(\lambda)$ will have k eigenvalues with negative real part and $n-k$ eigenvalues with positive real part. Problems of this type are difficult to solve numerically by shooting, because there are always unstable directions. This class of problems is reminiscent of hydrodyamic stability. Indeed, the Orr-Sommerfeld analysis of the Bickley jet leads precisely to a system of this form with $n = 4$. In fact the starting point for our development of a framework for computing stability exponents for solitary waves is *the compound matrix method*, which has been used to solve the Orr-Sommerfeld equation, cf. Bridges (1999), Allen & Bridges (2002). The general theory requires aspects of exterior algebra and the details of the theory are given in Bridges (1999), Allen (2001), Allen & Bridges (2002). This has been tested on a number of examples, Afendikov & Bridges (2001), Bridges, Derks & Gottwald (2002) and provides a robust general algorithm for the linear stability problem for model equations. The numerical framework relies principally on the fact that the spectral problem can be formulated as an ODE $\mathbf{v}_x = \mathbf{A}(x,\lambda)\mathbf{v}$ with $\mathbf{v}$ in a finite-dimensional space.

It would be of great interest to develop other and more general methods for numerical and analytical study of the stability of solitary waves of the water wave problem. On the numerical side, the results of Tanaka (1986) and Calvo & Akylas (2002) are encouraging, but there is an enourmous range of solitary waves of the water wave problem whose stability properties are not yet known.

REFERENCES

Afendikov, A.L. & Bridges, T.J. 2001 Instability of the Hocking-Stewartson pulse and its implications for three-dimensional Poiseuille flow, *Proc. R. Soc. Lond. A* **457**, 257–272.

Alexander, J., Gardner, R. & Jones, C.K.R.T. 1990 A topological invariant arising in the stability analysis of traveling waves, *J. Reine Angew. Math.* **410**, 167–212.

Allen, L. 2001 Modelling dolphin hydrodynamics: the numerical analysis and hydrodynamic stability of flow past compliant surfaces, *PhD Thesis, University of Surrey.*

ALLEN, L. & BRIDGES, T.J. 2002 Numerical exterior algebra and the compound matrix method, *Numerische Mathematik* (in press).

BENJAMIN, T.B. 1972 The stability of solitary waves, *Proc. Roy. Soc. Lond. A* **328**, 153–183.

BENJAMIN, T.B. 1984 Impulse, flow force and variational principles, IMA J. Applied Math. **32**, 3–68.

BONA, J.L. 1975 On the stability of solitary waves, *Proc. Roy. Soc. Lond. A* **344**, 363–374.

BONA, J.L., SOUGANIDIS, P.E. & STRAUSS, W.A. 1987 Stability and instability of solitary waves of KdV type, *Proc. Roy. Soc. Lond. A* **411** 395–411.

BRIDGES, T.J. 1996 Periodic patterns, linear instability, symplectic structure and mean-flow dynamics for three-dimensional surface waves, *Phil. Trans. Roy. Soc. Lond. A* **354**, 533–574.

BRIDGES, T.J. 1997 Multi-symplectic structures and wave propagation, *Math. Proc. Camb. Phil. Soc.* **121**, 147–190.

BRIDGES, T.J. 1999 The Orr-Sommerfeld equation on a manifold, *Proc. Roy. Soc. Lond. A* **455**, 3019–3040.

BRIDGES, T.J. 2000 Universal geometric condition for the transverse instability of solitary waves, *Phys. Rev. Lett.* **84**(12), 2614–2617.

BRIDGES, T.J. 2001 Transverse instability of solitary-wave states of the water-wave problem, *J. Fluid Mech.* **439**, 255–278.

BRIDGES, T.J. & DERKS, G. 1999 Unstable eigenvalues, and the linearisation about solitary waves and fronts with symmetry, *Proc. Roy. Soc. Lond. A* **455**, 2427–2469.

BRIDGES, T.J. & DERKS, G. 2001 The symplectic Evans matrix, and the linearization about solitary waves and fronts, *Arch. Rat. Mech. Anal.* **156**, 1–87.

BRIDGES, T.J. & DERKS, G. 2002 Linear instability of solitary wave solutions of the Kawahara equation and its generalizations, *SIAM J. Math. Anal.* (in press).

BRIDGES, T.J. & DERKS, G. 2002 Constructing the symplectic Evans matrix using maximally-analytic individual vectors, *Proc. Roy. Soc. Edin. A* (to appear).

BRIDGES, T.J., DERKS, G. & GOTTWALD, G. 2002 Stability and instability of solitary waves of the fifth-order KdV equation: a numerical framework, *Preprint, University of Surrey.*

CALVO, D.C. & AKYLAS, T.R. 2002 Stability of steep gravity-capillary solitary waves in deep water, *J. Fluid Mech.* **452** 123–143.

EVANS, J.W. 1975 Nerve axon equations IV. The stable and unstable impulse, *Indiana Univ. Math. J.* **24** 1169–1190.

GRILLAKIS, M., SHATAH, J. & STRAUSS, W.A. 1987 Stability theory of solitary waves in the presence of symmetry, I, *J. Func. Anal.* **74**, 160–197.

GRILLAKIS, M., SHATAH, J. & STRAUSS, W.A. 1990 Stability theory of solitary waves in the presence of symmetry, II, *J. Func. Anal.* **94**, 308–348.

KADOMTSEV, B.B. & PETVIASHVILI, V.I. 1970 On the stability of solitary waves in weakly dispersing media, *Sov. Phys. Dokl.* **15**, 539–541.

KIVSHAR, UU.S. & PELINOVSKY, D.E. 2000 Self-focusing and transverse instabilities of solitary waves, *Physics Reports* **331**, 117–195.

LONGUET-HIGGINS, M.S. & TANAKA, M. 1997 On the crest instabilities of steep surface waves, *J. Fluid Mech.* **336**, 51–68.

MADDOCKS, J.H. & SACHS, R.L. 1993 On the stability of KdV multi-solitons, *Comm. Pure Appl. Math.* **46** 867–901.

MIELKE, A. 2002 On the energetic stability of solitary water waves, Preprint, Universität Stuttgart.

PEGO, R.L. & WEINSTEIN, M.I. 1992 Eigenvalues, and instabilities of solitary waves, *Phil. Trans. Roy. Soc. Lond. A* **340**, 47–94.

Pelinovsky, D.E. 2002 Matrix stability theory for incoherent optical solitons, McMaster University Preprint.

Tanaka, M. 1986 The stability of solitary waves, *Phys. Fluids* **29**, 650–655.

Tanaka, M., Dold, J.W., Lewy, M. & Peregrine, D.H. 1987 Instability and breaking of a solitary wave, *J. Fluid Mech.* **185**, 235–248.

Kinematical Conservation Laws Applied to Study Geometrical Shapes of a Solitary Wave

S. Baskar and P. Prasad

Department of Mathematics
Indian Institute of Science
Bangalore 560 012.
Email: prasad@math.iisc.ernet.in.

ABSTRACT

Kinematical conservation laws (KCL), giving the successive positions of a moving curve in a plane, is used to describe all possible geometrical shapes of the crest-line (the line joining the highest points) of a curved solitary wave on a shallow water. The KCL is an under-determined system of two equations. We assume that the length of the curved solitary wave in the direction transverse to the direction of its propagation is very large compared to a length measuring the breadth of the solitary wave. This allows us to treat a section of the solitary wave by a plane perpendicular to the crest-line to be an one dimensional solitary wave and helps us to find an additional relation to close the KCL and solve the problem completely.

1. INTRODUCTION

Considerable amount of research has been done on the stability (both longitudinal and transverse) of solitary waves in the last 30 years (Bridges, 2001). In this paper, we have considered geometrical shapes of stable curved solitary waves in shallow water of constant depth. By a curved solitary wave, we mean a wave whose extent in the transverse direction (i.e., a direction perpendicular to the direction of propagation) is very large compared to its extent in normal direction (i.e., the direction of propagation) and the shape of the wave in a normal section is locally a solitary wave. This allows us to define at any time a crest-line Ω_t, the locus of the highest points of the local solitary waves in the normal sections. Our aim in this paper is to study the successive positions of Ω_t and its geometric shape starting from its initial configuration. When the amplitude of the wave on the crest-line vary slowly (compared to the variation of the amplitude in the local solitary wave), transverse waves are induced which propagate along the crest-line. The slow amplitude variation is coupled to the variation in the normal direction of the crest-line.

Propagation of waves (or shock waves) along the crest-line have been discussed by Ostrovsky and Shrira (1976), Miles (1977), Shrira (1980), Zakharov (1986) and Pederson (1994), some of these use Whitham's equations. Quite exhaustive results have been obtained by these authors. But one needs to fit each kink (shock - as usually called by these authors) on the crest-line individually since Whitham's equations are not in conservation form. Thus, long term solution of the problem becomes quite cumbersome. We shall write the basic equations for the propagation of the crest-line in physically realistic conservation forms, which are true for any propagating curve Ω_t in a plane. These are kinematical conservation laws (KCL), first derived by Morton, Prasad and Ravindran, 1992, for an appropriately defined velocity m of Ω_t, the Mach number of curve and angle θ which normal to Ω_t makes with the x-axis in terms of an appropriately defined ray coordinate system (ξ, τ). Here τ = constant give the successive positions of Ω_t (or Ω_τ) and ξ = constant are rays - in this case orthogonal to Ω_t. Let g be the metric associated with the variable ξ, then the KCL are

$$(g \sin\theta)_\tau + (m \cos\theta)_\xi = 0, \ \ (g \cos\theta)_\tau - (m \sin\theta)_\xi = 0 \tag{1.1}$$

Whitham's equations follow from these equations. The system (1.1) is an underdetermined system of equations for three dependent variables, the dynamics of the propagating curve appears in additional partial differential equations (Monica and Prasad, 2001) or an additional relation $g = g(m)$, where g is a known function. We shall use the local solitary wave solution to determine g in the form

$$g(m) = (m-1)^{-3/2}\, e^{-(3/2)(m-1)} \tag{1.2}$$

For $0 < m - 1 \ll 1$, to which our theory applies $g(m) = (m-1)^{-3/2}$, which is equivalent to the closure relation used by Miles (see his equations (2.3a,b)) and Ostrovskii and Shrira (1976). Miles deduced this relation from the expression for the solution of solitary wave in water wave problem in a channel of slowly varying breadth. We deduce (1.2) by a method which will be applicable to solitary waves in all physical systems in which an equation for curved solitary waves can be obtained.

Use of KCL provides a new understanding of the phenomenon of kinks - it shows that the kinks are basically geometric shocks in a ray coordinate system. KCL makes computation of the successive positions of the crest-line not only very easy but very robust - the method of numerical solution can be continued for a very long time even if there are more than one kink, which interact among themselves and also with more complicated solutions than simple wave solutions. This is because, the whole range of sophisticated methods of numerical solution of hyperbolic conservation laws (Prasad and Sangeeta 1999, and Monica and Prasad 2001) are applicable to KCL.

Baskar and Prasad (2001) have recently worked out existence and uniqueness of the Riemann problem for KCL with a general form of the metric function $g = G(m)$ and discussed all possible geometric shapes of the propagating curve Ω_t. We note that the function (1.2) satisfies their assumptions on $G(m)$. Thus, their results are applicable to the crest-line of a curved solitary waves in a shallow water.

2. Multi-dimensional KdV equation

First we briefly mention various length and time scales and non-dimensional quantities involved in discussion of waves in shallow water.
H = a length scale characterizing the depth of the undisturbed water,
λ = wave length of the waves,
A = a measure of the maximum height of the wave, it has the dimension of length.

$$\epsilon = (H/\lambda)^2 \quad , \quad \delta = A/H \quad , \quad \bar{h} = Hh \ , \ \bar{\eta} = H\delta\eta \tag{2.1}$$

$$\left.\begin{array}{rcl} (\bar{x}, \bar{y}, \bar{z}) & = & (\lambda x, \ \lambda y, \ Hz) \ , \ \bar{t} = \epsilon^{-1/2}(H/g)^{1/2}t \\ (\bar{u}, \bar{v}, \bar{w}) & = & \delta(gH)^{1/2}(u, v, \epsilon^{1/2}w) \ , \ \bar{p} = \bar{\rho}gHp \end{array}\right\} \tag{2.2}$$

where $(\bar{u}, \bar{v}, \bar{w})$ are components of the velocity in $\bar{x}, \bar{y}, \bar{z}$ directions, the undisturbed free surface is $\bar{z} = 0$, $\bar{z} = \bar{\eta}(\bar{x}, \bar{y})$ is the disturb free surface, $\bar{h}$ is the constant depth, $\bar{p}$ is the excess pressure above the atmospheric pressure and $\bar{\rho}$ is the constant density.

The water wave is dispersive, but in the long wave limit $\epsilon \to 0$ (without small amplitude assumption), it is well known that it supports non-dispersive waves and is governed by a system of hyperbolic equations (e.g., see Prasad and Ravindran, 1977, equations (2.28) - (2.30)). These are equations in variables $\eta, \tilde{u}$ and $\tilde{v}$ where $\tilde{u}$ and $\tilde{v}$ are defined as velocity components in x, y directions respectively upto order ϵ terms but after carefully removing z-dependent parts from the velocity components. This system supports nonlinear nondispersive curved waves with eigenvalues

$$c_1 = \sqrt{h + \delta\eta} - \delta(n_1\tilde{u} + n_2\tilde{v}), c_2 = \sqrt{h + \delta\eta}, c_3 = \sqrt{h + \delta\eta} + \delta(n_1\tilde{u} + n_2\tilde{v}) \tag{2.3}$$

where $n_1 = \cos\theta$ and $n_2 = \sin\theta$ are components of the unit normal to the wavefront. The system can have simple wave solution of third family propagating in a fixed direction (n_1, n_2) (section 3.1.3, Prasad 2001). The simple wave is a nonstationary phenomenon, it cannot reduce to a steady solution in any frame of reference. The stationary nature of a solitary wave results from the balance of the nonlinear effects in the simple wave by dispersion. The balance is beautifully described by the KdV equation under small amplitude assumption, however this balance takes place in certain situations - in one case when

$$\delta = \epsilon^{1/3} = \epsilon_1 \tag{2.4}$$

where ϵ_1 is a short length scale characterizing a neighbourhood of the wavefront over which the disturbance is concentrated (see Prasad and Ravindran, 1977, result (3.17)). Derivation of the KdV equation requires that in the small amplitude perturbations ($\delta \ll 1$) $\delta\eta$, $\delta\tilde{u}$ and $\delta\tilde{v}$ are related by

$$\tilde{u} = n_1\eta/\sqrt{h} \ , \ \tilde{v} = n_2\eta/\sqrt{h} \tag{2.5}$$

As is well known, the solitary wave is a solution of the KdV equation. The most important point to note is that the velocity of propagation of the solitary wave is not given by the eigenvalue c_3.

The plane KdV equation with wavefront normal in (n_1, n_2)-direction can be modified by taking (n_1, n_2) to very slowly over a length scale L such that

$$H \ll \lambda \ll L \tag{2.6}$$

Let $\phi(x, y, t)$ is the phase function such that $\phi = 0$ is the curved wavefront , and approximate the equations of surface water waves in propagation space (i.e., equations (2.31) - (2.33) of Prasad and Ravindran, 1977) in an ϵ_1 neighbourhood of $\phi = 0$. This leads to the modified KdV equation for the propagation of curved waves on a shallow water from the equation (4.5) of Prasad and Ravindran (1977) in the form

$$\frac{d\eta}{dt} - \sqrt{h}\eta\Omega + \frac{h}{6}\left\{(\phi_x^2 + \phi_y^2)\right\}^{3/2}\eta_{sss} = 0 \tag{2.7}$$

where

$$\frac{d}{dt} = \frac{\partial}{\partial t} + \left(\sqrt{h} + \frac{3}{2\sqrt{h}}\delta\eta\right)\left(n_1\frac{\partial}{\partial x} + n_2\frac{\partial}{\partial y}\right) \tag{2.8}$$

$$\Omega = -\frac{1}{2}\left(\frac{\partial n_1}{\partial x} + \frac{\partial n_2}{\partial y}\right) \tag{2.9}$$

and s is the fast variable defined by

$$s = \phi(x, y, t)/\epsilon \tag{2.10}$$

Note that we do not wish to replace ϕ by the linear phase function ϕ_0 as done in Prasad and Ravindran (1977) and do not wish to go up to the equation (4.17) there.

The function ϕ satisfies the eikonal equation $\phi_t + \left(\sqrt{h} + \frac{3}{2\sqrt{h}}\delta\eta\right)|\nabla\phi| = 0$, $\nabla = \left(\frac{\partial}{\partial x}, \frac{\partial}{\partial y}\right)$ and $(n_1, n_2) = \nabla\phi/|\nabla\phi|$. This means that the derivative $\frac{d}{dt}$ is along the paths given by the corresponding ray equations,

$$\frac{dx_\alpha}{dt} = \left(\sqrt{h} + \frac{3\delta\eta}{2\sqrt{h}}\right)n_\alpha\ , \quad \frac{dn_\alpha}{dt} = -\frac{3\delta}{2\sqrt{h}}L\eta \tag{2.11}$$

where

$$L = \nabla - \mathbf{n}\langle\mathbf{n}, \nabla\rangle. \tag{2.12}$$

The equations (2.7), and (2.11) are coupled together. Unlike the purely nondispersive problem (where the last term in (2.7) is absent) as in chapter 6 Prasad (2001), numerical solution of the system (2.7), (2.11) and (2.12) appears to be extremely complex. However, unlike the well known KP (Kadomstev and Petvaishvili) equation this system is a true generalization of the KdV equation to multi-dimensions for the propagation of a curved dispersive waves in high frequency approximation. Our aim in this paper is to study propagation not of a general dispersive wave but only a curved solitary wave. In the next section, we shall use the equation (2.7) to determine the function g which appears in (1.2).

3. Equation for the Average Flux of Energy

Consider now a point on the crest-line of a curved solitary wave such that at $t = 0$, the point coincides with a point $P(x_p, y_p)$. In a two-dimensional neighbourhood N_p in (x, y)-plane, of P of linear dimension ϵ_1, we can write $\phi = n_1(x-x_p)+n_2(y-y_p)-\sqrt{h}t+O(\epsilon_1^2)$. Let $\tilde{x} = (x-x_p)/\epsilon_1$, $\tilde{y} = (y-y_p)/\epsilon_1$, then $\tilde{x}$ and $\tilde{y}$ are of order one in N_p. Then $\tilde{\Omega} = -\frac{1}{2}\left(\frac{\partial n_1}{\partial \tilde{x}} + \frac{\partial n_2}{\partial \tilde{y}}\right) = -\frac{\epsilon}{2}\left(\frac{\partial n_1}{\partial x} + \frac{\partial n_2}{\partial y}\right) = \epsilon\Omega$, where Ω is of order one. Hence, when we consider the equation (2.7) locally in the neighbourhood N_p with

$$s = \tilde{s} = \{n_1(x - x_p) + n_2(x - y_p) - \sqrt{h}t\}/\epsilon_1 \tag{3.1}$$

the curvature term can be neglected and it reduces locally to the KdV equation

$$\eta_{t'} + \frac{3}{2\sqrt{h}}\eta\eta_{\tilde{s}} + \frac{h}{6}\eta_{\tilde{s}\tilde{s}\tilde{s}} = 0 \tag{3.2}$$

where

$$\frac{\partial}{\partial t'} = \frac{\partial}{\partial t} + \sqrt{h}\left(n_1\frac{\partial}{\partial x} + n_2\frac{\partial}{\partial y}\right) \tag{3.3}$$

The local solitary wave satisfying (3.2) is

$$\eta_0 = 3c\alpha \sec h^2\left(\frac{\sqrt{c}}{2}(s - \beta ct)/\beta\right) \tag{3.4}$$

where

$$\alpha = 2^{2/3}3^{-4/3}h^{4/3}\ , \ \beta = 6^{-1/3}h^{5/6} \tag{3.5}$$

(3.4) gives the amplitude A of the solitary wave $A = \delta \max_{\mathbb{R}} \eta_0 \equiv \delta\eta_c = 3c\alpha\delta$. Its normal velocity C (which is also the velocity of the crest-line), is given by $s - \beta ct = \text{constant}$ i.e., $C = \sqrt{h} + \beta c\delta$. We define the Mach number of the crest-line as $m = C/\sqrt{h}$ and hence the relation between m and the amplitude A at the point P is

$$m - 1 = \frac{\beta\delta_c}{\sqrt{h}} = \frac{\beta A}{3\alpha\sqrt{h}} = \frac{A}{2h} \tag{3.6}$$

As mentioned after the relation (2.5), the most important influence of the balance between the nonlinearity and dispersion in maintenance of the solitary wave shape is that the crest-line velocity is not the same as eigenvalue $c_3 = \sqrt{h} + \{3/(2\sqrt{h})\}\delta\eta_c$ but it is equal to

$$C = \sqrt{h} + \{\beta/(3\alpha)\}\,\delta\eta_c = \sqrt{h} + \left\{1/(2\sqrt{2})\right\}\delta\eta_c \tag{3.7}$$

which results in the relation (3.6).

The ray velocity of the solitary wave (i.e., also that of the crest-line) is $C(\cos\theta, \sin\theta)$. Hence the time rate of change, $\frac{d}{dT}$, when we move with the crest-line is

$$\frac{d}{dT} = \frac{\partial}{\partial t} + C\left(n_1\frac{\partial}{\partial x} + n_2\frac{\partial}{\partial y}\right) \tag{3.8}$$

Rearranging terms in the multi-dimensional KdV equation (2.7), we get

$$\frac{d\eta}{dT} - \sqrt{h}\eta\Omega + \left\{\frac{3}{2\sqrt{h}}\delta\eta - C\right\}\left(n_1\frac{\partial}{\partial x} + n_2\frac{\partial}{\partial y}\right)\eta + \frac{h}{6}\left\{h(\phi_x^2 + \phi_y^2)\right\}^{3/2}\eta_{sss} = 0 \tag{3.9}$$

Multiplying this by 2η we write another relation

$$\frac{d\eta^2}{dT} - 2\sqrt{h}\eta\Omega + \left\{\frac{3}{\sqrt{h}}\delta\eta^2 - \eta C\right\}\eta_s + \frac{h}{3}\{h(\phi_x^2 + \phi_x^2)\}^{3/2}\eta\eta_{sss} = 0 \tag{3.10}$$

Since C, n_1 and n_2 can be treated as constants in N_p, the operator $\frac{d}{dT}$ and integration with respect to s or $\zeta = \phi_0 - \beta\delta ct$ commute. Further, both integrations $\int_{-\infty}^{\infty}(\eta_0 - \eta_c\eta_0)\eta_{0s}ds$ and $\int_{-\infty}^{\infty}\eta_0\eta_{0sss}ds$ vanish. Hence, integrating (3.10) with respect to s from $-\infty$ to ∞, we get

$$\frac{dD_2}{dT} - 2\sqrt{h}\Omega D_2 = 0 \tag{3.11}$$

where

$$D_2 = \int_{-\infty}^{\infty}\eta_0^2(s')ds' \ , \ s' = s - \beta ct \tag{3.12}$$

(3.11) is a very important relation and implies that the product of D_2 and the ray tube area $\mathcal{A}$ (which for the propagating curve Ω_t in (x, y)-plane may be taken to be g) associated with the crest-line is constant (see (2.2.23) in Prasad, 2001 and also Whitham, 1974). Using the expressions (3.4) we get

$$D_2 = 24c^{3/2}\alpha^2\beta \tag{3.13}$$

so that (3.6) and (3.11) give

$$\frac{dm}{dT} = \frac{4}{3}\sqrt{h}\Omega(m-1) \tag{3.14}$$

Now, we define a new time variable $\tau = t/\sqrt{h}$ and use the ray coordinate system (ξ, τ) in which the time rate of change is denoted by the partial derivative $\frac{\partial}{\partial\tau}$, so that $\frac{d}{dT} = \sqrt{h}\frac{\partial}{\partial\tau}$. The ray equations for the crest-line are

$$x_\tau = m\cos\theta \ , \ y_\tau = m\sin\theta \tag{3.15}$$

From (1.1) we get the time rate of change of θ and g along the rays in the form

$$\theta_\tau = -\frac{1}{g}m_\xi \ , \ g_\tau = m\theta_\xi \tag{3.16}$$

(see also equation (3.3.15, 16 and 19) Prasad, 2001). Since $\Omega = -\frac{1}{2g}\frac{\partial\theta}{\partial\xi}$, elimination of θ from the above equations gives

$$m_\tau = -\{2(m-1)/(3mg)\}g_\tau \tag{3.17}$$

which finally leads to the expression (1.2) for g. Note that the constant of integration in (3.17) can be chosen to be one by suitable choice of ξ.

We have now a complete formulation of the problem. Given initial position

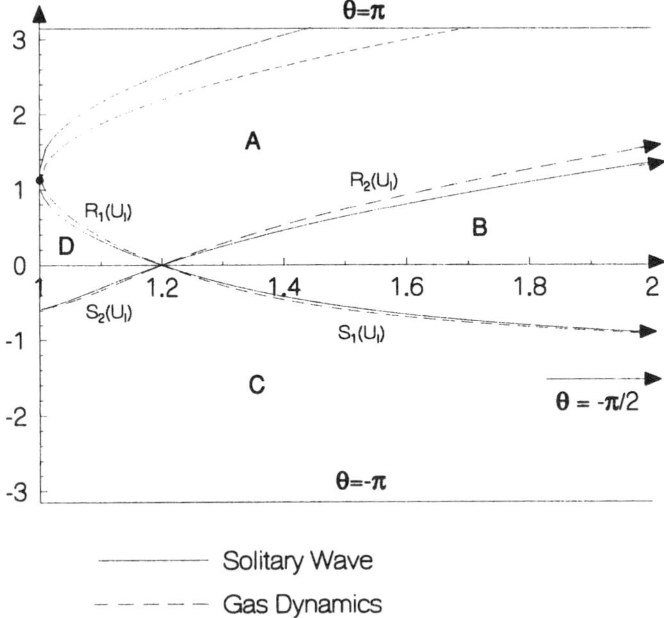

FIGURE 1. Domain A, B, C, D in (m,θ)-plane with $m_l = 1.2$ and $\theta_l = 0$. When (m_r, θ_r) belong to these domains, we get different geometrical shapes.

of the crest-line in terms of a parameter $\xi : x|_{\tau=0} = x_0(\xi)$, $y|_{\tau=0} = y_0(\xi)$ and amplitude distribution $m_0(\xi)$ on it, we determine initial value of g from (1.2). Then we first solve the system of conservation laws (1.1) and next we use the (3.15) to get the position of the crest-line at any time $t = \sqrt{h}\,\tau$ (for details, see chapter 6 in Prasad, 2001)

Since the local KdV has infinity of conservation laws, we can get as many transport equations as we wish, like (3.11). Infact, starting with equation (2.7), we shall get $g = (m-1)^{-1}e^{-(m-1)}$. But only one of these, namely (1.2) appears to be physically realistic. We first notice that $\delta\eta$ is proportional to the potential energy of the water per unit surface area of the water and $\frac{1}{2}\{(\delta\tilde{u})^2 + (\delta\tilde{v})^2\}$ is proportional to the corresponding kinetic energy. Therefore, to the first order in δ, the total energy density is $\delta\eta$ and the flux of the energy density crossing the lines parallel to the crest-line is proportional to

$$\{\delta(n_1\tilde{u} + n_2\tilde{v}\}\{\delta\eta\} + O(\delta^3) = (\delta^2/h)\eta^2 + O(\delta^3) \tag{3.18}$$

However, at the micro-scale of order $\epsilon_1, \eta^2\mathcal{A} = \eta^2 g$ is not constant due to the presence of the dispersion term in (3.2). D_2 is proportional to the integral of the square of the flux at the micro-scale. Thus, (3.11) is the physically realistic transport equation along the rays associated with the crest-line. Result (1.2) is valid for $0 < m - 1 \ll 1$, in which case we get $g(m) \simeq (m-1)^{-3/2}$. It is interesting to note that it agrees with the A-M relation used by Ostrovskii and Shrira (1976) and Miles (1977).

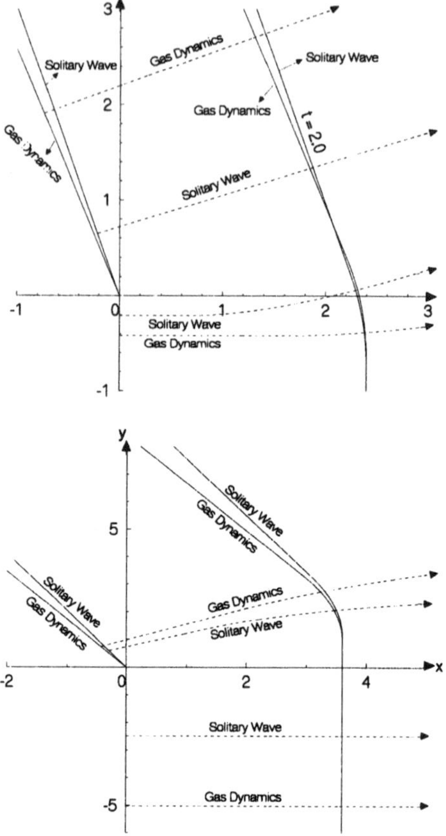

FIGURE 2. (a) Comparison of $\mathcal{R}_1$ elementary shape; and (b) Comparison of $\mathcal{R}_2$ elementary shape.

4. GEOMETRICAL SHAPES OF THE CREST-LINE

Consider the parametric representation of the initial wavefront in the (x, y)-plane to be

$$x(\xi, 0) = \begin{cases} 0 & \text{if} \quad \xi < 0 \\ -\xi g_r \sin\theta_r & \text{if} \quad \xi > 0 \end{cases}$$

$$y(\xi, 0) = \begin{cases} g_l \xi & \text{if} \quad \xi < 0 \\ \xi g_r \cos\theta_r & \text{if} \quad \xi > 0 \end{cases}$$

which is equivalent to the initial data

$$(m, \theta)(\xi, 0) = \begin{cases} (m_l, 0) & \text{if} \quad \xi < 0 \\ (m_r, \theta_r) & \text{if} \quad \xi > 0 \end{cases} \tag{4.1}$$

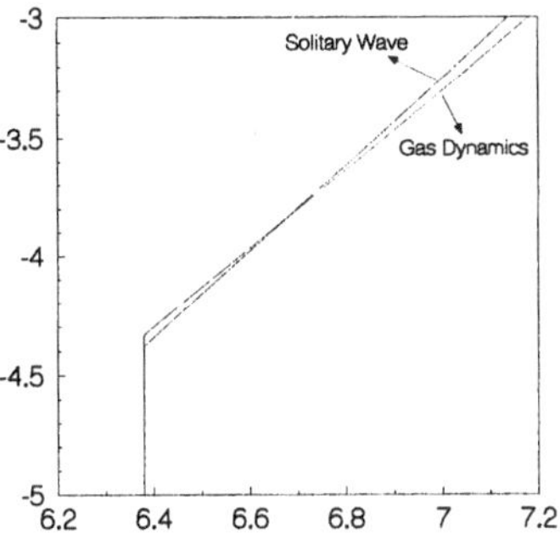

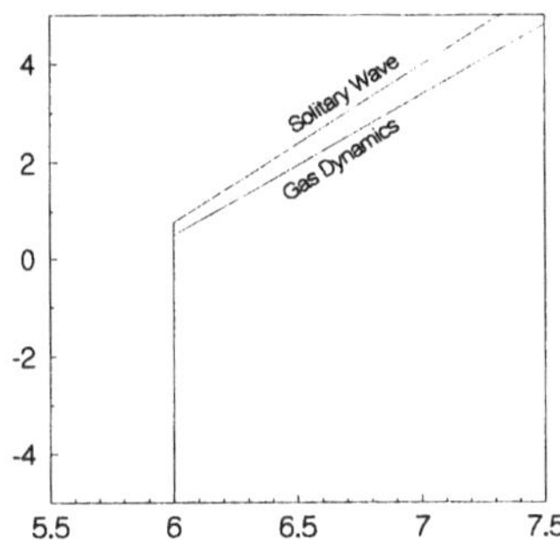

FIGURE 3. (a) Comparison of $\mathcal{K}_1$ elementary shape; and (b) Comparison of $\mathcal{K}_2$ elementary shape.

for the system (1.1) in the (ξ, t)-plane, where m_l, m_r and θ_r are constanta, $g_l = g(m_l)$ and $g_r = g(m_r)$ with g as defined in (1.2).

The initial value problem (1.1) together with the initial data (4.1) has been studied by Baskar and Prasad (2002) in the case of a general metric g satisfying a set of assumptions and shapes of the curve Ω_t for $t > 0$ has been computed for

$$g(m) = (m-1)^{-2}e^{-2(m-1)}$$

which appears for a front in weakly nonlinear ray theory in gas dynamics. As discussed in this work, the $(m > 1, \theta)$-plane is divided into four regions A, B, C and D (as shown in Fig: 1) by four curves $R_i(U_l), S_i(U_l), i = 1, 2$, which are given by

$$R_1(U_l) = \left\{ (m,\theta) | \begin{array}{l} 1 < m \leqslant m_l, \\ \theta = \sqrt{6(m_l - 1)} - \sqrt{6(m-1)} \end{array} \right\}$$

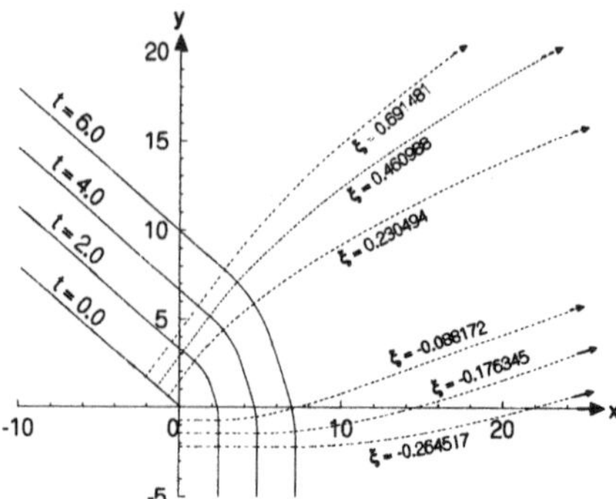

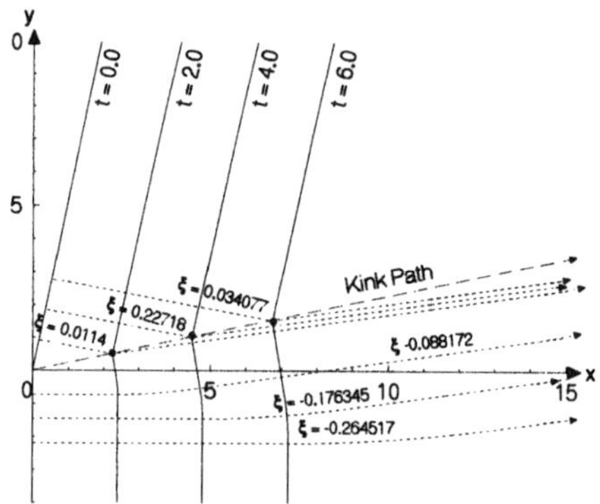

FIGURE 4. (a) Propagation of the crest line when $(m_r, \theta_r)\epsilon A$ with $m_l = 1.2$, $m_r = 1.3, \theta_r = 0.9$; (b) Propagation of the crest line when $(m_r, \theta_r)\epsilon B$ with $m_l = 1.2$, $m_r = 1.7, \theta_r = 0.5$.

$$R_2(U_l) = \left\{ (m,\theta)| \begin{array}{l} m_l \leqslant m < \infty \\ \theta = \sqrt{6(m-1)} - \sqrt{6(m_l-1)} \end{array} \right\}$$

$$S_1(U_l) = \left\{ (m,\theta)| \begin{array}{l} m_l \leqslant m < \infty \\ \theta = -\cos^{-1}\left(\frac{m_l g_l + m g(m)}{m g_l + m_l g(m)}\right) \end{array} \right\}$$

$$S_2(U_r) = \left\{ (m,\theta)| \begin{array}{l} 1 < m \leqslant m_l \\ \theta = -\cos^{-1}\left(\frac{m_l g_l + m g(m)}{m g_l + m_l g(m)}\right) \end{array} \right\},$$

the line $\theta = -\pi$ and the curve

$$\theta = \begin{cases} \sqrt{6(m_l-1)} + \sqrt{6(m-1)}, & \text{for } \sqrt{6(m_l-1)} < \theta < \pi \\ \pi, & \text{elsewhere} \end{cases}$$

where we have taken the positive determination of $\cos^{-1}$.

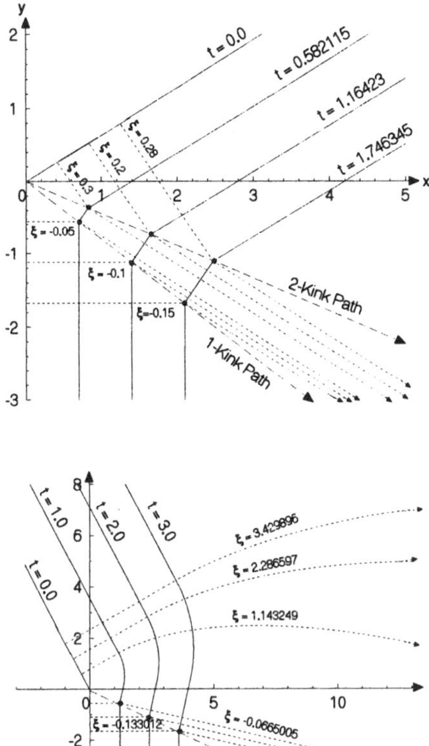

FIGURE 4. (c) Propagation of the crest line when $(m_r, \theta_r)\epsilon C$ with $m_l = 1.2$, $m_r = 1.3, \theta_r = -0.9$; and (d) Propagation of the crest line when $(m_r, \theta_r)\epsilon D$ with $m_l = 1.2$, $m_r = 1.05, \theta_r = -0.2$.

The existence and uniqueness of the intermediate state through which the left state can be joined to the right state has been discussed by Baskar and Prasad (2001). Their results with general function g show that qualitatively the shapes of Ω_t with *similar* initial conditions are the same. Since the curves R_1, R_2, S_1 and S_2 differ by a small quantity for two expressions for g, we do not expect much quatitative change also.

If the right state (m_r, θ_r) lies on $R_i(U_l), i = 1, 2$, then the singularity on the initial wavefront will be resolved and the wavefront becomes smooth for $t > 0$. This smooth part denoted as elementary shape $\mathcal{R}_i$ $(i = 1, 2)$, on the wavefront is the image of the rarefraction region in the (ξ, t)-plane. These are depicted in Fig: 2 along with the results for the nonlinear wavefront in gasdynamics.

If the right state (m_r, θ_r) lies on $S_i(U_l), i = 1, 2$ then the singularity on Ω_0 remains unresolved and propagates on Ω_t as a $\mathcal{K}_i$ $(i = 1, 2)$, elementary shape which is a kink. A comparison between the crest-line of the solitary wave and nonlinear wavefront in gas dynamics is shown in the figure 3.

In all the above cases the shape of a propagating crest-line is qualitatively same as that of a nonlinear wavefront in the gas dynamics. Infact, had we taken the initial position and initial shapes to be the same (which can be easily done) we would not have found much difference in the graphs. Therefore we omit the detailed discussion of different shapes except showing their graphs in Fig. 4. of the Ω_t in the present case when the right state (m_r, θ_r) lies in the regions A, B, C and D. For the full details, we refer to Baskar and Prasad, 2002.

ACKNOWLEDGEMENT: We Sincerely thank Professors D. H. Peregrine and V. I. Shrira for bringing to our notice the important developments on the subject in water waves. We also thank Prof. J. L. Bona for valuable discussion.

REFERENCES

BASKAR, S. & PRASAD, P. 2002 Riemann problem for kinematical conservation laws and geometrical features of nonlinear wavefront, Preprint, Department of Mathematics, Indian Institute of Science, Bangalore – 560 012.

BRIDGES, T. J. 2001 Transverse instability of solitary-wave states of the water-wave problem, *J Fluid Mech.*, **439**, pp. 255-278.

MORTON, K.W., PRASAD, P. & RAVINDRAN, R. 1992 Conservation forms of nonlinear ray equation, Tech. Rep.2, Dept. of Mathematics, Indian Institute of Science.

MILES, J. W. 1977 Diffraction of solitary waves, *J. Appl. Math and Phy.* (ZAMP), **28**, pp. 889-902.

MONICA, A. & PRASAD, P. 2001 Propagation of a curved weak shock, *J. Fluid Mech.*, **434**, pp. 119-151.

OSTROVSKY, L. A. & SHRIRA, V. I. 1976 Instability and self refraction of solitons, *Sov. Phys. JETP*, *44*, No.4, pp. 738-743.

PEDERSON, G. 1994 Nonlinear modulations of solitary waves, *J. Fluid. Mech.*, **267**, pp. 83-108.

PRASAD, P. 2001 Nonlinear hyperbolic waves *in multi-dimensions*, Chapman and Hall/CRC, Monographs and Surveys in Pure and Applied Mathematics - 121.

PRASAD, P. & RAVINDRAN, R. 1977 A theory of nonlinear waves in multi-dimensions with special reference to surface water waves, *J. Inst. Math. Applics.* **20**, pp. 9-20.

PRASAD, P. & SANGEETA, K. 1999 Numerical simulation of converging nonlinear wavefronts, *J. Fluid. Mech.*, **385**, pp. 1-20.

SHRIRA, V. I. 1980 Nonlinear refraction of solitons, *Sov. Phys. JETP*, **51**, No.1, pp. 44-49.

WHITHAM, G. 1974 *Linear and nonlinear waves*, Wiley.

ZAKHAROV, V. E. 1986 Shock waves propagated on solitons on the surface of a fluid, *Radiophys and Quantum Electronics*, **29**, pp 1073 - 1079.

Standing Waves in the Ocean

M.S. Longuet-Higgins
Institute for Nonlinear Science
University of California, San Diego
La Jolla, California 92093-0402, USA.

Abstract

Standing-wave spectra, namely those which have significant energy in opposite wavenumbers, are quite common in the North Atlantic and other parts of the open ocean. In addition they occur near the centres of hurricanes and in the wakes of moving atmospheric depressions; also near steep coastlines capable of reflecting significant amounts of wave energy. Standing waves generally can have steeper slopes and larger accelerations than progressive waves, and tend to throw up spray and droplets into the atmospheric boundary layer. Detection of standing-wave conditions at large distances is made possible by the elastic waves (microseismic) that they generate in the water and sea bed and in the adjacent land. Similar waves (microbaroms) in the atmosphere detected by a directional array of infrasound sensors in the Netherlands have been used to map standing-wave energy in the entire North Atlantic.

Laboratory studies of standing waves have mostly been focused on purely time-periodic waves. Here it is shown that the encounter between two opposite trains of progressive waves can produce energy densities exceeding the limit for periodic standing waves; hence the resulting motion must be aperiodic. Experiments are described in which a steep train of progressive wave meets a vertical wall, modelling the encounter of two opposite wave trains of equal amplitude. When the progressive-wave steepness exceeds a certain value it is found that the resulting motion develops a type of subharmonic instability in which every third wave (in time) is steeper than the preceding or following waves. The steep waves develop sharp crests and vertical jets. The other two waves are flat-topped or of intermediate form. This instability grows by a factor of about 2.2 for every three wave periods.

1. Introduction

Despite common belief among fluid dynamicists, standing waves are frequently found in the open ocean. By this we mean ocean wave fields whose directional spectra contain appreciable energy associated with components of the same frequency but opposite in direction. Figure 1, for example, shows a sequence of 28 directional ocean wave spectra, at 3-hourly intervals, for a point at 50° N, 40° W in the Labrador Sea, in March 1989 during the LEWEX trials. The spectra were hindcast by the WAM method, using estimated wind fields. Figure 2 shows some similar spectra derived from in situ observations of the sea state by a Wavescan directional buoy. In about one half of all these spectra one can discern some significant energy associated with diametrically opposite wavenumbers, indicating at least "partial" standing waves.

Similar types of directional spectra are to be expected at the centres of hurricanes and cyclones (for photographs see Charcot 1929); also in the wakes

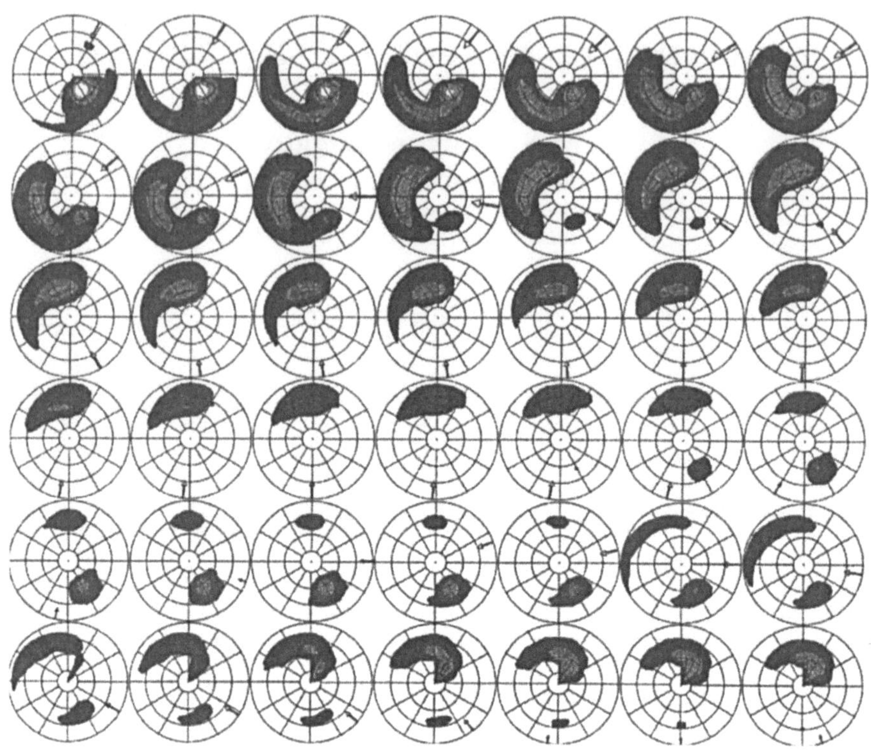

FIGURE 1. (from Gerling 1991, Figure 3). A 3-hourly sequence of hindcast wave spectra at 50° N, 45° W, beginning 12 March 1989 at 1500 UT.

of moving depressions, especially when the mean speed of the storm exceeds the group velocity of the waves (see Longuet-Higgins 1950).

It is also likely that in coastal regions close to steep cliffs such as the Pacific coastline of Oregon, U.S.A., or the Atlantic coast of northern Norway, reflection of waves from the coast will be sufficient to generate partial standing waves.

Standing or partly reflected waves can in general have steeper slopes and larger accelerations than pure progressive waves, and they tend to throw up jets and spray into the air, thereby contributing to the mechanics and thermodynamics of the atmospheric boundary layer. Yet compared to progressive waves, the literature on standing waves is very limited. The early theoretical studies by Penny & Price (1952), followed by Taylor's (1953) experimental investigation, were concerned only with *periodic* standing waves. As is known (Mercer & Roberts 1992) the energy of periodic standing waves of finite amplitude is rather strictly limited. Serious investigation of standing waves which are forced beyond their critical energy (see for example Jiang *et al.* 1998, Longuet-Higgins & Dommermuth 2001a,b) is quite recent, though there is now a growing interest in the subject.

In the present paper we shall first describe two interesting methods for remote-sensing of standing wave conditions in the open ocean, namely by

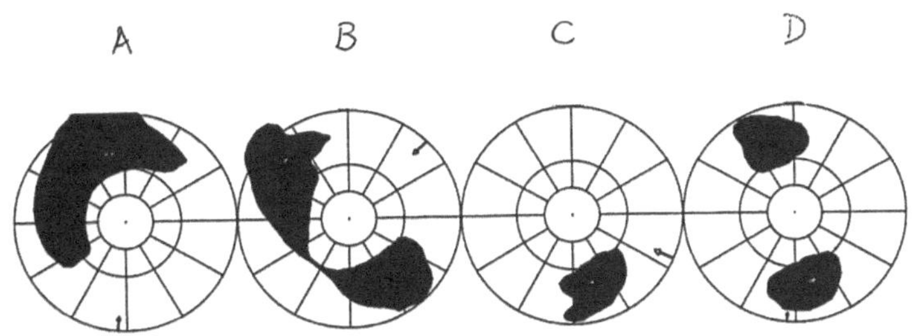

FIGURE 2. (from Zambresky 1991, Figure 7). Observed wave spectra measured by a Wavescan buoy at 50° N, 45° W. A. 16 March, 0000 UT; B. 16 March 1200 UT; C. 17 March 1200 UT: D. 17 March 1200 UT.

recording the propagating elastic waves that standing ocean waves have long been known to generate in the solid earth (microseisms) and in the atmosphere (microbaroms). These sensing methods can provide a synoptic picture of standing-wave fields over the North Atlantic and other parts of the ocean. We mention also, in Section 4, the occurrence of short-scale standing wave patterns in shear layers near a boundary current between two different water masses; these standing waves produce anomalous signals in radar scatterometers. In Section 5 we discuss the theoretical encounter of opposite trains of waves, of finite amplitude, in the open ocean, and show that there is a significant range of progressive-wave amplitudes in which the resulting wave pattern cannot be time-periodic. We then describe some new laboratory experiments designed to elucidate the actual behaviour of the water under these conditions. A new type of subharmonic instability affecting the waves is revealed.

2. DETECTION OF STANDING WAVES BY OCEANIC MICROSEISMS

Standing-wave conditions brought about by opposing swells can be detected at great distances through the nonlinear pressure oscillation which they produce. For, in any space-periodic motion of a uniform incompressible fluid, the oscillating part p' of the pressure at a point more than about half a wavelength from the free surface is given rigorously by the expression

$$p' = -\rho \frac{d^2}{dt^2} \left(\frac{1}{2} \overline{\eta^2} \right) \tag{2.1}$$

where ρ is the density and $\eta(\mathbf{x}, t)$ is the local surface elevation; see Longuet-Higgins (1950, 1953). $\mathbf{x}$ and t are a horizontal space coordinate and time, respectively, and an overbar denotes the horizontal average. The expression $\frac{1}{2}\overline{\eta^2}$ is clearly related to the potential energy of the waves, which in a progressive

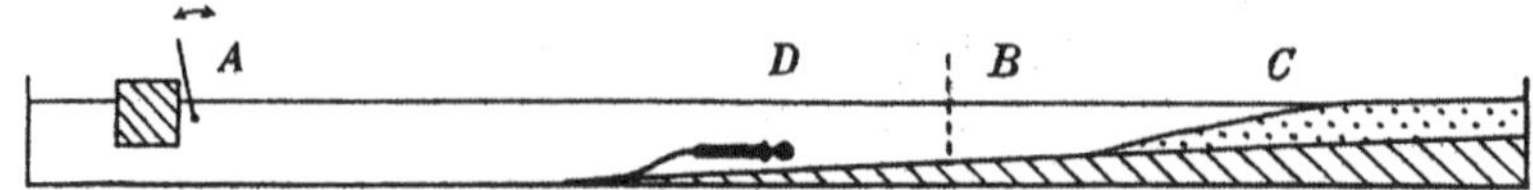

FIGURE 3. (after Cooper & Longuet-Higgins 1951). Sketch of the experimental wave tank, showing the relative positions of the wavemaker A, removable barrier B, absorbing beach C and hydrophone D. $AB = 5$ m, $BC = 3$ m, $DB = 1$ m. Total length of tank was 9.7 m; depth of water 41.5 cm.

wave does not change but in a standing wave fluctuates in time. In fact if we substitute for η a general expression describing (to first order) two waves of amplitude a_1 and a_2 having the same radian frequency σ but travelling in opposite directions we find, after a little algebra, that

$$p' = -2\rho\sigma^2 a_1 a_2 \cos 2\sigma t \qquad (2.2)$$

representing a pressure oscillation which has twice the frequency of either wave and is independent of the vertical coordinate z. For pure progressive waves, when either a_1 or a_2 vanishes, so also does p'. For pure standing waves, when $a_1 = a_2 = \frac{1}{2}a$, the term was first discovered by Miche (1944) and was generalised by Longuet-Higgins & Ursell (1948). It was then confirmed by laboratory experiments (Cooper & Longuet-Higgins 1951).

These experiments are summarised briefly in Figures 3 and 4 below. Figure 3 shows a sketch of the wave tank used, which had a width of 24 cm and an overall length of 9.7 m. A wavemaker at A generated waves of period about 0.5 s which, if no barrier were present at B, dissipated on the beach at C. Subsurface pressure fluctuations were measured by a hydrophone at D. In Figure 4, which shows the pressure versus time t, the wavemaker was switched on at point O. Travelling with the group-velocity, the waves were detectable first at A, then the main energy arrived at around G. In Figures 4a and 4b there was no vertical barrier. These show that as the depth z of the hydrophone was increased from 8.8 cm to 17.3 cm so the pressure fluctuations diminished according to the exponential law $e^{-\sigma^2 z/g}$. Then (Figure 4c) a vertical barrier was inserted at B, Figure 1, and when the reflected wave energy from the barrier arrived at the hydrophone, that is at time G' in Figure 4c, a double-frequency oscillation was evident. Lastly in Figure 4d the depth z of the hydrophone was increased to 37 cm. The first-order pressure fluctuation was then negligible, but the amplitude of the second-order pressure fluctuation, well given by equation (2.2), remained the same. For further details of the experiments, and for applications to the measurement of reflection coefficients, the reader is referred to the original (1951) paper.

For ocean waves, the physical importance of the pressure oscillation (2.1) is that it is responsible for the generation of the acoustical waves in the ocean and the sea bed known as *oceanic microseisms*, that is to say small displacements of the earth's surface easily detectable by a sensitive seismograph. For some years microseims were actually used for tracking hurricanes over the Caribbean Sea. A quantitive theory for microseism generation was first given by Longuet-Higgins (1950, 1953), and has been confirmed by many field observations; see espe-

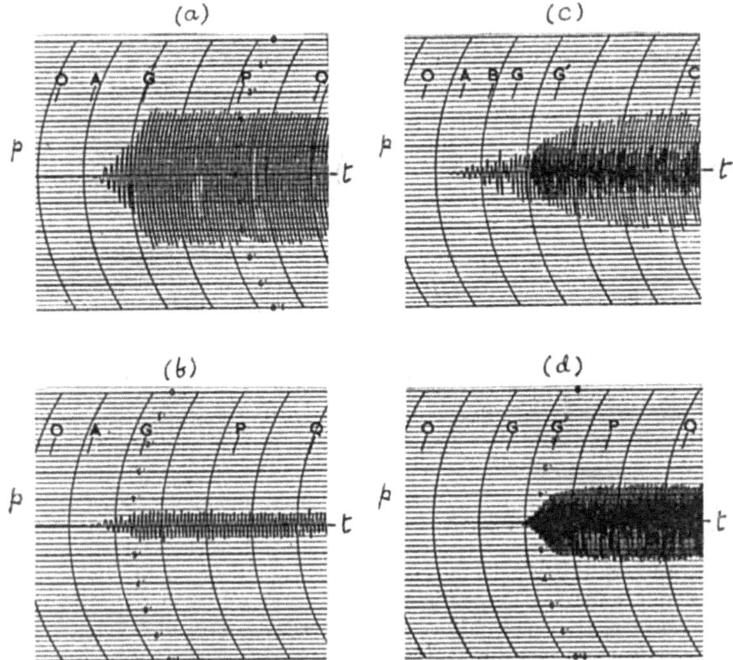

FIGURE 4. (after Cooper & Longuet-Higgins 1951). Pressure vs time t as recorded by the hydrophone at D in Figure 3. (a) Progressive wave with no vertical barrier. Hydrophone depth $z = 8.8$ cm, (b) As (a) but with $z = 17.3$ cm, (c) As (b) but with reflecting barrier at B, (d) As (c) but with $z = 37.0$ cm.

cially Orcutt *et al.* (1993). The theory involved the generalisation of equations (2.1) and (2.2) to the representation of the surface by a two-dimensional wave spectrum (the first use of such a representation in fact). Effects of the compressibility of the water and the sea-bed, which lead to the outward propagation of elastic waves from the generating area, were also taken into account.

3. Detection of standing ocean waves by infrasound

A related method of detecting standing-wave conditions in the ocean has recently been demonstrated by Evers & Haak (2001). This uses a directional array of sensitive microbarographs (first installed for monitoring nuclear tests in the atmosphere) which can detect low-frequency sound waves in the frequency range 3 - 10 Hz, the same as oceanic microseisms. Such waves in the atmosphere are known as *microbaroms*. Observations by Benioff & Gutenberg (1940) in California and by Baird & Banwell (1940) in New Zealand first associated microbaroms with oceanic microseisms. Clearly microbaroms could conceivably be generated by the same mechanism as microseisms. For, equation (2.1) shows that it is essentially the displacements $\eta(\mathbf{x}, t)$ of the air-water interface which give rise to the pressure oscillations p' below the water surface. Similar pressure oscillations will therefore be present in the atmosphere, but with the density ρ'

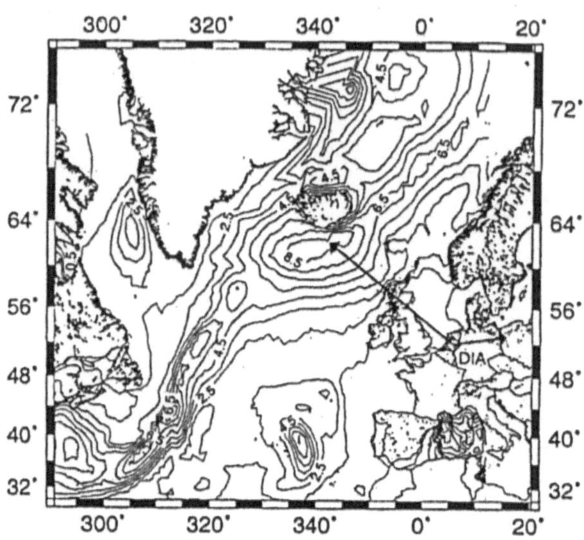

FIGURE 5. (after Evers & Haak, 2001). Contours of the nominal "standing-wave pressure" $a^2\sigma^2$, in $\mathrm{m}^2/\mathrm{s}^2$ over the North Atlantic, as measured by the Deelen Infrasound Array (DIA) at Deelen, Netherlands on 8 November 1999.

of the air replacing the density ρ of the water. The amplitude of the pressure oscillations will of course be modified by the presence of winds and by the radial spreading of the sound rays in the layered atmosphere. A quantitative theory of the generation of microbaroms by this mechanism has been given by Posmentier (1967).

Figure 5, taken from Evers & Haak (2001) shows the estimated equivalent standing-wave source $a^2\sigma^2$ of ultrasound, calculated from an observational array at Deelan, Netherlands. The maximum intensity south of Iceland corresponds to a standing-wave period of 9.8 s, generating microbaroms of half this period, that is a frequency of 0.2 Hz. The direction of the microbaroms at this frequency corresponds rather well with the position of the maximum of $a^2\sigma^2$ as shown by the contours of intensity. (The eastwards vector corresponds to the direction of a meteor trail detected by the array.)

4. SHORT-SCALE STANDING WAVES

On smaller scales, short standing waves with wavelengths of order 10 cm are sometimes observed near the boundary of a shearing current. In Figure 6, for example, the roughest zone is near the boundary of an estuarine current in the far zone flowing to the left. The standing waves appear to be due to the bending of the incident ray by the current. Such an increase in surface roughness, together with the jets and droplets thrown upwards by the standing waves are of interest to the remote-sensing of the sea surface by radar, since the back-scattered radar return is not related to the local wind in the usual way (Trizna 2000).

FIGURE 6. (from Trizna 2000). Standing waves on a shearing current separating two different water masses.

FIGURE 7. Sketch of two opposite trains of progressive waves.

5. Steep, aperiodic standing waves

Most experimental or theoretical investigations, including those of Taylor (1953), have been concerned with standing waves that are periodic both in space and time – a natural first assumption. But as shown by Mercer & Roberts (1992) there is an upper bound to the energy of time-periodic standing waves of a given wavelength. In dimensionless units where the density ρ, the acceleration of gravity g and the wavenumber k are taken as unity, the maximum energy density E per unit horizontal distance (for 2-dimensional waves in deep water)

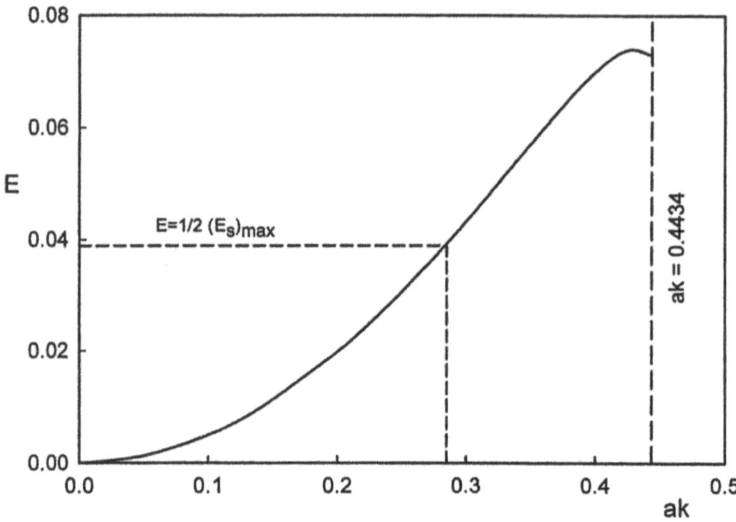

FIGURE 8. The energy density E_p of a progressive wave on deep water, as a function of the wave steepness parameter ak.

is given by

$$(E_s)_{\max} = 0.07774. \tag{5.1}$$

What happens if more energy than this is supplied to a space-periodic wave motion? Numerical studies by Longuet-Higgins & Dommermuth (2001a,b) have shown that if a standing with an initially narrow circular trough is released from rest (so that the initial kinetic energy vanishes) then a jet arises out of the trough which first becomes flat-topped and then breaks to either side. If on the other hand the surface is initially flat (so that the potential energy vanishes) and a vertical velocity, sinusoidal in the horizontal coordinate x, is imparted impulsively at the free surface, then a vertical jet again arises, which becomes thinner and more sharp-pointed before falling back into the wave trough where it may produce a circular cavity. These numerical experiments are summarised by Longuet-Higgins (2001).

Here we shall be concerned with the physically more realistic but still ideal situation when two groups of waves of the same frequency and amplitude but travelling in opposite directions meet and interact in deep water, as in Figure 7. If the waves are of low steepness, then according to linear theory they will combine to form a standing wave of energy

$$E_s = 2E_p \tag{5.2}$$

where E_p denotes the energy-density of each progressive wave. However, if the waves are sufficiently steep, and if we assume that their energies are added to produce a periodic standing wave of energy E_s, then we must have

$$E_p = \frac{1}{2}E_s \leqslant 0.03882 \tag{5.3}$$

by equation (5.1). Now a plot of the energy density E_p of a progressive wave

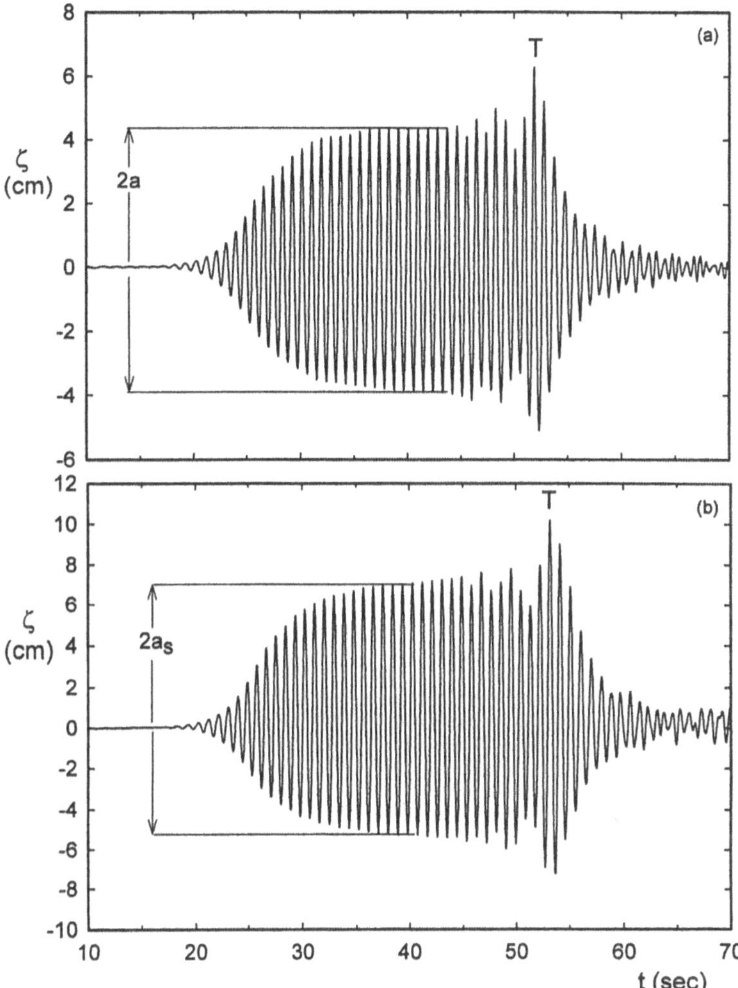

FIGURE 9. Record of the surface elevation η at the point $x = 15.35$ cm as a function of the time t when $G = 1.6$, (a) with no barrier at B, and (b) with the barrier in place.

against its steepness parameter ak (Figure 8) shows that a progressive wave may have an energy as great as 0.0745, which is achieved when $ak = 0.429$. The maximum value of ak is 0.443. So from Figure 8 we see there is a certain range of progressive-wave steepnesses, namely

$$0.285 < ak < 0.443 \tag{5.4}$$

where the resulting motion cannot be a perfectly periodic standing wave. The resulting motion must be non-periodic or chaotic in some way, perhaps leading to breaking.

Not all progressive waves for which $ak < 0.285$ will necessarily combine to

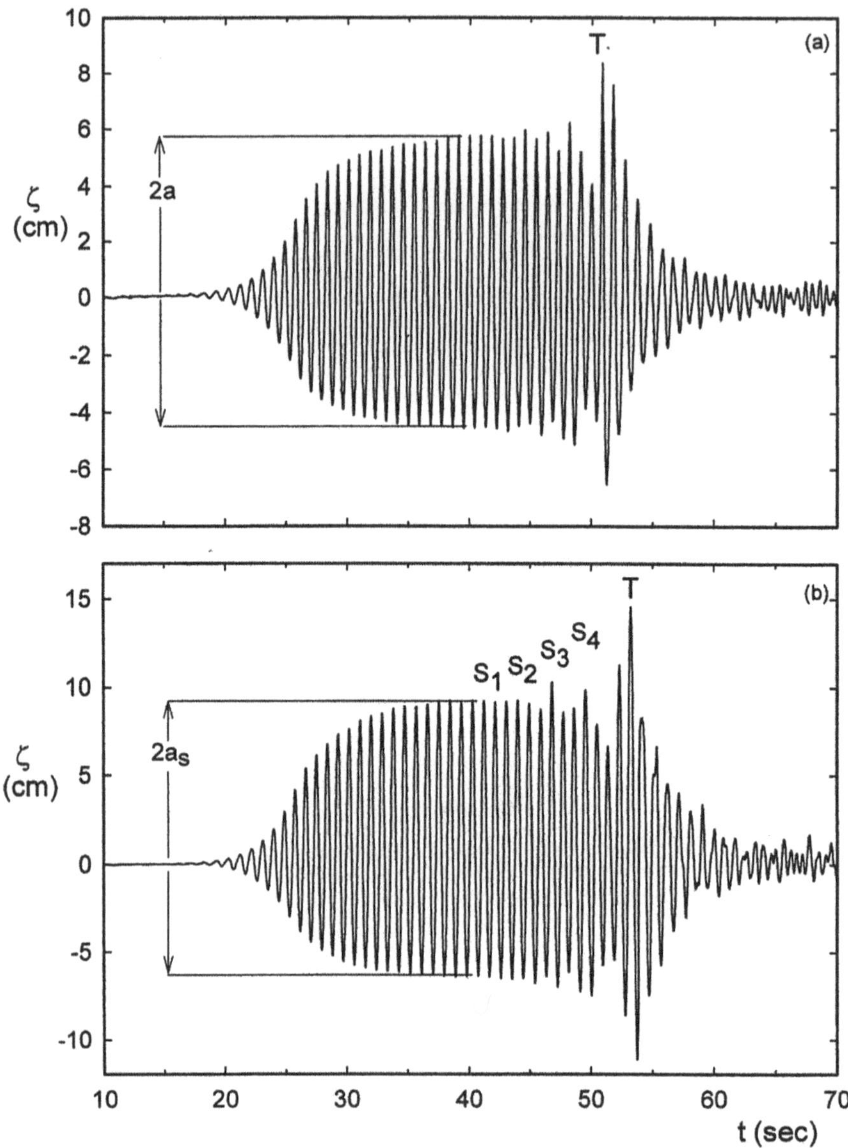

FIGURE 10. As Figure 9 but with $G = 2.0$.

produce perfectly periodic standing waves. To investigate their actual behaviour we have resorted to experiment.

6. Experimental procedure

It was assumed that the encounter between two opposite trains of progressive waves could be simulated well enough by the reflection of a wave-train from a

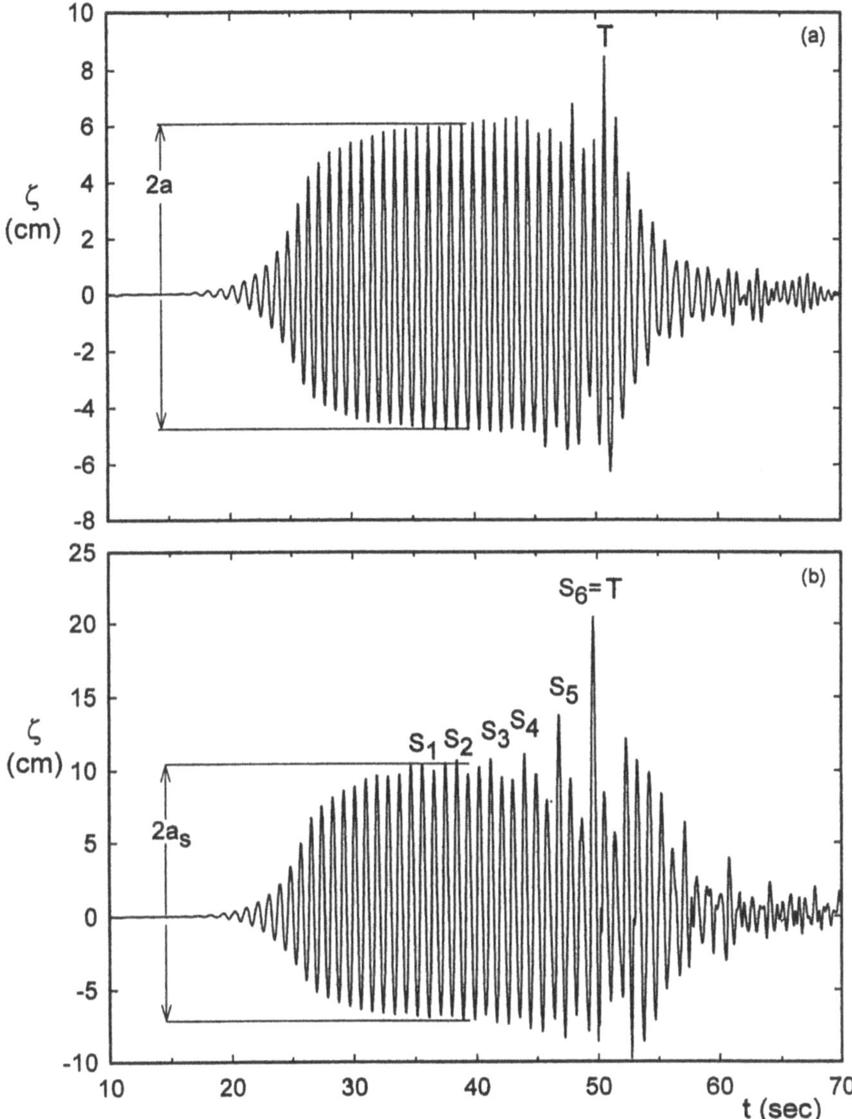

FIGURE 11. As Figure 9 but with $G = 2.2$.

smooth vertical barrier, as in Section 2 above. Experiments to determine the pressure exerted on a vertical wall by the impact of steep gravity waves have previously been conducted by Bagnold (1939) and by Chan & Melville (1988) among others. Here we are more concerned with the form of the waves and with their sequence in time.

The following experiments were carried out in a glass-sided wave channel in the Hydraulics Laboratory at the Scripps Institution of Oceanography. We

G	a_o (cm)	a (cm)	$a\sigma^2/g$	ak	a_s (cm)
1.6	4.53	4.22	0.206	0.200	7.50
1.8	5.19	4.63	0.226	0.216	6.95
2.0	5.52	5.66	0.251	0.236	7.87
2.2	5.82	5.51	0.268	0.252	8.49
2.4	6.61	5.85	0.285	0.266	9.28
2.6	7.52	6.36	0.310	0.285	10.89

TABLE 1. Range of experiments with $\lambda = 0.1, N = 40$

give here a summary; for further details, see the paper by Longuet-Higgins & Drazen (2002).

The experimental arrangement was similar to Figure 3 above, with a controlled wavemaker at A, a solid plane beach (gradient 1:10) at C and a removeable plane barrier at B. In this case the distances AB and AC were 15.65 m and 26.96 m respectively. The width of W of the tank and the stillwater depth h were both 0.50 m. The surface elevation η on the mid-line of the tank was recorded by a resistance-wire wave gauge on the up-wave side of B, and sometimes also close to the wavemaker at A. A video recording of the free surface, at a rate of 30 frames/s was made from a position slightly above the mean water level and to the left of the barrier. The frequency of the waves was 1.1 Hz, so that the waves were effectively in deep water, while the most pronounced cross-wave instabilities were avoided.

To avoid the Fresnel envelope pattern associated with sudden starting or stopping of a train of waves of constant amplitude (Miles 1962) the wavemaker was started gradually from rest with a horizontal displacement given by

$$\zeta = \left\{ \begin{array}{ll} 0, & t < 0 \\ C \tanh \lambda t \sin 2\pi f t, & t > 0 \end{array} \right\} \tag{6.1}$$

where C and λ were constants, $\lambda = 0.1$. On the other hand the wavemaker was switched off suddenly after an integral number N of wave cycles, producing a typical Fresnel envelope at the end of the wave group, as seen in Figure 9, for example, having a maximum elevation at time T.

The input voltage to the wavemaker was governed by a gain factor G, which it is useful to use as a parameter. Runs were made with G in the range 1.6 to 2.6; see Table 1 for the corresponding wave parameters. At each value of G the surface elevation η at B was first recorded without the barrier in place, so that the waves passed by and were largely dissipated on the beach at C. Then the barrier was put in place and the surface elevation at B was again recorded. Simultaneously a video sequence of the water surface was made as described above.

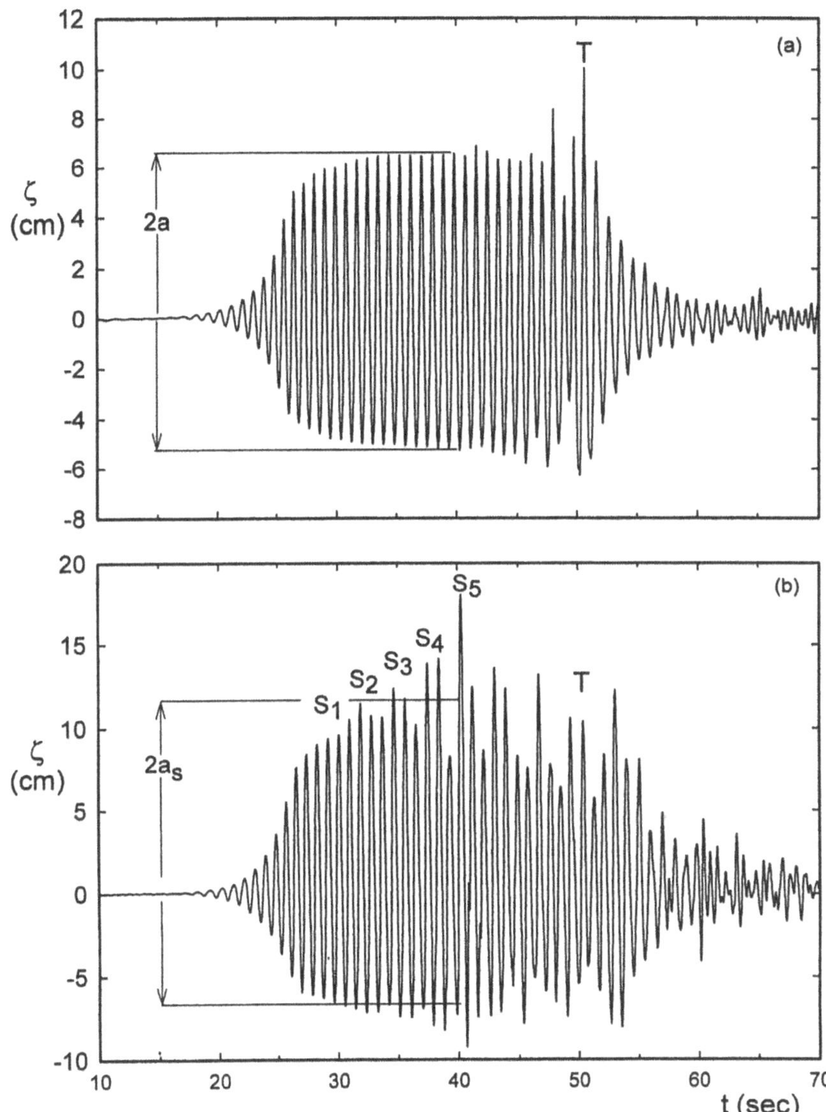

FIGURE 12. As Figure 9 but with $G = 2.4$.

7. EXPERIMENTAL RESULTS

Figure 9 shows the case $G = 1.6$. In Figure 9a, without the barrier, the steady wave amplitude a is 4.2 cm, attained at about time $t = 35$ s. It remains constant until about $t = 42$ s, after which it is affected by the Fresnel pattern from the wave cut-off. When $t > 30$ s there are slight indications of a Benjamin-Feir instability, but these are small compared to the Fresnel oscillations.

Figure 9b is as 9a but with the barrier in place at B. The wave amplitude

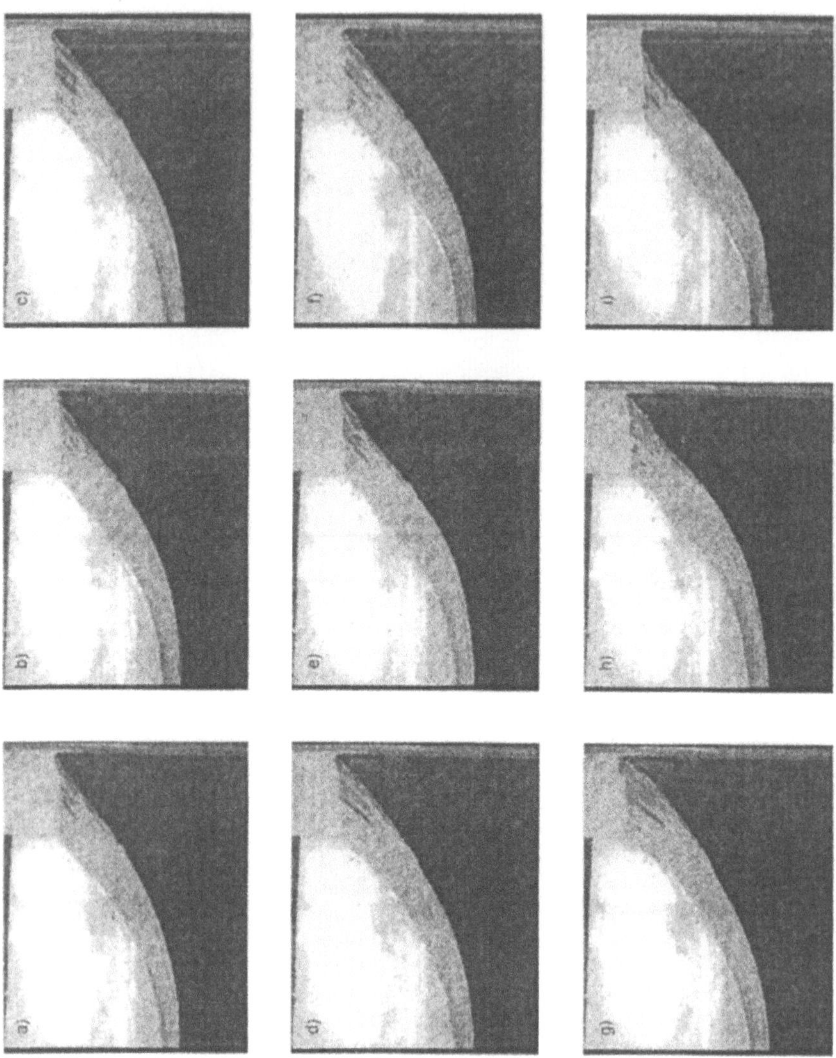

FIGURE 13. A sequence of frames from the video corresponding to Figure 12b ($G = 2.4$) shown consecutive wave crests between $t = 28$s and $t = 42$s. The timing of each frame is as close as possible to a maximum of the surface elevation shown in Figure 12b, that is within 1/60s. Each frame in the left-hand column corresponds to one of the maxima S_i in Figure 12b.

a_s is approximately equal to $2a$ (see Table 1) as one would expect on linear theory.

In the case $G = 1.8$ the corresponding records were quite similar to $G = 1.6$, but with the amplitudes increased; see Table 1. However when $G = 2.0$ some new features appeared. In Figure 10a, taken without the barrier at B, we can see the usual Fresnel envelope for a progressive wave, with a maximum at T.

Shortly before $t = 40$ s there is a slight modulation of the envelope due either to a Benjamin-Feir instabililty or to some three-dimensionality of the motion. However, Figure 10b, taken with the barrier in place, shows that between $t = 40$ s and $t = 50$ s there is apparently a new instability in which every third wave, marked with the symbol S_i ($i = 1$ to 4) is higher than its two neighbours.

This is confirmed by Figures 11a and 11b, taken when $G = 2.2$. In Figure 11b, which shows the surface elevation in the reflected wave, the three-fold pattern now extends as far as from $t = 35$ s to $t = 50$ s. It appears to have overwhelmed the Fresnel pattern even as far as the maximum T.

Figure 12b, corresponding to $G = 2.4$, shows the same pattern extending as far back as $t = 30$ s, but by $t = 45$ s the waves have become chaotic and the Fresnel pattern is quite ragged. A similar phenomenon was apparent when $G = 2.6$ except that the pattern began and broke down even earlier.

An examination of the photographic record, see Figure 13 for the case $G = 2.4$, reveals that the highest peaks in each triplet are always sharp-pointed. The lower peaks are either round-crested on flat-topped or sometimes have profiles that are intermediate between flat-topped and sharp-crested. After the crest S_5, at $t = 40$ s, the motion becomes markedly three-dimensional, which contributes to the chaotic appearance of the surface elevation.

8. The triple instability; discussion

A rough measure of the amplitude of the instability noted in Figures 10 to 12 is the difference $\Delta\zeta$ in crest height between the highest and lowest waves of each triplet. In Figure 14, $\Delta\zeta$ has been plotted against the suffix i in S_i on a log-linear scale, for each value of G, except that when $G = 2.6$, i has been increased by 2 to bring the plots closer together. (This does not of course affect the proportional rate of increase of $\Delta\zeta$.) It will be seen that in every case except one, namely $G = 2.0$ and $i = 1$, the plots lie close to the same straight line. This indicates an increase in $\Delta\zeta$ by a factor of about 2.2 for every 3 wave cycles, that is an increase of 1.3 per wave cycle. The exceptional plotted point (x) corresponds to a very small value of $\Delta\zeta$, lying within the noise-level of the experiment.

Thus we have detected a subharmonic instability which tends to occur at values of G greater than about 2.0, that is to say incident wave steepnesses $ak \geqslant 0.236$ (see Table 1). The observed rate of growth is about 1.3 per wave cycle, practically independent of the incident wave amplitude.

The above instability is probably related dynamically to the "period tripling" phenomenon observed by Jiang *et al.* (1998) in forced standing waves. However, in Jiang *et al.* (1998) standing waves were forced subharmonically by oscillating the wave tank vertically at a frequency twice that of the resulting surface waves. Such a method of excitation is of course unlikely to be found in nature. Also, in their experiments a quasi-steady state was achieved by balancing the input of wave energy from the vertical forcing against the loss of energy due to wave breaking. In our experiments there was no energy input due to vertical motion of the bottom, the loss of energy due to wave breaking was negligible, and the instability grew in time.

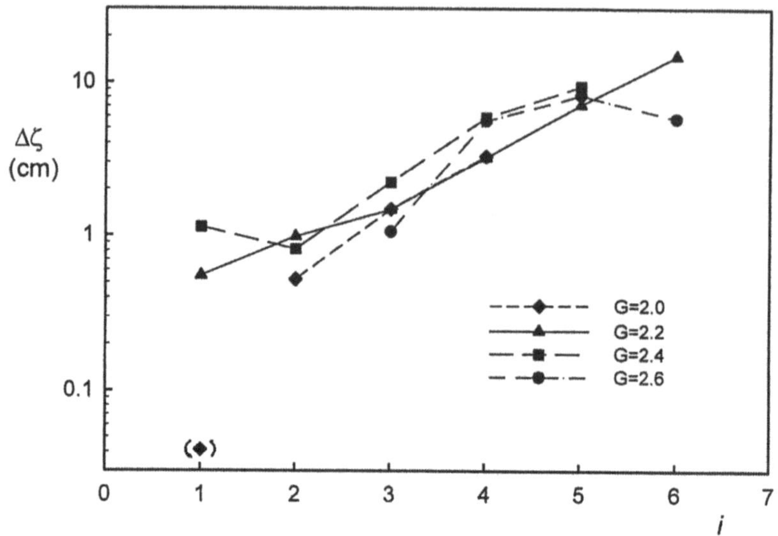

FIGURE 14. Growth of the triple-period instability, as measured by the difference $\Delta\zeta$ in crest-elevation between the highest and lowest waves of a triplet.

Note that some instabilities of periodic standing waves that are subharmonic in space were found analytically by Mercer & Roberts (1992). Those described here, however, were subharmonic in time.

It is worth noting that flat-topped standing waves are also seen in the open ocean, especially at the centres of hurricanes. Charcot (1929) shows an example from a storm south of Greenland. A similar wave, closer to land, is shown in Figure 15.

9. CONCLUSIONS

Standing-wave or partial standing-wave spectra are quite common in the open ocean. They are observed to throw up water droplets and spray into the atmospheric boundary-layer. They are clearly detectable on land by the double-frequency oscillations that they generate in the ground and in the atmosphere. If a standing wave of given length is to be precisely periodic it cannot contain more than a certain energy. Hence when two trains of progressive waves having the same frequency and amplitude encounter one another the resulting motion cannot be perfectly periodic if the steepness ak of each wave-train exceeds 0.285. Laboratory experiments show that if $ak \geqslant 0.236$ a new type of instability occurs in which every third crest of the standing wave pattern becomes steeper than its predecessor or successor, leading to the formation of upward jets. This instability grows by a factor of about 2.2 in every three wave periods.

FIGURE 15. Photograph of a flat-topped standing wave, taken by Mr. Fukumi Kuriyama at Katasurahama Beach on Shikoku Island, Japan during a typhoon. This photograph, using a long-focus lens, was awarded a prize by the Nikkor Club, and appeared in their Year Book for 1973. Reproduced by kind permission.

ACKNOWLEDGEMENT: I am indebted to Mr Fukumi Kurijama and to the Nikkor Club for permission to reproduce the photograph in Figure 15.

REFERENCES

BAGNOLD, R.A. 1939 Interim report on wave pressure research. *J. Inst. Civil Engrs.*, **12**, 201-226.

BAIRD, H.F. & BANWELL, C.J. 1940 Recording of air-pressure oscillations associated with microseisms at Christchurich, *N.Z. J. Sci. Tech.*, **21B**, 314-329.

BENIOFF, H. & GUTENBERG, B. 1939 Waves and currents recorded by electromagnetic barographs. *Bull. Amer. Meteor. Soc.*, **20**, 421-426.

CHAN, E.S. & MELVILLE, W.K. 1988 Deep-water plunging wave pressures on a vertical wall. *Proc. R. Soc. Lond.* A, **417**, 95-131.

CHARCOT, J. 1929 La mer du Groenland, croisièies du "Pourquoi Pas?" Paris-Bruges, Desclée, de Brouwer & Cie, 207 pp; see p. 162.

COOPER, R.I.B. & LONGUET-HIGGINS, M.S. 1951 An experimental study of the pressure variations in standing water waves. *Proc. R. Soc. Lond.* A, **206**, 424-435.

EVERS, L.G. & HAAK, H.W. 2001 Listening to sounds from an exploding meteor and ocean waves. *Geophys. Res. Lett.*, **28**, 41-44.

GERLING, T.W. 1991 A comparative anatomy of the LEWEX wave systems. pp. 182-

193 in *Directional Ocean Wave Spectra*, ed. R.C. Beal, Baltimore, The Johns Hopkins Univ. Press, 218 pp.

JIANG, L., PERLIN, M. & SCHULTZ, W.W. 1998 Periodic tripling and energy dissipation of breaking standing waves. *J. Fluid Mech.*, **369**, 273-299.

LONGUET-HIGGINS, M.S. & URSELL, F. 1948 Sea waves and microseisms. *Nature, Lond.*, **162**, 700.

LONGUET-HIGGINS, M.S. 1950 A theory of the origin of microseisms. *Phil. Trans. R. Soc. Lond.*, A **243**, 1-35.

LONGUET-HIGGINS, M.S. 1953 Can sea waves cause microseisms? *Proc. Symposium on Microseisms.* Washington, D.C., *U.S. Nat. Acad. Sci. Publ.*, **306**, 74-93.

LONGUET-HIGGINS, M.S. & DOMMERMUTH, D.G. 2001a Vertical jets from standing waves. II. *Proc. R. Soc. Lond.*, A **457**, 2137-2149.

LONGUET-HIGGINS, M.S. & DOMMERMUTH, D.G. 2001b. On the breaking of waves by falling jets. *Phys. Fluids*, **13**, 1652-1659.

LONGUET-HIGGINS, M.S. & DRAZEN, D. 2002 On steep gravity waves meeting a vertical wall: a triple instability. *J. Fluid Mech.*, **466**, 305-318.

LONGUET-HIGGINS, M.S. 2001 Vertical jets from standing waves in deep water: The bazooka effect. pp. 195-204 in *IUTAM Symp. on Free Surface Flows*, ed. Y.D. Shikhmarzaev, Kluwer Acad. Publ., 360 pp.

MERCER, G.N. & ROBERTS, A.J. 1992 Standing waves in water: Their stability and extreme form. *Phys. Fluids* A, **4**, 259-269.

MICHE, M. 1944 Mouvements ondulatoires de la mer en profondeur constante on décroissante. *Ann. Ponts et Chaussées*, **114**, 25-87, 131-164, 270-292 and 396-406.

MILES, J.W. 1962 Transient gravity wave response to an oscillating pressure. *J. Fluid Mech.*, **13**, 145-150.

ORCUTT, J.A., COX, C.S., KIBBLEWHITE, A.C., KUPERMAN, W.A. & SCHMIDT, H. 1993 Observations and causes of ocean and seafloor noise at ultra-low and very-low frequencies. pp. 203-232 in *Natural Physical Causes of Underwater Sound:* Sea Surface Sound (2), ed. B.R. Kerman, Dordrecht, Kluwer Acad. Publ., 750 pp.

PENNY, W.G. & PRICE, A.T. 1952 Finite periodic stationary gravity waves in a perfect fluid, Part 2. *Phil. Trans. R. Soc. Lond.* A, **244**, 254-284.

POSMENTIER, E.S. 1967 A theory of microbaroms. *Geophys. J. R. Astronom. Soc.*, **13**, 487-501.

TAYLOR, G.I. 1953 An experimental study of standing waves. *Proc. R. Soc. Lond* A, **218**, 44-59.

TRIZNA, D.B. 2000 Oceanic fronts as a source of high air-sea flux due to breaking of standing waves of intermediate scale. *Proc. IGARSS 2000* Honolulu, Hawaii, July 2000, p. 117.

ZAMBRESKY, L.F. 1991 An evaluation of two WAM hindcasts for LEWEX. pp. 167-172 in Directional Ocean Wave Spectra, ed. R.C. Beal, Baltimore, The Johns Hopkins Univ. Press, 218 pp.

Discretising Barrick's Equations

J.J. Green
Department of Applied Mathematics, University of Sheffield, UK

Abstract

Barrick's equations, derived by Barrick, Weber and others on the 1970's, describe the interaction of radar with the dynamic ocean surface, and so provide a mathematical basis for oceanic remote sensing. We describe a general discretisation scheme for these equations based on the observation that the linearised equations can be viewed as a local weighted projection transform, and demonstrate that techniques from the global unweighted case, in particular from transmission tomography, can be applied quite naturally. The generality of the discretisation allows it to be applied to a number of inversion algorithms, and we demonstrate implementation with the popular tomographic inversion algorithm ART, which we find able to perform real-time inversion of measured backscatter on modest hardware.

1. Introduction

The use of the Doppler spectrum of HF radar backscatter for the remote sensing of the ocean's surface is now well-established, with a large literature and numerous deployments for commercial and scientific applications. Part of the attraction is the versatility of the technology: ocean currents can be monitored over a large coastal area with relatively inexpensive radars and small computational cost; the same signal can be used to estimate high-frequency ocean waves, and so the strength and direction of prevailing winds. The theoretical extraction of the full directional spectrum of ocean waves from HF backscatter, first established by Barrick in the 1970s, has been slower to develop due to its need for high quality measured backscatter and substantial computational resources, barriers that have all but disappeared in recent years.

In this article we describe a general discretisation scheme for Barrick's equations based on the observation that the linearised equations can be viewed as a local weighted projection transform, and demonstrate that techniques from the global unweighted case, in particular from transmission tomography, can be applied quite naturally. The generality of the discretisation allows it to be applied to a number of inversion algorithms, and we demonstrate implementation with the popular tomographic inversion algorithm ART, which we find able to perform real-time inversion of measured backscatter on modest hardware.

2. Background

In this section we give a brief overview of Barrick's equations modelling the Doppler spectrum of HF (high frequency) radar echos from the sea's surface. Readers seeking the derivation of these are referred to the articles of Barrick

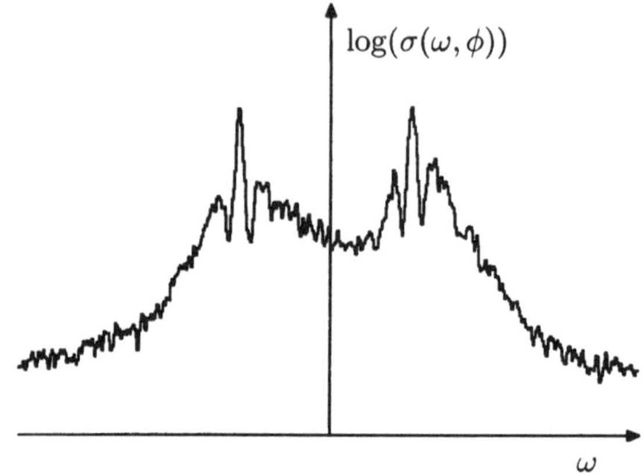

FIGURE 1. An example of measured Doppler spectra

(1972), Weber & Barrick (1977), Barrick & Weber (1977), and to the more recent expositions of Holden & Wyatt (1992) and Hisaki (1996).

In the interests of notational brevity we will treat the version of Barrick's equations for water of infinite depth — the extension to the shallow water case is straightforward.

2.1. *Barrick's equations*

Barrick's equations relate σ, the Doppler spectrum of the backscattered radar signal, to S, the ocean wavenumber spectrum in the absence of current. The equations are derived by a perturbation analysis leading to an expansion $\sigma = \sigma_1 + \sigma_2 + \cdots$, whose first term is given by

$$\sigma_1(\omega, \phi) = 2^6 \pi k_0^4 \sum_{m=\pm 1} S(-m\mathbf{k}_0)\delta(\omega - m\omega_\mathrm{b}), \tag{2.1}$$

where ω is the angular frequency of the Doppler shift, $\mathbf{k}_0$ the wavevector of the incident HF radar beam with wavenumber k_0 and direction ϕ, S the ocean wavevector spectrum and ω_b the frequency of the Bragg matched waves, given by the linear dispersion relationship

$$\omega_\mathrm{b} = \sqrt{2gk_0}.$$

The distributional equation (2.1) defines impulses in the Doppler spectrum σ at $\pm\omega_\mathrm{b}$, features which can clearly be identified in measured spectra, for example in Figure 1.

The second order term, σ_2, derived in the series of papers by Barrick *et al.* mentioned above, may be written

$$\sigma_2(\omega, \phi) = 2^6 \pi k_0^4 \sum_{m,\, m'=\pm 1} \int_{\mathbf{R}^2} |\Gamma|^2 \, S(m\mathbf{k})S(m'\mathbf{k}')\delta(\omega - m\sqrt{gk} - m'\sqrt{gk'}) \, \mathrm{d}\mathbf{p}. \tag{2.2}$$

Here the wave vectors $\mathbf{k}$ and $\mathbf{k}'$ satisfy the Bragg resonance condition $\mathbf{k} + \mathbf{k}' = -2\mathbf{k}_0$, and are related to the variable of integration $\mathbf{p}$ by

$$\mathbf{k} + \mathbf{k}_0 = \mathbf{p}, \quad \mathbf{k}' + \mathbf{k}_0 = -\mathbf{p}$$

as is illustrated in Figure 2.

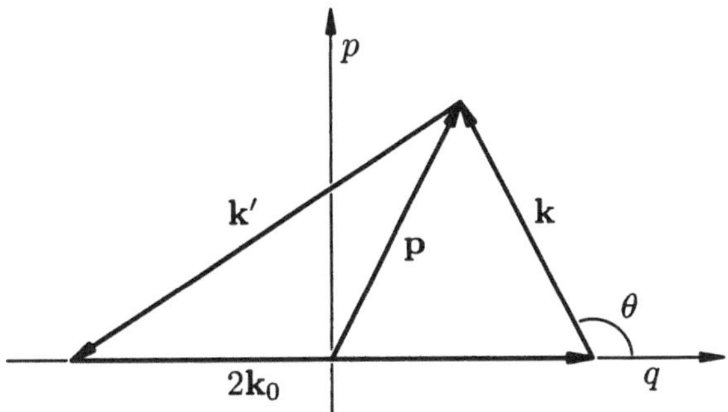

FIGURE 2. Geometry of the wavevectors at second order

The kernel $|\Gamma|^2$ in (2.2) is determined by the *coupling coefficient*,

$$\Gamma = \Gamma(\omega, k, \theta, mm'),$$

which is the sum of contributions accounting for nonlinear hydrodynamic effects (direct backscatter from second-order, Stokes, waves) and for nonlinear electromagnetic effects (indirect backscatter from pairs of first-order waves). The reader is referred to Weber & Barrick (1977), Barrick & Weber (1977) and Barrick & Lipa (1986) for details on the kernel, as we will only need the fact that it is symmetric about the natural coordinate axes of the integration variable $\mathbf{p} = (p, q)$ illustrated in Figure 2.

It is convenient to write Barrick's equations in terms of the dimensionless variables $\eta = \omega/\omega_\text{b}$, $\mathbf{K} = \mathbf{k}/2k_0$, $\mathbf{K}' = \mathbf{k}'/2k_0$. The normalised directional spectrum $Z = S/(2k_0)^4$ is then related to the normalised Doppler spectrum $\sigma(\eta, \phi) \equiv \omega_\text{b}\sigma(\omega, \phi)$ to first order by

$$\sigma_1(\eta, \phi) = \sum_{m=\pm1} Z(-2m\hat{\mathbf{k}}_0)\delta(\eta - m) \tag{2.3}$$

where $\hat{\mathbf{k}}_0 = \mathbf{k}_0/k_0$, and to second order by

$$\sigma_2(\eta, \phi) = 8\pi \sum_{m,\, m'=\pm1} \int_{\mathbf{R}^2} |\Gamma|^2\, Z(m\mathbf{K})Z(m'\mathbf{K}')\delta(\eta - m\sqrt{K} - m'\sqrt{K'})\,\mathrm{d}\mathbf{p}. \tag{2.4}$$

For each Doppler frequency $|\eta| < 1$, the Dirac constraint ensures that there are exactly two nonzero summands in (2.4), integrals over a curves which are reflections in the q-axis. On the other hand, for $|\eta| > 1$ there is only one nonzero summand in (2.4), and this is an integral over a curve (or curves, for $1 < |\eta| < \sqrt{2}$) symmetric about the q-axis. It follows that we can write (2.4) as

$$\sigma_2(\eta, \phi) = 16\pi \int_{K<K'} |\Gamma|^2\, Z(m\mathbf{K})Z(m'\mathbf{K}')\delta(|\eta| - m\sqrt{K} - m'\sqrt{K'})\,\mathrm{d}\mathbf{p}. \tag{2.5}$$

where m and m' depend on η,

$$m = \begin{cases} 1 & \eta > 0 \\ -1 & \eta < 0 \end{cases}, \qquad m' = \begin{cases} 1 & |\eta| < 1 \\ -1 & |\eta| > 1 \end{cases}.$$

The curves of integration in (2.5) are illustrated in Figure 3.

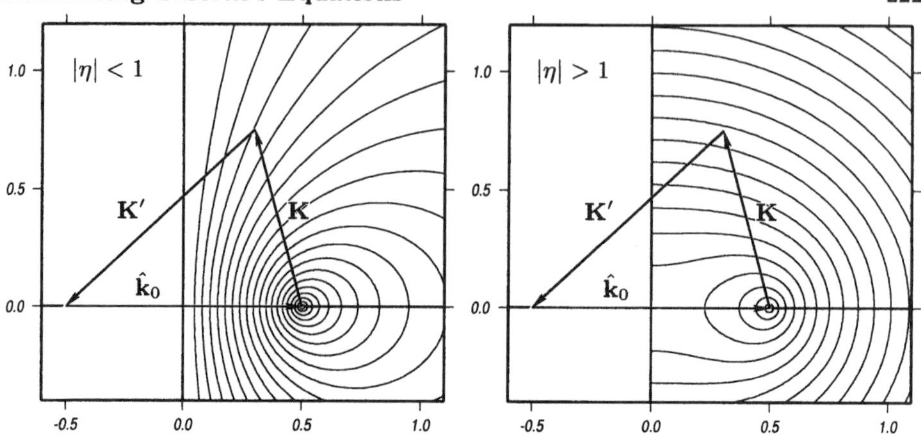

FIGURE 3. Integration curves for the normalised second-order equation

2.2. *Linearisation*

The inversion of equation (2.5), i. e., finding the directional spectrum Z from Doppler spectral data σ, is complicated by nonlinearity. Fortunately the inversion is particularly amenable to linearisation in practice — for radars of suitable frequency the larger wave vector $\mathbf{K}'$ corresponds to higher frequency ocean waves which quickly align to prevailing winds, and whose directional spectrum is well approximated by empirical models such as Pierson & Moskowitz (1964). Indeed, it is practical to fit such a model to the ocean spectral data provided by the inversion of the first-order equation (2.1), i. e., from the Bragg peaks. See, for example, the discussion in Wyatt, Ledgard & Anderson (1997).

Writing $G = G(\mathbf{K})$ for the estimate of the directional spectrum obtained from such a model, solving (2.5) for the Dirac constraint and making θ, the direction of $\mathbf{K}$ relative to $\mathbf{k}_0$ (as in Figure 2), the variable of integration, we obtain the linear approximation

$$\sigma(\eta, \phi) = \int_{-\pi}^{\pi} \Lambda(\mathbf{K}, \phi) Z(m\mathbf{K})\, \mathrm{d}\theta \tag{2.6}$$

where the kernel Λ has absorbed that in (2.5), the model estimate of $Z(m'\mathbf{K}')$, the Jacobian of the change of integration variable and the integration constraint

$$\Lambda = \begin{cases} 16\pi \left|\Gamma\right|^2 G(m'\mathbf{K}') \frac{\partial|\omega|}{\partial K} & K < K' \\ 0 & \text{otherwise} \end{cases}.$$

We henceforth consider only the linearised equation (2.6), but much of what is described below could be extended to work with the fully nonlinear equation, albeit with additional computational and notational complexity.

2.3. *Availability of data*

Although the equations (2.3) and (2.6) provide rather good agreement with measured HF radar backscatter, there are a number of limitations on the measurement of Doppler spectra which should be borne in mind when attempting the inversion of (2.6) with real data.

In measured backscatter, e. g., Figure 1, the Bragg peaks, although evident, are smeared in (Doppler) frequency. Such effects could be caused by variation in the ocean current in the observed area, or by the measurement of the signal with (necessarily) band-limited equipment. This leads to a problem, yet to be satisfactorily solved, in determining which part of the signal is first-order, and which second. In practice this leads to a bound on the quantity $|\eta| - 1$ for second order data σ, corresponding to a lower bound on K in the directional spectrum Z.

The inevitable presence of noise in the measured signal is also apparent in Figure 1, since for any physically reasonable exact data Z we would expect σ to be a smooth function of η, barring ± 1, decaying rapidly with increasing η and as $\eta \to \pm 1$. Thus measured data with reasonable signal to noise ratio is restricted to Doppler frequencies for which $||\eta| - 1|$ is in some positive interval, and so restricts the wavenumbers K at which we can reconstruct the directional spectrum Z.

Finally, radars capable of measuring second-order Doppler spectra to the accuracy required for inversion are rather expensive. Two radars are needed to resolve left-right ambiguities, but for purely economic reasons an inversion technique requiring more than two would find limited application.

3. Discretisation for inversion

The problem of inverting Barrick's equation (2.4) is that of recovering a function from integrals over curves in its domain. In this respect the problem is one of *integral geometry*, who's most studied special case is the inversion of the x-ray and Radon transforms (which are the same in two dimensions). See, for example, Helgason (1999). Such inversions arise in a number of problems in image processing, electron microscopy, astronomy, and in particular, medical and process tomography. The reader is referred to Natterer & Wübbeling (2001) for an extensive overview of this field and its literature.

One popular method for inversion of the x-ray transform is to evaluate numerically the inversion formula derived by Radon in 1937 (see §5.1 of Natterer & Wübbeling (2001)). This allows the expression of the problem in terms of the Fourier transform, allowing solution by fast Fourier methods. Such techniques, however, depend rather delicately on the geometry of the lines along which integrals are taken, and so would seem difficult to extend to more general curves.

Another widely studied approach for these problems relies on the fact that they can be sparsely discretised. If the unknown function is represented by a basis of functions with small support (for example the characteristic functions of a grid of pixels), then each line over the function's domain will intersect only a few of the supports. The resultant sparse discretisations can then be solved efficiently for the coefficients, often by iterative row-action methods as described by Censor (1981). For a discretised problem $y = Ax$ and initial estimate $x^{(0)}$ one obtains iteratively improved estimates $x^{(i+1)}$ from the calculated value of $A_j x^{(i)}$ and y_j, where A_j is the jth row of A and the index j cycles through the row-indices of A.

The seminal example of a row-action method, discovered by Kaczmarz in 1917, is the algebraic reconstruction technique (ART), (see Gordon, Bender & Hermann (1970)) implemented by Hounsfield in the EMI computerised tomography scanner (see Hounsfield (1973)). For the ART, the iterative step is

$$x^{i+1} \leftarrow x^i + \lambda \frac{(y_j - A_j x^{(i)})}{\|A_j\|_2^2} A_j^{\mathrm{T}}$$

where $\lambda \in (0,2)$ is the relaxation parameter. For $\lambda = 1$ each step of the ART is the projection of $x^{(i)}$ onto the hyperplane defined by jth row of the problem, $A_j x = y_j$.

The ART and its many variants are, of course, easy to implement in a manner which exploits sparsity, and in tomographic applications surprisingly few iterations are needed, typically 4–15 cycles through the rows of the matrix (Matej *et al.* (1994)). Indeed, it is shown in Flemming (1980) that early termination of the ART regularises (in the sense of Tikhonov) the ill-posed nature of the problem that has been discretised.

In the remainder of this note we discuss the discretisation of Barrick's equation for inversion by such row-action methods.

3.1. *Barrick's equation*

Suppose that the normalised directional spectrum Z in Barrick's linearised second order equation (2.6) is to be represented as a linear combination of basis functions b_i,

$$Z(\mathbf{K}) = \sum_i \xi_i b_i(\mathbf{K}). \tag{3.1}$$

Seeking a sparse but general discretisation, we initially assume only that each b_i has a small support. Measured Doppler spectra are typically available as binned data,

$$\sigma_j = \frac{1}{\eta_{j+1} - \eta_j} \int_{\eta_j}^{\eta_{j+1}} \sigma(\eta)\, \mathrm{d}\eta, \tag{3.2}$$

where $\Delta\eta \equiv \eta_{j+1} - \eta_j$ is henceforth assumed constant. Thus each inversion datum σ_j corresponds to a *strip* integral in the wavevector domain of Z. Since these strips narrow quadratically as η_j approaches ± 1, the representation in (3.1) should scale similarly if it seeks to discretise uniformly with respect to the measured data. This scaling is absorbed into the problem by introducing the variable $\mathbf{y}$, the vector of length $\sqrt{K}$ in the direction of $\mathbf{K}$, so that (2.6) becomes

$$\sigma(\eta, \phi) = \int_{-\pi}^{\pi} \Lambda(\mathbf{y}, \phi) Z(m\mathbf{y})\, \mathrm{d}\theta \tag{3.3}$$

where

$$\Lambda = \begin{cases} 16\pi\, |\Gamma|^2\, G(m'\mathbf{K}') \frac{\partial|\omega|}{\partial K} \frac{\partial K}{\partial y} & K < K' \\ 0 & \text{otherwise} \end{cases}$$

has absorbed the Jacobian of the transformation. The integration curves in the y-θ domain are illustrated in Figure 4.

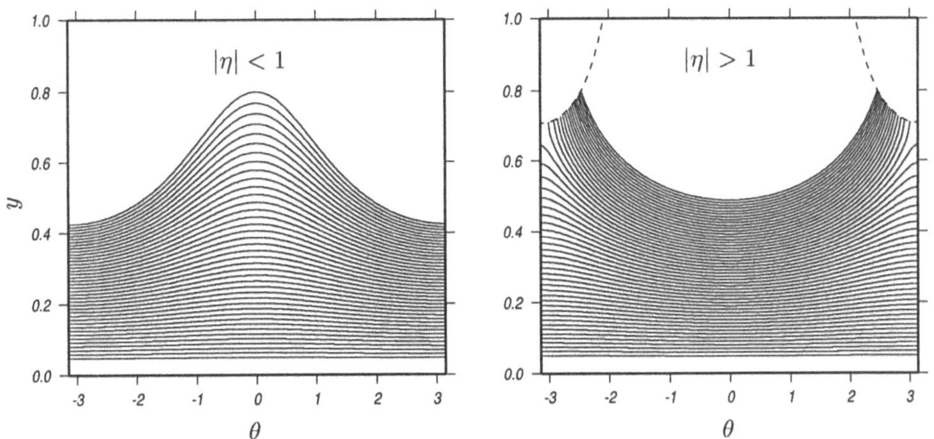

FIGURE 4. Integration curves for the normalised second-order equation. Points $\mathbf{y} = (y, \theta)$ above the dashed line correspond to normalised wave-vectors $\mathbf{K}$ with $\mathbf{K} > \mathbf{K}'$.

Combining the representations (3.1) and (3.2) with the formulation of Barrick's equation in the $\mathbf{y}$ domain gives

$$\sigma_j = \sum_i \frac{1}{\Delta\eta} \xi_i \int_{\eta_j}^{\eta_{j+1}} \int_{-\pi}^{\pi} \Lambda(\mathbf{y}, \psi) b_i(m\mathbf{y}) \, \mathrm{d}\theta \, \mathrm{d}\eta. \tag{3.4}$$

Taking $\mathbf{y}_i$ to be some point in the support of b_i, and supposing that this support is so small that Λ is well approximated on it by $\Lambda_i \equiv \Lambda(\mathbf{y}_i)$, we find that (3.4) reduces to the linear system

$$\sigma_j = \frac{1}{\Delta\eta} \sum_i \Lambda_i B_{i,j} \xi_i, \tag{3.5}$$

where

$$B_{i,j} \equiv \int_{\eta_j}^{\eta_{j+1}} \int_{-\pi}^{\pi} b_i(m\mathbf{y}) \, \mathrm{d}\theta \, \mathrm{d}\eta \tag{3.6}$$

is the jth strip integral over the ith basis function. Writing C_j for the strip

$$\{\mathbf{y}(\eta, \theta) : \eta_j \le \eta \le \eta_{j+1}\}$$

we have

$$B_{i,j} = \int_{C_j} b_i(m\mathbf{y}) \frac{\partial \eta}{\partial y}(m\mathbf{y}) \, \mathbf{dy} \approx \frac{\partial \eta}{\partial y}(m\mathbf{y}_i) \int_{C_j} b_i(m\mathbf{y}) \, \mathbf{dy} \tag{3.7}$$

provided, again, that the support of b_i is sufficiently small.

The final integral in (3.7) can be evaluated approximately by introducing orthogonal coordinates (s, t), based at $\mathbf{y}_i$ and with s in the direction of increasing η. Then by Taylor

$$\eta(s, t) \approx \eta(\mathbf{y}_i) + |\nabla\eta(\mathbf{y}_i)| \, s \tag{3.8}$$

so that

$$\int_{C_j} b_i(m\mathbf{y})\,\mathrm{d}\mathbf{y} \approx \int_{s_{i,j}}^{s_{i,j+1}} \int_{-t_i}^{t_i} b_i(m\mathbf{y}(s,t))\,\mathrm{d}s\,\mathrm{d}t, \tag{3.9}$$

where the t_i are chosen so that all points (s,t) in the support of b_i have $-t_i < t < t_i$ and where, from (3.8), the

$$s_{i,j} = \frac{\eta_j - \eta(\mathbf{y}_i)}{|\nabla\eta(\mathbf{y}_i)|} \tag{3.10}$$

are the limits of integration of the *projection* of the basis function b_i in the direction of the coordinate t. The point of the approximation of a curved strip integral by a projection, as illustrated in Figure 5, is that for a suitable choice of basis function b_i, projections can be calculated explicity.

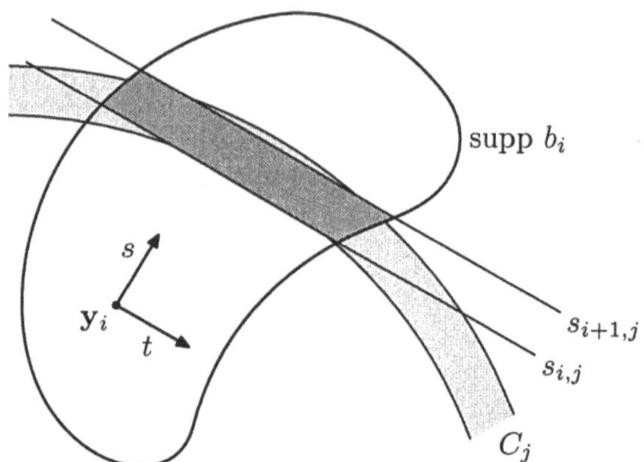

FIGURE 5. Geometry of the strip integral approximation

For completeness, we mention that the partial derivatives in (3.7) and (3.8), conveniently written with the dual variable $\mathbf{y}'$ in the direction of $\mathbf{K}'$ and of length $y' = \sqrt{K'}$, are

$$\frac{\partial\eta}{\partial y} = mF(y,D) + m'\frac{y(y^2+\cos\theta)}{(y')^3}F(y',D), \qquad \frac{\partial\eta}{\partial\theta} = -m'\frac{y^2\sin\theta}{2(y')^3}F(y',D) \tag{3.11}$$

with the dependence on the normalised water depth $D \equiv d/2k_0$ given by

$$F(y,D) = (1 + 2y^2 D \operatorname{cosech} 2y^2 D)\sqrt{\frac{\tanh y^2 D}{\tanh D}},$$

which can taken as 1 in the deep water case.

3.2. *Basis functions*

The strip integral approximation (3.7) aims to treat the transformation defined by Barrick's equation (3.4) as being, locally, a weighted projection transform. This approach makes available the store of techniques and results on projection transforms, provided they are amenable to localisation. In this section we discuss the recent research of Lewitt and co-workers on functional representation for projection transforms (Lewitt (1990), Matej *et al.* (1994), Matej (1996)) and

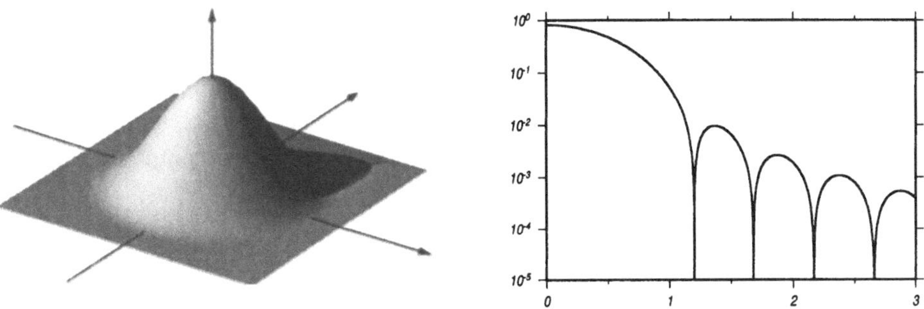

FIGURE 6. Lewitt's basis function Ψ for $m = 2$, $\alpha = 12$ (left) and the absolute value of the radial dependence of its Fourier transform (right)

describe how these can be applied to Barrick's equation in a straightforward manner.

Lewitt has observed that, in a representation of a function by a linear combination of basis functions, it is desirable that the basis be effectively band-limited, i. e., that the Fourier transforms of the basis functions be localised around zero, since this imposes constraints on the smoothness of the function so represented (Lewitt (1990)). For an inversion problem that is ill-posed (in the sense of Hadamard), such as the inversion of the x-ray and Radon transforms (see Chapter 4 of Natterer & Wübbeling (2001)), such a constraint act to regularise the iterative solution. In particular, for the smoothness assumption suppresses the recovery in inversion of *ghosts*, functions in the null-space of the corresponding finite-data transform (see Louis (1984)). Noting that the usual pixel representation (i. e., with respect to a basis of characteristic functions of pixels) has a slow decay in the Fourier transform of its basis, Lewitt proposes representation with respect to a basis of translates of a single, radially symmetric, function

$$b_i(\mathbf{y}) = \Psi(\mathbf{y} - \mathbf{y}_i) = \psi(\|\mathbf{y} - \mathbf{y}_i\|),$$

whose *window* function, ψ, is a generalisation of the Kaiser-Bessel window used in signal processing (and reduces to it for $m = 0$)

$$\psi(r) = \begin{cases} (1 - (r/a)^2)^{m/2} I_m(\alpha\sqrt{1 - (r/a)^2})/I_m(\alpha) & (r < a) \\ 0 & \text{otherwise} \end{cases}, \tag{3.12}$$

where I_m is the modified Bessel function of order m, as in Watson (1944).

In (3.12), the parameter a is the radius of the support of Ψ, α controls the localisation of Ψ about zero, while m determines its smoothness of $\Psi(\mathbf{y})$ for $\|\mathbf{y}\| = a$. An example of the basis function in dimension two is shown in Figure 6, and one can see the reason for Lewitt's choice of the nickname "blobs" for these functions.

Lewitt's basis functions possess a number of attractive features, both theoretically and computationally:

- The radial dependence of the Fourier transform $\hat{\Psi}(\|\mathbf{y}\|)$ (calculated explicitly in Lewitt (1990)) has asymptotic decay $O(1/\|\mathbf{y}\|^{m+1})$, and thus the degree of localisation of the Fourier transform about zero can be controlled by a single parameter;

• A number of interesting quantities associated with Ψ: its gradient, Fourier transform and Laplacian, may be calculated explicitly. In particular the projection p of Ψ, i.e., the integral over the line whose closest point to zero is a distance s from it, is

$$p(s) = 2\int_o^\infty \psi_m(\sqrt{s^2+t^2})\,\mathrm{d}t = a\frac{I_{m+1/2}(\alpha)}{I_m(\alpha)}\left(\frac{2\pi}{\alpha}\right)^{1/2}\psi_{m+1/2}(s) \qquad (3.13)$$

where the dependence of ψ on m is here indicated by its subscript;

• The freedom in the choice of the parameters may be used to ensure that the resulting basis has desirable properties (a small error in representing constant functions, invertibility of the interpolation matrix $[\Psi(\mathbf{y}_i - \mathbf{y}_j)]$ and so on). These matters are discussed in Matej (1996) and Green (2001);

• The modified Bessel function I_ν can be calculated simply and efficiently for ν a half-integer, as is described in Thompson (1997).

A systematic evaluation of the application of Lewitt's basis for a number of inversion methods (including ART) in positron emission tomography has found a clear advantage over pixel (or voxel) based discretisations (see Matej *et al.* (1994)). It is our hope that these benefits will transfer to the discretisation of Barrick's equation.

3.3. *Implementation for Barrick's equation*

The combination of Lewitt's basis with the discretisation of Barrick's equation (3.5) is achieved simply by substituting the expression for the projection of Lewitt's basis (3.13) into that for the localised projection (3.9). Thus the coefficients $B_{i,j}$ can be obtained by the following procedure applied for each grid-node $\mathbf{y}_i$ and for each chosen sideband (i.e., for each of the corresponding choices of m and m'):

- calculate $\frac{\partial\xi}{\partial y}(\mathbf{y}_i)$ and $|\nabla\xi|(\mathbf{y}_i)$ using (3.11);
- calculate the integration limits $s_{i,j}$ from (3.10);
- for each pair $(s_{i,j}, s_{i,j+1})$:
  - calculate the strip-integral approximation (3.9) using the formula (3.13);
  - evaluate $B_{i,j}$ from (3.6).

The main computational cost in this procedure is the (numerical) integration of the projection p of the basis function, but this can be greatly reduced by creating an appropriate approximation of the indefinite integral.

For practical discretisation a number of minor complications, ignored in the above for the sake of brevity, must be incorporated in the calculations.

For Doppler spectra obtained from shallow-water observations, the formulae for the kernel and integration contours need to be modified to account for the shallow-water dispersion relation. Details on the required changes can be found in Holden & Wyatt (1992).

Combining the discretisations for multiple radar systems may present a problem if the radars are operating at different frequencies, for then the y variables from the discretisations are incommensurate. This is solved by a suitable scaling, but then one might as well discretise in the k domains, and perform the normalisation as required (or use the corresponding unnormalised formulae).

Finally, the $\mathbf{y}$ domain as described above is rather narrow (see Figure 4), so discretisation with a basis function whose support is a disc leads to undersampling in the y variable (or oversampling in angle). Consequently, a scaling of the y variable (as mentioned in Lewitt (1990)) is needed, forcing a few trivial modifications.

3.4. *Review of discretisation methods*

We here provide a brief overview of some of the discretisation schemes that have been used in the inversion of Barrick's equation.

The first inversion method, described in Lipa (1977), makes reductions in the complexity of the inversion by assuming that the directional spectrum is separable, $S(\mathbf{k}) = g(k)h(\theta)$. Such an assumption, often reasonable, allows the separation of the discretisation and reduces the number of unknowns to such an extent that the resulting linear system is overdetermined.

The discretisation adopted by Lipa & Barrick (1980), and later by Howell & Walsh (1993), replaces separability by the less restrictive assumption that the direction spectrum is well approximated by a truncated Fourier series in angle

$$S(\mathbf{k}) = \sum_{n=1}^{N} (a_n(k) \cos n\theta + b_n(k) \sin n\theta).$$

As with Lipa's approach, this leads to a lightweight discretisation since only the $2N$ one-dimensional functions a_n and b_n need be determined, and satisfactory results are reported for $N = 2$ or 3.

The inversion method of Wyatt (1990), a two-dimensional version of the iterative scheme described in § 7.6 of Twomey (1977), is rather unusual in that the discretisation of the directional spectrum is along the contours of integration defined by the Dirac constraint (as in Figure 4). This approach has the advantage that the solution to the forward problem (needed in each iteration) is rapid, albeit at the expense of finding nearest-neighbours for each discretisation point, needed for the smoothing which stabilises the inversion.

Finally we mention the more traditional grid-based discretisations associated with the nonlinear inversion method of Hisaki (1996), and the Bayesian inversion of Hashimoto & Tokuda (1999). These authors represent the directional spectrum as piecewise-constant on pixels which are uniform in θ, but exponential in wavenumber.

4. Preliminary results

The discretisation described above has been implemented as part of an experimental replacement inversion kernel for the Sheffield wave inversion toolset. The toolset implements the results of research into the inversion problem by L. R. Wyatt and co-workers over 15 years (Holden & Wyatt (1992), Wyatt (1986), Wyatt (1990), Wyatt, Ledgard & Anderson (1997)), and was recently deployed as part of the EuroROSE project described in Wyatt *et al.*.

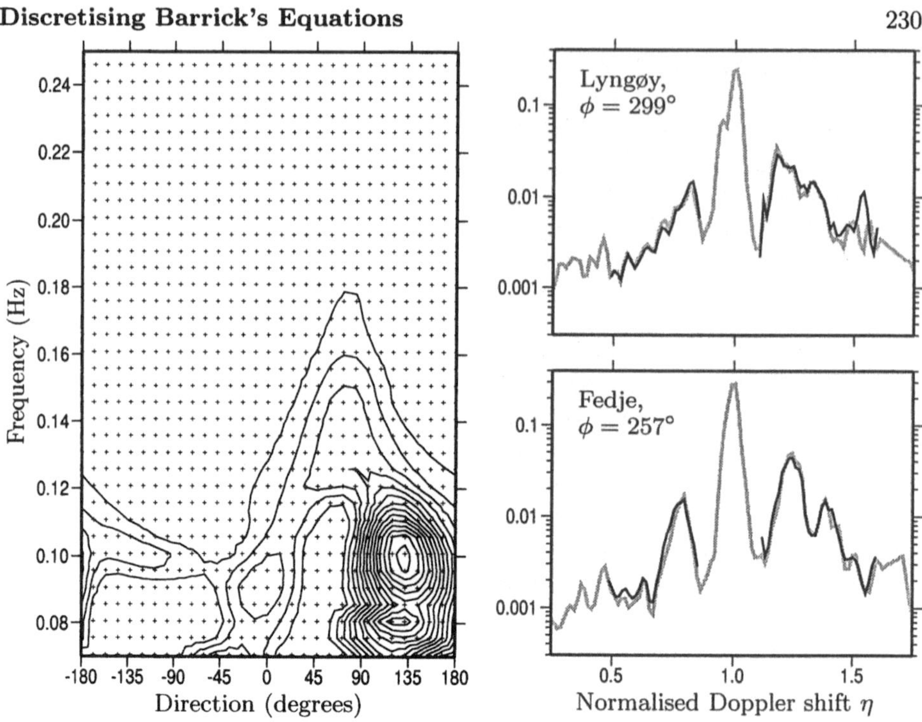

FIGURE 7. Sample directional spectrum (left) as inverted from measured Doppler spectra, detail around the positive Bragg peak shown in grey (right).

An important design goal for the implementation was modularity, since we wish the discretisation to be usable with any inversion method which can be recast as a row-action method (in particular with Wyatt's inversion method). The discretisation, generic row-action method and ART modules were implemented in around 20,000 lines of C, chosen for reasons of efficiency and portability. The initial implementation was found to have a performance close to that of the (highly optimised) Wyatt algorithm used in the EUROROSE project, with an inversion performed in an amortised half-second on a 750MHz PIII, around four times as fast as was needed for real-time operation during EUROROSE (where 200–350 inversions were performed every 10 minutes).

Figure 7 shows an example of inversion of data acquired during the EUROROSE Fedje deployment. A pair of Doppler spectra (right, in grey) were inverted using the ART, with 25 iterations and a relaxation parameter of 0.1. The discretisation for the inversion used the radial Lewitt basis arranged on a regular 64×30 cylindrical grid in the y-θ domain. The directional spectrum (left) is dominated by low frequency waves at around 0.1 Hz heading south eastward, typical for the exposed location. The estimate of the Doppler spectra derived from the directional spectrum *via* Barrick's second-order equation, which drive the inversion, is shown superimposed (in black) on the measured Doppler spectra (left).

A detailed comparison of row-action methods for use with our discretisation will be reported elsewhere.

ACKNOWLEDGEMENT: The author wishes to thank L. R. Wyatt for support and advice on the research leading to this article, and acknowledges the financial support provided through the EUROROSE project (EU contract MAS3CT9801 68).

REFERENCES

BARRICK, D. E. 1972 First-order theory and analysis of MF/HF/VHF scatter from the sea. *IEEE Trans. Antennas Propogat.* **AP-20**, 2–10.

BARRICK, D. E. & LIPA, B. J. 1986 The second-order shallow-water hydrodynamic coupling coefficient in interpretation of HF radar sea echo. *IEEE J. Oceanic Eng.* **OE-11 (2)**, 310–315.

BARRICK, D. E. & WEBER, B. L. 1977 On the nonlinear theory for gravity waves on the ocean's surface. Part II: Interpretation and applications. *J. Phys. Oceanog.* **7**, 11–21.

CENSOR, Y. 1981 Row-action methods for huge and sparse systems and their applications. *SIAM Review* **23 (4)**, 444–466.

FLEMING, H. E. 1980 Equivalence of regularization and truncated iteration in the solution of ill-posed image reconstruction problems. *Linear Alg. Applic.* **130**, 133–150.

GORDON, R., BENDER, R. & HERMANN, G. T. 1970 Algebraic reconstruction techniques (ART) for three-dimensional electron microscopy and x-ray photography. *J. Theoret. Biol.* **29**, 471–481.

GREEN, J. J. 2001 Approximation with the radial basis functions of Lewitt. To appear in *Proceedings of Algorithms for Approximation 4*.

HASHIMOTO, N. & TOKUDA, M. A Bayesian approach for estimating directional spectra with HF radar. *Coastal Eng. J.* **41 (2)**, 137–149.

HELGASON, S. 1999 *The Radon transform.* Progress in Mathematics, vol. 5, Birkhäuser.

HISAKI, Y. 1996 Nonlinear inversion of the integral equation to estimate ocean wave spectra from HF radar. *Radio Science* **31 (1)**, 25–39.

HOLDEN, G. J. & WYATT, L. R. 1992 Extraction of sea state in shallow-water using HF radar. *IEE Proc. F Radar Sig. Proc.* **139**, 175–181.

HOUNSFIELD, G. N. 1973 Computerized transverse axial scanning (tomography), I: Description of system. *Brit. J. Radiol.* **46**, 1016–1022.

HOWELL, R. & WALSH, J. 1993 Measurement of ocean wave spectra using narrow-beam HF radar. *IEEE J. Oceanic Eng.* **18**, 296–305.

LEWITT, R. M. 1990 Multidimensional digital image representations using generalized Kaiser-Bessel window functions. *J. Opt. Soc. Am. A* **7 (10)**, 1834–1846.

LIPA, B. 1977 Derivation of directional ocean-wave spectra by integral inversion of second-order radar echos. *Radio Science* **12 (3)**, 425–434.

LIPA, B. & BARRICK, D. 1980 Methods for the extraction of long-period ocean-wave parameters from narrow beam HF radar sea-echo. *Radio Science* **15 (4)**, 843–853.

LOUIS, A. K 1984 Orthogonal function series expansion and the null space of the Radon transform. *SIAM J. Math. Anal.* **15 (3)**, 621–633.

MATEJ, S., HERMAN, G. T., NARAYAN, T. K., FURUIE, S. S., LEWITT, R. M. & KINAHAN, P. E. 1994 Evaluation of task-oriented performance of several fully 3d PET reconstruction algorithms. *Phys. Med. Biol.* **39**, 355–367.

MATEJ, S. & LEWITT, R. M. 1996 Practical considerations for 3-d image reconstructions using spherically symmetric volume elements. *IEEE Transactions on Medical Imaging* **15 (1)**, 68–78.

NATTERER, F. & WÜBBELING, F. 2001 *Mathematical Methods in Image Reconstruction* SIAM Monographs on Mathematical Modeling and Computation, vol. 5, SIAM.

PIERSON JR, W. J. & MOSKOWITZ, L. 1964 A proposed spectral form for fully developed wind seas based on the similarity theory of S. A. Kitaigorodskii. *J. Geophys. Res.*, 5181–5190.

THOMPSON, W. J. 1997 *Atlas for computing mathematical functions.* John Wiley & Sons Inc.

TWOMEY, S. 1977 *Introduction to the Mathematics of Inversion in Remote Sensing and Indirect Measurements.* Dover.

WATSON, G. N. 1944 *A Treatise on the Theory of Bessel Functions.* Cambridge.

WEBER, B. L. & BARRICK, D. E. 1977 On the nonlinear theory for gravity waves on the ocean's surface. Part I: Derivations. *J. Phys. Oceanog.* **7**, 3–10.

WYATT, L. R. 1986 The measurement of the ocean wave directional spectrum from H.F. radar Doppler spectra. *Radio Science* **21**, 473–485.

WYATT, L. R. 1990 A relaxation method for integral inversion applied to HF radar measurement of the ocean wave directional spectrum. *Int. J. Remote Sensing* **11**, 1481–1494.

WYATT, L, R., GREEN, J. J., GURGEL, K.-W., NIETO BORGE, J. C., REICHERT, K., GÜNTHER, H., ROSENTHAL, W., SAETRA, O. & REISTAD, M. (in press) Comparisons of wave measurements from the EuroROSE Fedje experiment.

WYATT, L. R., LEDGARD, L. J. & ANDERSON, C. W. 1997 Maximum-likelihood estimation of the directional distribution of 0.53-Hz ocean waves. *J. Atmos. Oceanic Tech.* **14**, 591–603.

Zeitfracht Medien GmbH
Ferdinand-Jühlke-Straße 7
99095 Erfurt, Deutschland
produktsicherheit@kolibri360.de